T0141911

Advances in Intelligent Systems and Computing

Volume 846

Series editor

Janusz Kacprzyk, Polish Academy of Sciences, Warsaw, Poland
e-mail: kacprzyk@ibspan.waw.pl

The series "Advances in Intelligent Systems and Computing" contains publications on theory, applications, and design methods of Intelligent Systems and Intelligent Computing. Virtually all disciplines such as engineering, natural sciences, computer and information science, ICT, economics, business, e-commerce, environment, healthcare, life science are covered. The list of topics spans all the areas of modern intelligent systems and computing such as: computational intelligence, soft computing including neural networks, fuzzy systems, evolutionary computing and the fusion of these paradigms, social intelligence, ambient intelligence, computational neuroscience, artificial life, virtual worlds and society, cognitive science and systems, Perception and Vision, DNA and immune based systems, self-organizing and adaptive systems, e-Learning and teaching, human-centered and human-centric computing, recommender systems, intelligent control, robotics and mechatronics including human-machine teaming, knowledge-based paradigms, learning paradigms, machine ethics, intelligent data analysis, knowledge management, intelligent agents, intelligent decision making and support, intelligent network security, trust management, interactive entertainment, Web intelligence and multimedia.

The publications within "Advances in Intelligent Systems and Computing" are primarily proceedings of important conferences, symposia and congresses. They cover significant recent developments in the field, both of a foundational and applicable character. An important characteristic feature of the series is the short publication time and world-wide distribution. This permits a rapid and broad dissemination of research results.

Advisory Board

Chairman

Nikhil R. Pal, Indian Statistical Institute, Kolkata, India
e-mail: nikhil@isical.ac.in

Members

Rafael Bello Perez, Universidad Central "Marta Abreu" de Las Villas, Santa Clara, Cuba
e-mail: rbellop@uclv.edu.cu

Emilio S. Corchado, University of Salamanca, Salamanca, Spain
e-mail: escorchado@usal.es

Hani Hagras, University of Essex, Colchester, UK
e-mail: hani@essex.ac.uk

László T. Kóczy, Széchenyi István University, Győr, Hungary
e-mail: koczy@sze.hu

Vladik Kreinovich, University of Texas at El Paso, El Paso, USA
e-mail: vladik@utep.edu

Chin-Teng Lin, National Chiao Tung University, Hsinchu, Taiwan
e-mail: ctlin@mail.nctu.edu.tw

Jie Lu, University of Technology, Sydney, Australia
e-mail: Jie.Lu@uts.edu.au

Patricia Melin, Tijuana Institute of Technology, Tijuana, Mexico
e-mail: epmelin@hafsamx.org

Nadia Nedjah, State University of Rio de Janeiro, Rio de Janeiro, Brazil
e-mail: nadia@eng.uerj.br

Ngoc Thanh Nguyen, Wroclaw University of Technology, Wroclaw, Poland
e-mail: Ngoc-Thanh.Nguyen@pwr.edu.pl

Jun Wang, The Chinese University of Hong Kong, Shatin, Hong Kong
e-mail: jwang@mae.cuhk.edu.hk

More information about this series at http://www.springer.com/series/11156

M. Arun Bhaskar · Subhransu Sekhar Dash
Swagatam Das · Bijaya Ketan Panigrahi
Editors

International Conference on Intelligent Computing and Applications

Proceedings of ICICA 2018

 Springer

Editors
M. Arun Bhaskar
Department of Electrical
    and Electronics Engineering
Velammal Engineering College
Chennai, Tamil Nadu, India

Subhransu Sekhar Dash
Department of Electrical
    and Electronics Engineering
Government College of Engineering
Keonjhar, Odisha, India

Swagatam Das
Electronics and Communication
    Sciences Unit
Indian Statistical Institute
Kolkata, West Bengal, India

Bijaya Ketan Panigrahi
Department of Electrical Engineering
Indian Institute of Technology Delhi
New Delhi, Delhi, India

ISSN 2194-5357   ISSN 2194-5365   (electronic)
Advances in Intelligent Systems and Computing
ISBN 978-981-13-2181-8   ISBN 978-981-13-2182-5   (eBook)
https://doi.org/10.1007/978-981-13-2182-5

Library of Congress Control Number: 2018950951

© Springer Nature Singapore Pte Ltd. 2019
This work is subject to copyright. All rights are reserved by the Publisher, whether the whole or part of the material is concerned, specifically the rights of translation, reprinting, reuse of illustrations, recitation, broadcasting, reproduction on microfilms or in any other physical way, and transmission or information storage and retrieval, electronic adaptation, computer software, or by similar or dissimilar methodology now known or hereafter developed.
The use of general descriptive names, registered names, trademarks, service marks, etc. in this publication does not imply, even in the absence of a specific statement, that such names are exempt from the relevant protective laws and regulations and therefore free for general use.
The publisher, the authors and the editors are safe to assume that the advice and information in this book are believed to be true and accurate at the date of publication. Neither the publisher nor the authors or the editors give a warranty, express or implied, with respect to the material contained herein or for any errors or omissions that may have been made. The publisher remains neutral with regard to jurisdictional claims in published maps and institutional affiliations.

This Springer imprint is published by the registered company Springer Nature Singapore Pte Ltd.
The registered company address is: 152 Beach Road, #21-01/04 Gateway East, Singapore 189721, Singapore

Preface

This AISC volume contains the papers presented at the *International Conference on Intelligent Computing and Applications* (ICICA) held on February 2 and 3, 2018, at Velammal Engineering College, Chennai, India. ICICA 2018 received 286 paper submissions from various countries across the globe. After a rigorous peer review process, 126 full-length articles were accepted for oral presentation at the conference. This corresponds to an acceptance rate of 44% and is intended for maintaining the high standards of the conference proceedings. The papers included in this LNAISC volume cover a wide range of topics on genetic algorithms, evolutionary programming, and evolution strategies such as AIS, DE, PSO, ACO, BFA, HS, SFLA, artificial bees and fireflies algorithm, parallel computation, membrane, grid, cloud, DNA, mobile computing, computer networks and security, data structures and algorithms, data compression, data encryption, data mining, digital signal processing, digital image processing, watermarking, security and cryptography, AI methods in telemedicine and eHealth, document classification and information retrieval, optimization techniques, and their applications for solving problems in these areas.

In the conference, separate sessions were arranged for delivering the keynote address by eminent members from various academic institutions and industries. Four keynote lectures were given on February 2 and 3, 2018. On the first day, Dr. Ivan Zelinka gave a talk on "Recent Advances and Progress in Evolutionary Algorithms and its Dynamics," and in the post-lunch session, Dr. Swagatam Das, Assistant Professor, Electronics and Communication Sciences Unit, Indian Statistical Institute, Kolkata, gave his lecture on "Deep Machine Learning" and clarified the queries raised by the participants. On the second day, Dr. Deepti Prasad Mukherjee, Professor and Head, Electronics and Communication Sciences Unit, Indian Statistical Institute, Kolkata, gave his talk on "What is this object?" an insight into research ideas, product and patent applications, and in the post-lunch session, Dr. Kyung Tae Kim, Professor Emeritus, Hannam University, Republic of Korea, gave a lecture on "4th Industrial Revolution and what do we have to prepare?" All these lectures generated great interest among the participants of ICICA 2018 in paying more attention to these important topics in their research work.

We take this opportunity to thank the authors of all the submitted papers for their hard work, adherence to the deadlines, and suitably incorporating the changes suggested by the reviewers. The quality of a refereed volume depends mainly on the expertise and dedication of the reviewers. We are indebted to the Program Committee members for their guidance and coordination in organizing the review process.

We are indebted to the Chairman, Chief Executive Officer, Advisor, Principal, faculty members, and administrative personnel of Velammal Engineering College, Chennai, for encouraging us to organize the conference on a grand scale. We would like to thank all the participants for their interest and enthusiastic involvement. Finally, we would like to thank all the volunteers whose tireless efforts in meeting the deadlines and arranging every detail meticulously made sure that the conference could run smoothly. We hope the readers of these proceedings find the papers useful, inspiring, and enjoyable.

Chennai, India   M. Arun Bhaskar
Keonjhar, India   Subhransu Sekhar Dash
Kolkata, India   Swagatam Das
New Delhi, India   Bijaya Ketan Panigrahi
May 2018

Organizing Committee

Chief Patrons

Thiru. M. V. Muthuramalingam, Chairman, VET
Thiru. M. V. M. Velmurugan, CEO, VET

Patrons

Dr. M. Balasubramanian, Advisor, VEC
Dr. N. Duraipandian, Principal, VEC
Dr. B. Venkatalakshmi, Vice Principal, VEC

General Chairs

Dr. Bijay Ketan Panigrahi, IIT Delhi
Dr. Swagatam Das, ISI, Kolkata

Program Chairs

Dr. S. S. Dash, Government College of Engineering, Keonjhar, Odisha
Dr. M. Arun Bhaskar, VEC

Organizing Chairs

Dr. S. Srinath, HOD/EEE, VEC
Dr. Ramazan Bayindir, Gazi University, Turkey
Dr. S. Premalatha, VEC

Special Session Chairs

Dr. P. N. Suganthan, NTU, Singapore
Prof. Ing. Ivan Zelinka, VSB-TUO, Czech Republic
Dr. Bolanle Tolulope Abe, TUT, South Africa

Publication Chairs

Dr. Suresh Chandra Satapathy, ANITS, India
Dr. C. Subramani, SRM University, Chennai
Dr. Adisa Jimoh, TUT, South Africa

International Advisory Committee

Dr. Ahmed Faheem Zobaa, BU, UK
Dr. Akhtar Kalam, VU, Australia
Dr. Alfredo Vaccaro, US, Italy
Dr. David Yu, UWM, US
Dr. Dmitri Vinnikov, TUT, Estonia
Dr. Gorazd Štumberger, UM, Slovenia
Dr. Hussain Shareef, UKM, Malaysia
Dr. Joseph Olorunfemi Ojo, TTU, USA
Dr. Junitamohamad-Saleh, US, Malaysia
Dr. Mohamed A. Zohdy, OU, MI
Dr. Murad Al-Shibli, Head, EMET, Abu Dhabi
Dr. Nesimi Ertugrul, UA, Australia
Dr. Omar Abdel-Baqi, UWM, USA
Dr. Richard Blanchard, LBU, UK
Dr. Shashi Paul, DM, UK
Dr. Zhao Xu, HKPU, Hong Kong

National Advisory Committee

Dr. Kumar Pradhan, IIT Kharagpur
Dr. Bhim Singh, IIT Delhi
Dr. Ganapati Panda, IIT Bhubaneswar
Dr. S. Jeevananthan, PEC, Puducherry, India
Dr. C. Christober Asir Rajan, PEC, Puducherry, India
Dr. Debashish Jena, NITK, India
Dr. N. P. Padhy, IIT Roorkee
Dr. M. Somasundaram, AU, Chennai, India
Dr. Sidhartha Panda, VSSUT EC, Burla
Dr. K. Vijayakumar, SRM University, Chennai
Dr. N. Veerappan, AVIT, Chennai
Dr. S. K. Patnaik, Anna University
Dr. G. Uma, Anna University
Dr. M. Mohanthy, SOA University, Odisha
Dr. S. N. Bhuyan, SOA University, Odisha

Contents

About the Editors

Dr. M. Arun Bhaskar is presently working as an assistant professor in the Department of Electrical and Electronics Engineering, Velammal Engineering College, Chennai. He has received his PDF from the Tshwane University of Technology, South Africa, and Ph.D. from SRM University, Chennai. He is having more than 12 years of research and teaching experience. He has published more than 30 papers in national conferences, 36 papers in international conferences, 12 papers in IEEE CPS, 26 in international journals. His research areas are modeling of FACTS controller, voltage stability improvement, design of nonlinear controllers for FACTS devices. He is a life member of Indian Society for Technical Education and a member of Institute of Electrical and Electronics Engineering.

Dr. Subhransu Sekhar Dash is presently working as a professor in the Department of Electrical and Electronics Engineering, Government College of Engineering, Keonjhar, Odisha, India. He received his Ph.D. from College of Engineering, Anna University, Guindy. He has more than 20 years of research and teaching experience. His research areas are power electronics and drives, modeling of FACTS controller, power quality, power system stability, and smart grid. He is a visiting professor at the Francois Rabelais University, POLYTECH, France. He is the chief editor of International Journal on Advanced Electrical and Computer Engineering.

Dr. Swagatam Das received B.E. Tel. E., M.E. Tel. E. (control engineering), and Ph.D., all from Jadavpur University, India, in 2003, 2005, and 2009, respectively. Currently, he is serving as an associate professor at the Electronics and Communication Sciences Unit of the Indian Statistical Institute, Kolkata. His research interests include evolutionary computing, pattern recognition, multi-agent systems, and wireless communication. He has published one research monograph, one edited volume, and more than 150 research articles in peer-reviewed journals and international conferences. He is the founding co-editor-in-chief of "Swarm and Evolutionary Computation," an international journal from Elsevier. He serves as associate editor of the IEEE Transactions on Systems, Man, and Cybernetics:

Systems and Information Sciences (Elsevier). He is an editorial board member of Progress in Artificial Intelligence (Springer), Mathematical Problems in Engineering, International Journal of Artificial Intelligence and Soft Computing, and International Journal of Autonomous and Adaptive Communications Systems. He is the recipient of the 2012 Young Engineer Award from the Indian National Academy of Engineering (INAE) and also the recipient of the 2015 Thomson Reuters Research Excellence India Citation Award as the highest cited researcher from India in engineering and computer science category between 2010 and 2014.

Dr. Bijaya Ketan Panigrahi is a professor of Electrical and Electronics Engineering Department at the Indian Institute of Technology Delhi, India. He received his Ph.D. from Sambalpur University. He is serving as a chief editor to the International Journal of Power and Energy Conversion. His interests focus on power quality, FACTS devices, power system protection, and AI application to power system.

A Novel IDS to Detect Multiple DoS Attacks with Network Lifetime Estimation Based on Learning-Based Energy Prediction Algorithm for Hierarchical WSN

N. Dharini, N. Duraipandian and Jeevaa Katiravan

Abstract Lifetime of sensor network plays a crucial part in designing any WSN-based application. On the other hand, securing them against malicious activity is also important. Security and lifetime should work hand in hand to protect and conserve the network. For example, imparting any security solution should not decrease WSN's lifetime. Thus, a novel energy prediction algorithm is developed and deployed in WSN structure that can actively look and report for any adversaries present within the network and can also report the network's lifetime stating the remaining amount of time the network can work perfectly without losing its energy. The key idea is that by retrieving the residual energy of the nodes at an instance of time, the proposed algorithm can predict its energy consumption at various other time instances; in such a way how long the nodes can stay alive is predicted. Moreover, we observe that different types of DoS attack consume abnormal amount of energy; by predicting its energy consumption at various instances and comparing with its actual energy consumption, attacks can be identified and classified. This novel mechanism can look for and detect 5 types of DoS attacks at a time. By formulating the network's lifetime, the design of WSN can be optimized according to the application requirements. Detailed performance analysis is done using NS-2 Mannasim framework. Simulation studies are done under various scenarios to prove its efficiency and accuracy.

Keywords WSN · Network lifetime · Energy prediction · IDS DoS

N. Dharini (✉)
Anna University, Chennai, Tamil Nadu, India
e-mail: dharini1990@gmail.com

N. Duraipandian · J. Katiravan
Velammal Engineering College, Chennai, Tamil Nadu, India

© Springer Nature Singapore Pte Ltd. 2019
M. A. Bhaskar et al. (eds.), *International Conference on Intelligent Computing and Applications*, Advances in Intelligent Systems and Computing 846,
https://doi.org/10.1007/978-981-13-2182-5_1

1

1   Introduction

Hierarchical routing in WSN involves multi-hop communication within a cluster of
nodes followed by data aggregation, thereby achieving energy efficiency among
sensor nodes. Geographical area of nodes is partitioned into cluster. Each cluster
gets one node elected as cluster heads (CH), usually, high energy nodes are chosen
as CH nodes within the cluster sense and report the data to their respective CH.
Ilyas and Mahgoub in their handbook [1] and Sharma and Trivedi [2] mentioned
that hierarchical routing is the two layers routing where one layer is used to select
the cluster heads and the other layer is used for routing. Various hierarchical routing
protocols have been developed as fundamental hierarchical routing algorithms
employed for wireless sensor network communication. Broadcast wireless communication nature of sensor nodes makes them prone to various attacks. According
to [3], WSN security challenges are different and difficult as apart from other
wireless networks. Particularly, cluster heads become intruder's ideal targets. If any
one of the cluster head is captured and made malicious, the entire cluster gets
affected. Thus, an IDS is very essential for cluster-based WSN.

2   Related Works

Trust management and encryption schemes were developed in the state-of-the-art
approaches. One such technique known as spontaneous dogs [4] involves local and
global agents to overhear the communications taking place among nodes. Global
agents found within each cluster can overhear relayed and normal packets. If any
node tries to drop or selectively forward the packets, the global agents can detect
those using spontaneous watchdogs. **Drawbacks**: The drawback of this approach is
that overhearing all the packets is hardly possible by the global agents, due to the
randomness of the selection process.

Group-based IDS proposed involves grouping algorithm where groups are
formed only if there are spatially close enough to each other and their sensed values
are similar within the nodes present inside the group. Groups are further divided
into subgroups. Sensor's sensed data, packet dropping, sending receiving
packet mismatch, and sending power were collected, from which fabrication,
jamming, sinkhole, gray hole, and flooding attacks can be detected. The eHIP [5]
system proposes authentication-based intrusion prevention subsystem and
collaboration-based intrusion detection subsystem to secure cluster-based WSN.
Both the subsystems introduced 2 different authentication mechanisms to verify
control messages and the sensed data to defend external attacks. The simulation
results prove it to be an energy-efficient secure system in a highly hostile area.

Drawbacks: Neighbor nodes need to be monitored continuously in order to
detect attacks. Moreover, monitoring the sensor nodes for any packet loss does not
effectively detect the presence of attacks. WSN communication infrastructure is

unstable, packet loss may occur due to various reasons in wireless communication such as due to fading or attenuation of the signal, etc., thus completely relying on the monitoring characteristics of wireless communication does not confirm the presence of attacks. Advanced techniques need to be devised.

The main idea of IDSEP [6] is to detect malicious nodes based on energy consumption of sensor nodes. Abnormal energy consuming nodes are detected using Markov chain based prediction algorithm. Malicious nodes launching gray hole attack will consume less energy than the other legitimate node. Those launching flooding attack will consume maximum energy. Those launching wormhole will consume twice the amount of energy than the legitimate node. Those of which launching black hole attack will consume higher energy than wormhole, gray hole and Sybil. Finally, Sybil nodes consume energy as much as the number of their fake identities. **Drawbacks**: Drawback of this approach is that it can only detect malicious cluster heads and a highly computationally complex algorithm is used.

Nodes having highest residual energy are selected for the packet routing process [7]. Ant colony based algorithm is proposed to predict the energy consumption of nodes. Nodes' energy consumption is predicted ahead of their activity based on their previous activity pattern. This work does not consider the security aspect of the network.

The work proposed in [8] detected black hole and selective forward attacks by means of local information obtained from neighbors. **Drawbacks**: Neighboring nodes do not have the overall view of the network, which is vital for intrusion detection design. Xiao et al. [9] proposed a lightweight security scheme called CHEMAS (Check point-based Multi-hop Acknowledgement Scheme) to detect selective forward attacks. Certain nodes are selected randomly to generate acknowledgements along the packet forwarding path. Random selection of nodes increases the resilience against attacks. Suspected nodes are identified for packet loss if the intermediate nodes do not receive enough acknowledgements from the checkpoints. Simulation study and theoretical analysis were carried out. The proposed work achieves high detection rate, even in harsh radio conditions. Communication overhead is also within reasonable bounds. **Drawbacks**: There are possibilities for misalerts during upstream packet transfer. Because during such process having known the default route from source node to base station, packet loss might happen due topology change which may be wrongly alerted as malicious dropper.

Adaptive security design proposed in [10] developed authentication-based security modules to improve secure communication in hierarchical WSNs. Trust evaluation was also carried out. **Drawbacks**: In general, prevention techniques like authentication and trust model does not detect internal compromised nodes. This will greatly affect the performance of the network.

UnMask: Utilizing Neighbor Monitoring for Attacks Mitigation for multi-hop WSN [11] detects, diagnose and isolates malicious nodes. Neighbor nodes communication is an overhead in UnMask. Based on the gathered information, the authors build a secure routing protocol that provides additional protection against malicious nodes by supporting multiple node-disjoint paths. **Drawbacks**: Overhearing neighbor nodes' communication increases the overhead and communication cost. Multiple attacks were not addressed.

The work by [12] proposed a hierarchical IDS scheme where the network is divided into clusters where each cluster has a cluster head which is responsible for the nodes' sensed data transportation to the base station. Each cluster head also acts a centralized intrusion detection agent in such a way the network is covered by reasonable number of intrusion detection agents. **Drawbacks**: There was no simulation or experimental study conducted to prove the proposed work.

3   Attack Taxonomy

3.1   Denial of Service Attacks

1. Black hole attack

Malicious nodes falsify the route information to attract the neighbor nodes and to capture and drop the packet. All packets routed via these nodes are dropped.

2. Wormhole attack

A tunnel is formed by any two malicious nodes. These nodes advertise to its neighbor nodes as they have the shortest routes to the destinations by falsifying the routing data. Thus, all the nodes' packets are attracted towards their tunnel where anything can be done to the captured data.

3. Gray hole attack

Gray hole attack is a kind of black hole attack where the data packets are dropped or selective forwarded in a statistical manner.

4. Sybil attack

This kind of attacker will spoof their identity either randomly or by using legitimate node's identity to other nodes within the network. This will cause a lot of packets to be routed towards the fake identity nodes which makes severe attacks.

5. Flooding attack

It is a kind of Denial of Service attack where legitimate nodes will be denied of getting their resources at the right time. Network will be preoccupied and flooded by unwanted data or control packets. If the packet flooded is a data packet it is called data flooding; if the packet flooded is a control packet say for example RREQ, it is called RREQ flooding.

4 Proposed Work

4.1 Assumptions

- Homogeneous WSN is deployed in a static manner, so that all the nodes will have the same transmit, receive, and idle power
- Initial energy of sensor node—10 J
- Initial energy of cluster head node—50 J
- Initial energy of sink node—100 J
- Initially all the nodes are assumed to be legitimate. They are compromised externally only after certain amount of time
- Sensor nodes cannot lie about their energy consumption.

4.2 Energy Model

Sensor nodes operate in five different states such as sleep, idle, transmit, receive and calculate. Out of which transmit and receive mode consume high amount of energy. We adopt the energy model proposed by Heinzelman [14] to obtain energy consumption of normal nodes. When k-bit message is transmitted, the energy consumption by the transmitting node is given in Eq. (1)

$$E_{Tx} = E_{elec} \times k + \varepsilon_{amp} \times k \times d^2 \qquad (1)$$

where E_{Tx} is the transmitting energy, E_{elec} is the transmitter and receiver electronics, i.e., energy consumption for sending and receiving each bit, d is the distance between the sender and receiver, and ε_{amp} is the energy consumption exponent. Energy consumption of each sensor node to receive a message is given in Eq. (2)

$$E_{Rx} = E_{elec} \times k \qquad (2)$$

where E_{Rx} is the receiving energy

$$E_{c(k)} = \left(n_{s(k)} \times E_{Tx} \right) + \left(n_{r(k)} \times E_{Rx} \right) + E_{sensing} + E_{idle} + E_{calculate} \qquad (3)$$

where $E_{c(k)}$ is the total energy consumption of sensor node at time k; $n_{s(k)}$ is the number of bits transmitted at time k; $n_{r(k)}$ is the number of bits received at time k; $E_{sensing}$-Energy consumption during sensing; E_{idle}-Energy consumption in idle mode; $E_{calculate}$-Energy consumption during processing of information.

4.3 Modeling of DoS Attacks

Gray hole, Black hole, Sybil, Wormhole, and Flooding attacks are implemented using the steps given below. Usually, the nodes having maximum residual energy are chosen as cluster head.

1. A simple sensor node(N_i) works according to the AODV routing protocol. According to AODV routing protocol, the source node which requires the route to the destination will propagate a route request message (RREQ). Neighboring nodes either forward the RREQ to its own neighbors else if it has a valid active route to the destination, it transmits a route reply message (RREP). Freshness of the route is maintained by destination sequence numbers. If there is any path break, a route error (RERR) message is propagated along the upstream path with infinite hop count to the source node.
2. Nodes launching Flooding attack unwantedly flood the network with large number of route request (RREQ) packets.
3. Nodes launching black hole attack works as follows

   - Send fake route reply (RREP) packets with large sequence numbers
   - Disable the route error (RERR) messages regarding the fake packets to the neighbor nodes
   - Neighbor nodes get falsified route information and thus they forward their packets
   - Malicious nodes receive the packets and drop the packets.

4. Nodes launching gray hole or selective forward attack is a kind of black hole except that the malicious nodes drop the packets either only for a particular interval of time or from a particular source node. So the above steps are repeated for a certain amount of time interval.
5. Nodes launching wormhole attack works as follows

   - Send fake route reply (RREP) packets with large sequence numbers
   - Disable the route error (RERR) messages regarding the fake packets to the neighbor nodes
   - Neighbor nodes get falsified route information and thus they forward their packets to malicious wormhole node
   - Packets are thus tunneled to the unknown node (Wormhole receiver).

6. The nodes launching Sybil attack

   - Capture any one node and falsifies its node id;
   - Receives all the packet by spoofing its identity to neighbors.

4.4 Learning-Based Energy Prediction Algorithm

Nodes are programmed to unicast their residual energy after a particular interval of time (say 50 s) to their CH. Upon receiving the residual energy, the actual energy consumed by the node is calculated as follows in (4). The algorithm is already devised in [13] by myself. This is the extended work.

$$\text{Actual energy } (E_1(v)) = \text{Initial energy} - \text{Residual energy} \quad (4)$$

where $E_1(v)$ is the energy consumed by the node v in the first interval

$$E_0(v) = 0 \quad (5)$$

where $E_0(v)$ in Eq. (5) is the energy consumed by the sensor node at simulation start time of 0 s. At time instant 0, the network will just be formed, there will not be any sensing transmitting or receiving process taking place, so at that time instant, the energy consumption of sensor nodes is zero.

$$E_K(v) = e_k \quad (6)$$

where e_k is the energy consumed by the node v during $T_{k-1,k}$ as shown in (3). e_k is the energy, the cluster heads obtain from all nodes at kth interval(say 100 s). Upon having E_0 and E_k, where $k = 1, 2, 3, \ldots$; the predicted energy consumption is calculated as follows in (7)

$$\text{Predicted Energy}(E_{k+1}(v)) = e_k + \emptyset (E_k(v) - E_{k-1}(v)) \quad (7)$$

$\emptyset$ (Weight Factor) is a parameter used to balance "past" and "current" energy consumption included in the prediction energy consumption. Small value of $\emptyset$ indicates past energy consumption emphasis is higher than current energy consumption. Conversely, large value of $\emptyset$ emphasis current energy consumption. Specifically, if we take $\emptyset = 1$, no past energy consumption contributes to $E(v) k$. The above process is briefly explained in the form of algorithm below

Algorithm
For(i=1;i<N;i++) /*N=Total number of sensor nodes within the cluster*/
    {
    Let k=0, 1, 2, 3... be the time instants;
    Let $E_{k(i)}$ be the energy consumed at time instant "k" by node i;
    Let $E_{0(i)}$= 0;
    $E_{k(i)}=e_{k(i)}$;
    Predicted Energy($Ep_{k+1(i)}$)=$E_{k(i)}$+$\emptyset$($E_{k(i)}$–$E_{k-1(i)}$);
    }

Usually, learning-based prediction is accompanied by prediction error. Weight factors need to be adjusted accordingly to reduce the error between the desired and predicted output. Initially by default 0.5 is set as weight factor Ø and prediction is carried out. Prediction error is calculated using the formula given below

$$\text{Error}_{k(i)} = \left| \text{Actual Energy Consumed}_{k(i)} - \text{Predicted Energy}_{k(i)} \right| \qquad (8)$$

In order to bring the prediction error minimal, the weight factor is calculated using the formula given below, thus from Eq. (7)

$$\emptyset = \left(\text{Actual Energy}_{k+1(i)} - F_{k(i)} \right) / \left(E_{k(i)} - E_{k-1(i)} \right) \qquad (9)$$

where Ø is the weight factor, $E_{k(i)}$ be the energy consumed at time instant "k" by node i, and $E_{k-1(i)}$ be the energy consumed at time instant "$k - 1$" by node i. From Eq. (9), weight factor is tuned according to the actual energy consumption. Thus, the weight factor calculation is carried out for all the sensor nodes. Simulations are carried out to adjust the weight factor under various scenarios and tabulated in the simulated results section. The algorithm is lightweight in nature because it requires only the past two inputs to predict the future output at various time instants. Prediction error can also be maintained within an optimum range by suitably adjusting the weight factor. Prediction accuracy also depends upon the time interval chosen. The simulations were carried out at an interval of 20 s. Thus, for every 20 s, prediction of nodes' energy consumption takes place.

4.5 Detection of Malicious Nodes

Routing tables maintained by the cluster heads are exchanged and updated every time a new CH gets elected. Initial energy E_{i1} of the cluster member nodes is known to the corresponding cluster heads. Cluster member nodes will send a packet indicating their residual energy E_{r1} at the end of first time interval, from which the actual energy consumption is calculated as follows

$$E_{c1} = E_{i1} - E_{r1} \qquad (10)$$

CH predicts sensor nodes' energy consumption for the next interval (50–100 s), denoted as E_{p1} by using the proposed learning-based energy prediction algorithm. Energy prediction for the further consecutive intervals $E_{p(i)}$ is calculated and placed in a routing table. After receiving the residual energy $E'_{r(i)}$ from all sensor nodes for the consecutive intervals, the actual energy consumption is

$$E_{c(i)} = E_{r(i)} - E'_{r(i)} \qquad (11)$$

A comparison between predicted and actual energy consumption is carried out, if there is a mismatch between $E_{p(i)}$ and $E_{c(i)}$ then the node is regarded as a malicious node and the type of attack is differentiated as follows. When this scheme detects abnormal energy consumption of a sensor node, the cluster head node identifies the node id launching the attack and isolates it from the network. The same process is carried out by the sink node to identify the malicious cluster heads. The above energy comparison is made among the cluster head. If any cluster head is found with abnormal energy consumption, it is marked as an attacker node and isolated from the network.

4.6   Categorization of Attacks

The difference between actual and predicted energy consumptions (Prediction Error) of every node is calculated using the equation given below.

$$E_i = \text{Actual Energy Consumed}_i - \text{Predicted Energy}_i \tag{12}$$

$$T1 = \min\left(\text{Error}_{k(i)}\right) \tag{13}$$

$$T2 = \text{average}\left(\text{Error}_{k(i)}\right) \tag{14}$$

$$T3 = \max\left(\text{Error}_{k(i)}\right) \tag{15}$$

$$T4 = 2\left(\text{Error}_{k(i)}\right) \tag{16}$$

$$T5 = M\left(\text{Error}_{k(i)}\right) \tag{17}$$

In Eqs. 13, 14 15, 16, and 17, $\text{Error}_{k(i)}$ refers to Eq. 8. Error values of all the nodes are aggregated and minimum, maximum, and average of all the nodes' error are formulated and set as thresholds $T1$, $T2$, $T3$, $T4$, $T5$. Attackers are classified by comparing every node's error value with the minimum, maximum, and average error values as follows.

If $0 < E_i \leq T1$, then sensor node i is regarded as a malicious one launching a gray hole attack. In gray hole attack, the attacker node selectively drops certain number of packets, so its transmission becomes less. The prediction error is very less than the expected minimum value hence it is a gray hole node which consumes energy less than the legal one.

If $T1 < E_i \leq T2$, then the sensor node i is the legal one. The prediction error is greater than the minimum and maintained within the average value.

If $T2 < E_i \leq T3$, then sensor node i is regarded as a malicious one launching a flooding attack. The prediction error is greater than the average value and more or less equal to maximum error value thus contributing high energy consumption in the network. Unwanted data or control packets are transmitted with high transmission energy. Thus, flooding node consumes higher energy than the legitimate node.

If $T2 < E_i \leq T4$, the sensor node i is maliciously attacked by wormhole attacker as it almost consumes twice the amount of energy as that of legal node; since in wormhole attack the node has to forward the packet it received to the indented neighbor and also to the malicious wormhole receiver, thus consuming twice the amount of energy as that of legal one.

If $T4 < E_i \leq T5$, then the sensor node i is attacked by Sybil attacker. In Sybil attacks, malicious nodes will have multiple M identities. Thus, the node may consume M times of energy as that of legal one.

If $T5 < E_i \leq T3$, then the sensor node i is regarded as the one attacked by the black hole attacker, where the error value lies somewhere between Sybil node and flooding node's error value. Thus, the nodes detected as malicious are added in the blacklist of the routing table by the cluster heads and sink and thereby they are removed from the network. No routing process takes place via malicious nodes.

4.7   Network Lifetime Estimation

Having the knowledge of above energy prediction algorithm, the lifetime of sensor nodes can also be predicted as follows. As stated above in the learning-based energy prediction algorithm

$$\text{Predicted Energy}(E_{k+1}(v)) = e_k + \emptyset(E_k(v) - E_{k-1}(v)) \tag{18}$$

From the above equation, the energy consumption of nodes at various intervals can be predicted; let us consider two consecutive intervals such as K and $K + 1$. The difference between these energy values gives the actual energy consumed by the node i at that particular interval.

$$H_{(k)i} = E_{(K+1)i} - E_{(K)i} \tag{19}$$

Let us assume the iteration interval of this estimation is 50 s. Thus, for every 50 s, node consumes $H(K)$ amount of energy. By iterating the intervals, node may drain its energy.

$$\text{Initial Energy} = H_{(k)i} * \text{Iteration interval} \tag{20}$$

Total available initial energy of the sensor node is calculated by multiplying the energy consumed at say (50–100) s by iteration interval. Knowing the values of $H_{(k)i}$ and initial energy, the interval at which the node deplete its total energy is formulated

$$\text{Iteration interval} = \text{Initial Energy}/H_{(k)i} \tag{21}$$

Lifetime of the node (in secs) = (Iteration interval $*$ Preset interval (in sec)) (22)

This difference multiplied by the iteration interval equals the total available energy value of the sensor node. For better understanding, lifetime obtained in seconds is converted into minutes

Lifetime of the node (in mins) = (Iteration interval $*$ Preset interval (in sec))/60 (23)

Network Estimation Algorithm
for(i = 1;i < N;i++) /*N = Total number of nodes within the cluster*/
{
$H_{(k)i} = E_{(K+1)i}-E_{(K)Ii}$
Iteration interval = Initial Energy /$H_{(k)i}$ /* Determination of the time interval at which it may drain its energy completely*/
Lifetime of the node (in mins) = (Iteration interval*Preset interval(in sec))/60
}

5 Simulation Results

NS-2 Mannasim Framework [15] is used to simulate the proposed work. Network of 10 and 20 number of sensor nodes with one sink, access point, and 2 cluster heads is deployed. AWK scripts were used to analyze the trace files. Script generator tool of Mannasim framework is used to configure the network parameters as stated in Table 1. Simulation configuration parameters are in Table 1.

Table 1 Simulation parameters

Parameters	Values	Parameters	Values
No. of nodes	10, 20	Initial energy (sink and access point)	100 J
No. of sink and access point	Each 1	Initial energy (cluster head)	50 J
No. of cluster heads	2	Initial energy (nodes)	10 J
No. of attackers	Totally 14; Sybil-3; Wormhole-2; Black hole-3; Flooding-3; Gray hole-3.	Initial energy (sensor node)	5 J
Routing protocol	AODV	Sensing interval	5 s
MAC layer	MAC/802.11	Disseminating interval	20 s
Physical layer	Phy/ wirelessphy-mica2	Simulation time	150 s

5.1 Energy Prediction Results

Based on the proposed energy prediction algorithm, appropriate weight factors are formulated according to the equation under various scenarios. Ideal, black hole, gray hole, wormhole, Sybil and flooding attacks were simulated. Under each scenario, networks with 10, 20, 30, 40, 50 nodes were deployed and attackers were modeled. Weight factors were obtained from each scenarios and their average value is taken. Thus, obtained weight factors are tuned according to each attack scenario (Table 2).

Prediction accuracy is based on weight factor determination. By fixing the above values as weight factor for the proposed energy prediction algorithm, the prediction accuracy is very high with a minimal error. Accuracy is proved by comparing the actual energy consumed with predicted energy consumption of nodes using the weight factor obtained. Under each scenario, comparison is made and the error between actual and predicted energy consumed is minimal in the range of 0.02.

5.2 Attack Detection

Thus by using the proposed system, the five different types of attacks are classified according to their error value. Sample terminal output is shown in Figs. 1 and 2.

5.3 Performance Analysis of Hierarchical WSN with Flooding, Black Hole, Gray Hole, Wormhole, Sybil, Proposed IDS, Proposed Lifetime Prediction Algorithm and Without Attack

(1) **Effective Throughput**: Throughput refers to successful transmission and delivery of packets. It is measured in bits/second. Throughput of the network is reduced (greatly affected) under black hole attack scenario. Other attacks too reduce

Table 2 Weight factor determination

No. of nodes	Ideal scenario	Black hole	Gray hole	Flooding	Sybil	Wormhole
10	0.7	0.6	0.68	0.9	0.5	0.87
20	0.5	0.5	0.5	0.99	0.54	0.74
30	0.5	0.8	0.8	0.98	0.46	0.56
40	0.8	0.8	0.8	0.95	0.67	0.67
50	0.8	0.9	0.8	0.98	0.59	0.6
AVG	0.66	0.72	0.73	0.96	0.55	0.68

Fig. 1 Detection of flooder node in terminal

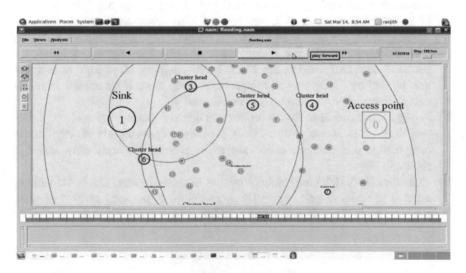

Fig. 2 Detection of flooding node in Nam window

the throughput but the significance is less. Proposed algorithms also do not degrade the performance of the throughput thus proving its efficiency as shown in Fig. 3.

(2) **Packet delivery ratio (PDR)**: PDR is the ratio between number of packets received to the number of packets sent. Ideally, PDR is 100%. PDR of different attacks are shown in Fig. 4, similar to throughput, PDR of black hole becomes

Fig. 3 Throughput

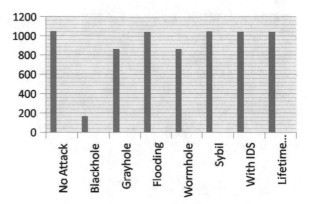

very less. Ideally, the PDR is maintained even under the proposed algorithms. Proposed work does not affect the PDR. Out of the other attacks implemented, Sybil and wormhole reduce the PDR significantly after black hole attack.

(3) **Average end-to-end delay**: The overall time taken by the node to transmit its data and control packets is called average end-to-end delay. It is the sum of transmitting, forwarding, and queuing delays. Delay is expected to be minimum in a network. From Fig. 5 Delay is very high under Sybil, black hole and flooding attack scenario, where flooded packets blocks the resources and delay the transmission, and Sybil creates multiple false identities and reroutes the packets unwontedly and under black hole, dropping of packets occurs frequently and causes routing table updates thereby these three attacks increases delay. Proposed work incurs delay considerably.

(4) **Energy Consumption**: Sensor nodes are usually battery operated; thus they may drain off their energy soon. Thus, the energy consumption of nodes due to computation and communication are to be monitored periodically and it is shown in Fig. 6.

(5) **Detection Ratio(DR) and False Positive Detection ratio**: DR is the ratio of number of nodes detected correctly as malicious to the total number of malicious nodes in the network. False positive is used to describe the number of

Fig. 4 PDR

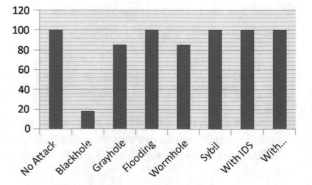

Fig. 5 Delay

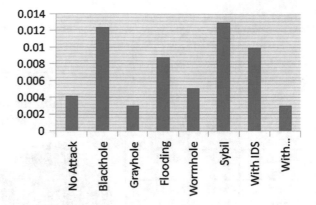

Fig. 6 Energy consumption

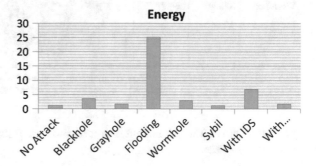

innocent sensor nodes incorrectly identified as malicious nodes. Randomly chosen 5 and 10 number of nodes are made to initiate the attack and their corresponding detection ratio and false positive were analyzed below. Upon using the energy prediction algorithm, malicious nodes are identified successfully in the network with a maximum detection ratio of 4:5 and falseof 0.1.

Thus among the other routing layer attacks, flooding attack affects the network's lifetime severely. Network lifetime computation algorithm proposed does not affect the network performance.

5.4 Network Lifetime Computation

Lifetime of the sensor nodes is predicted ahead. How long each node will stay alive under each attack scenario is obtained from the terminal output. Iteration interval value is obtained for each from the proposed algorithm and tabulated as follows. Iteration value is obtained for all nodes. Only the average value of all the nodes is

Table 3 Lifetime of attacker nodes

	Iteration value	Network lifetime (min)	
Ideal nodes	627.45	522.87	
Black hole nodes	423.51	352.92	
Gray hole nodes	432.53	360.54	
Flooding nodes	12.53	10.44	
Sybil nodes	367.87	306.55	
Wormhole nodes	320.26	266.88	

Thus from the above graph lifetime of sensor nodes are predicted and it is observed that apart from ideal nodes (legitimate) gray hole nodes have a longer lifetime than the other attacker nodes

presented below. The calculation is preceded by assuming initial energy of a sensor node as 5 J (Table 3).

6 Conclusion

Five different types of DoS routing layer attacks are detected successfully using the proposed energy prediction algorithm. Lightweight learning-based energy prediction is implemented and a comparison between the predicted and actual energy is carried out. Prediction is accurate with determined weight factors thereby detection accuracy is reasonable in the range of 4:5. Also, the network lifetime computation algorithm predicts the lifetime of different attacker and legitimate nodes well ahead thus giving a template for the designers of various WSN applications.

References

1. Ilvas, M & Mahgoub, I: Handbook of Sensor Networks: Compact Wireless and Wired Sensing Systems, CRC Press, (2004).
2. Sharma, P & Trivedi, A: An Approach to Defend Against Wormhole Attack in Ad Hoc Network Using Digital Signature. In: Proceedings of IEEE International Conference on Communication software and Networks, pp. 307–31(2011).

3. Abduvaliyev, A, Pathan, AK, Zhou, J, Roman, R & Wong, WC: On the Vital Areas of Intrusion Detection Systems in Wireless Sensor Networks. In: Proceedings of IEEE Communication Surveys and Tutorials 15(3),1223–1237(2013).
4. Mohi, M, Movaghar, A & Zadeh, PM: A Bayesian game approach for preventing DoS attacks in wireless sensor networks. In: Proceedings of WRI International Conference on Communications and Mobile Computing, pp. 507–511(2009).
5. Su, WT, Chang, KM & Kuo, YH: eHIP: an energy-efficient hybrid intrusion prohibition system for cluster-based wireless sensor networks. Computer Networks, 51(4), 1151–1168 (2006).
6. Han, G, Jiang, J, Shen, W, Shu, L & Rodrigues, J: IDSEP: a novel intrusion detection scheme based on energy prediction in cluster-based wireless sensor networks. IET Information Security journal, 7(2), 97–105(2013).
7. Zahariadis, T, Leligou, HC, Trakadas, P& Voliotis, S: Trust management in wireless sensor networks. European Transactions on Telecommunication, 21(4),386–395(2010).
8. Shen, ZW & Zhu, YH: An Ant Colony System Based Energy Prediction Routing Algorithms for Wireless Sensor Networks. In: Proceedings of 4th International Conference on Wireless communication, networking and Mobile Computing, pp 1–4(2008).
9. Xiao, B, Yu, B&Cao, C: CHEMAS: identify suspect nodes in selective forwarding attacks. Journal of parallel distributed computing, 66(11),1218–1230(2007).
10. Hsieh, MY, Huang, YM &Chao, HC: Adaptive security design with malicious node detection in cluster-based sensor networks, Computer Communications, 30(11), 2385–2400(2007).
11. Khalil, I, Bagchi, S& Rotaru, CN: UnMask: utilizing neighbor monitoring for attack mitigation in multi hop wireless sensor networks. Ad Hoc Networks, 8(2), 148–164(2009).
12. Strikos, AA: A full approach for intrusion detection in wireless sensor networks, Wireless and Mobile network architectures, School of Information and Communication Technology, 2007.
13. Dharini, N, Ranjith, B, Pravin Renold, A: Distributed Detection of Flooding and Grayhole attack in hierarchical Wireless sensor networks. In: Proceedings of IEEE International Conference on Smart Technologies and Management for Computing, Communication, Controls, Energy & Materials, pp 192–198(2015).
14. Heinzelman, A, Chandrakasan, H, Balakrishman: Energy-efficient communication protocol for wireless microsensor networks. In: Proceedings of Hawaii International Conference of System Sciences, pp. 3005–3014 (2000).
15. NS-2 Mannasim framework information available from: http://www.mannasim.dcc.ufmg.br/ msg.htm and http://www.nsnam.com.

On Static Control of the Differential Evolution

Lukas Tomaszek and Ivan Zelinka

Abstract In this article, we review how to capture the swarm and evolutionary algorithm dynamic into a network, and we show how we can cover the dynamical network into a coupled map lattices. The main part of this article focuses on the control of the algorithm via the coupled map lattices. Mainly, we concentrate on the statical control and the differential evolution algorithm. In the experimental part, we show that we can successfully lower or raise the control property. All the experiments are done on well-known CEC 2015 benchmark functions.

Keywords Differential evolution · Networks · Coupled map lattices Control

1 Introduction

Networks provide us with a powerful tool for a data analysis. We can use them for analyzing the world wide web [1], brain connectivity [12], social interactions [14, 19], and for many other datasets.

In our research, we are attempting to capture a run of swarm or evolutionary algorithm (SEA) into a network. Network vertices represent individuals (particles, fireflies), and edges capture interactions between them. As an interaction example, we can mention leader-individuals migration in the self-organizing migrating algorithm or crossing between individuals in the differential evolution (DE). More about the conversions, you can find in [21].

Given networks may be analyzed by a well-developed theory about networks [2, 3, 9, 10]. For example, we can calculate global properties like diameter, average

L. Tomaszek · I. Zelinka (✉)
Department of Computer Science, FEI VSB Technical University of Ostrava, Ostrava, Czech Republic
e-mail: ivan.zelinka@vsb.cz

L. Tomaszek
e-mail: lukas.toamszek@vsb.cz

© Springer Nature Singapore Pte Ltd. 2019
M. A. Bhaskar et al. (eds.), *International Conference on Intelligent Computing and Applications*, Advances in Intelligent Systems and Computing 846,
https://doi.org/10.1007/978-981-13-2182-5_2

19

path length or clustering coefficient to determine which network, and consequently which run of an algorithm, is better. Alternatively, we can count metrics like degree centrality, closeness centrality, or eigenvector centrality to differentiate between good and bad individuals.

With the information about the network, we can make the algorithm better and faster. For example, we can improve the algorithm by replacing or removing the individuals from the population, or by changing the algorithm parameters. Some improvements were presented in [8, 17].

Also, we can convert the network into a coupled map lattices (CML) [13]. Each row of the CML is understood as a time development of a vertex from a network. We can analyze and study given CMLs for a different kind of behavior (deterministic and chaotic regimes, intermittency) or we may attempt to control it [20, 22].

In this article, we want to focus on the phase of static control of the DE algorithm. The main aim is to describe how we can convert the DE algorithm into the CML and show how the CML look like and demonstrate that it is possible to change the dynamic of the algorithm by controlling it.

2   Differential Evolution

The DE is an optimization algorithm, which belongs to the class of algorithms called evolutionary algorithms. It was presented in 1995 by Storn and Price [15, 16]. The canonical DE consists of four steps: initialization, mutation, crossover, and selection. Initialization is done only once at the beginning. Next three steps are repeated in loops called the generations.

2.1   Initialization

In this step, we have to generate an initial population. This population consists of NP individuals, and these individuals are represented by vectors. Each vector (individual) represents one possible solution. To be able to differentiate between individuals in various generations, the following notation is used $\vec{x}_i^g = \left(x_{i,1}^g, x_{i,2}^g, \ldots, x_{i,D}^g\right) \cdot \vec{x}_i^g$ is individual I in generation g. Also, the parameters j of one solution of the optimization problem. $x_{i,j}^g$ is parameter j of individual i in generation g. Also, parameter j of one solution of the optimization problem. D is the dimension of the optimization problem. Also, the size of the individual.

Each optimization problem may have some requirements for input parameters. For example, length and weight are always positive. With regards to this, each optimization problem has two boundaries represented by vectors: minimum $\vec{x}_{\min} = \left(x_{\min,1}, x_{\min,2}, \ldots, x_{\min,D}\right)$ and maximum $\vec{x}_{\max} = \left(x_{\max,1}, x_{\max,2}, \ldots, x_{\max,D}\right) \cdot$

$\vec{x}_{min}$ ($\vec{x}_{max}$) is the minimum (maximum) boundary of the optimization problem, and $x_{min,j}$ ($x_{max,j}$) is the minimum (maximum) value of parameter j.

When we know the boundaries, we can initialize all individuals according to

$$x_{i,j}^0 = x_{min,j} + rand_{i,j}[0, 1] \cdot \left(x_{max,j} - x_{min,j}\right). \tag{1}$$

$x_{i,j}^0$ is parameter j of individual i at the beginning of the algorithm. $rand_{i,j}[0, 1]$ is a uniformly distributed random number lying between 0 and 1. This number is different for each individual and each parameter.

2.2 Mutation

After the initialization is done, the generation loops start. In each loop, DE creates a mutant vector for each individual in the population. There are several strategies, how we can create this vector. The most common strategies, we can see below:

$$rand/1 : \vec{v}_i^g = \vec{x}_{R_1^i}^g + F \cdot \left(\vec{x}_{R_2^i}^g - \vec{x}_{R_3^i}^g\right), \tag{2}$$

$$best/1 : \vec{v}_i^g = \vec{x}_{best}^g + F \cdot \left(\vec{x}_{R_1^i}^g - \vec{x}_{R_2^i}^g\right), \tag{3}$$

$$current\text{-}to\text{-}best/1 : \vec{v}_i^g = \vec{x}_i^g + F \cdot \left(\vec{x}_{best}^g - \vec{x}_i^g\right) + F \cdot \left(\vec{x}_{R_1^i}^g - \vec{x}_{R_2^i}^g\right), \tag{4}$$

$$best/2 : \vec{v}_i^g = \vec{x}_{best}^g + F \cdot \left(\vec{x}_{R_1^i}^g - \vec{x}_{R_2^i}^g\right) + F \cdot \left(\vec{x}_{R_3^i}^g - \vec{x}_{R_4^i}^g\right), \tag{5}$$

$$rand/2 : \vec{v}_i^g = \vec{x}_{R_1^i}^g + F \cdot \left(\vec{x}_{R_2^i}^g - \vec{x}_{R_3^i}^g\right) + F \cdot \left(\vec{x}_{R_4^i}^g - \vec{x}_{R_5^i}^g\right). \tag{6}$$

$\vec{v}_i^g$ is the mutual vector for individual i. $\vec{x}_{R_1^i}, \vec{x}_{R_2^i}, \ldots, \vec{x}_{R_5^i}$ are randomly selected individuals from the population, which are mutually exclusive. $R_1^i \neq R_2^i \neq \cdots \neq R_5^i \neq i$. $\vec{x}_{best}^g$ is the best individual in the population. The individual with the best fitness (with the lowest cost function value for a minimization problem). F is the control parameter called scaling factor.

2.3   Crossover

Next step is named crossover. In this phase, we create a trial vector $\vec{u}_i^g$ for each population member. This new vector (offspring) is created from the original individual $\vec{x}_i^g$ and the mutant vector $\vec{v}_i^g$ by mixing its components. The DE algorithm commonly uses two crossover schemes: binomial and exponential.

For the binomial crossover, we generate a random number between 0 and 1 for each parameter of the individual. If the generated number is lower than or equal to the parameter CR (crossover rate), we insert into the offspring the parameter from the trial vector. Otherwise, we use the parameter from the original individual. The binomial scheme can be expressed according to:

$$\vec{u}_{i,j}^g = \begin{cases} \vec{v}_{i,j}^g & \text{if } j = K \text{ or rand}_{i,j}[0,1] \leq \text{CR}, \\ \vec{x}_{i,j}^g & \text{otherwise.} \end{cases} \tag{7}$$

$\vec{u}_i^g$ is the trial vector for individual i. K is a randomly chosen integer number in $\{1,2,\ldots,D\} \cdot \text{rand}_{i,j}[0,1]$ is a uniformly distributed random number lying between 0 and 1. This number is different for each individual and each parameter. CR is the control parameter called crossover rate.

The exponential crossover is determined by two integer numbers: n and L. The first number n determines the index of the first parameter we choose from the trial vector. This number is chosen randomly from $\{1,2,\ldots,D\}$. The second number L denotes how many parameters we pick from this vector, and we obtain this number by running next pseudo-code:

$$L = 0;$$
do
$$L = L+1;$$
while$(rand[0,1] \leq CR) \wedge (L < D)$

Finally, when we know the numbers n and L, the exponential crossover can be described like this

$$\vec{u}_{i,j}^g = \begin{cases} \vec{v}_{i,j}^g & \text{if } j = n \text{modd}, (n+1)\text{modd}, \ldots, (n+L-1)\text{modd}, \\ \vec{x}_{i,j}^g & \text{otherwise.} \end{cases} \tag{8}$$

2.4 Selection

Last part of each generation is the selection. In the selection process, we choose whether the original individual or the offspring survives into the next generation. The selection process is described by the next equation:

$$\vec{x}_i^{(g+1)} = \begin{cases} \vec{u}_i^g & \text{if } F(\vec{u}_i^g) \geq F(\vec{x}_i^g), \\ \vec{x}_i^g & \text{otherwise,} \end{cases} \qquad (9)$$

where function F returns the fitness of the individual.

Above described version of the DE, the algorithm is canonical. Since 1995 many researchers have presented new strategies and versions with the parameter adaptations. You can find the summary of the developments of the DE algorithm in [5].

3 Networks

Networks [3, 9] are a very powerful tool for a data analysis. A network, or a graph $G = (V, E)$, is a collection of n vertices V (also called nodes or actors) joined by m edges E (also called links, ties or arcs). According to an edge type, we can distinguish several types of networks:

- **Undirected**. In these networks, we do not distinguish the direction of the edges. We cannot determine from which and to which vertex the edge flows.
- **Directed**. These networks are the opposite of undirected networks. We can distinguish from and to which vertex the edge flows.
- **Weighted**. In weighted networks, each edge is assigned a real number w, which denotes the strength of the bonding between connected vertices.

3.1 Network Model for the DE Algorithm

As we have mentioned above, the networks are a very powerful tool for a data analysis. Networks can help us to understand the behavioral of an algorithm. We can discover the weaknesses and strength of the algorithm. We can find out why some strategies work and other not. Also, according to the obtained data from the analysis, we can create more powerful and robustness algorithm.

For our experiments, we proposed a network model [23], which is based on an ant behavioral [6]. Ants possess unique pheromones, and they release them during the traveling for food or materials for an anthill. The pheromones form paths which cross, and the paths create a network. In the network, crossroads are represented by the vertices, and the edges connect the vertices (crossroads), which are joined by a road.

When many ants travel on the same road, this road will contain significantly more pheromones according to other roads, so each road is as big (as strong) as how many ants travel on it. Also, the pheromones are vaporized by the wind, so when ants stop walking on some road, this road disappears. The strength of the road, we represent as an edge weight. This model, we can use for capturing the dynamic of the DE algorithm, but also for other SEA [18].

In the network which captures the dynamic of the DE algorithm, the vertices are individuals in the population, the edges represent the interactions between these individuals, and the edge weights captures the strength of these interactions. If we want to create a network, from this algorithm, we have to follow two simple rules:

1. **Movement of ants, improvement**. If in the selection process, we replace original individual $\vec{x}_i$ by the offspring $\vec{u}_i$, we raise the edges weights from vertices, which represent the individuals used in mutation $\vec{x}_{\text{best}}, \vec{x}_{R_j}$, to the vertex, which represents the offspring $\vec{u}_i$. Note that, in the network, the offspring replace the original individual, so it is represented by the same vertex.
2. **Vaporization of the pheromones**. After each generation loop, we reduce all edges weights by $N\%$.

Given network will be dynamical, weighted and directed. The network will change during the run of the algorithm. Also, we can determine, who were used for improvement, and how strong were the interactions between selected individuals. The obtained networks can be analyzed by well-developed theory about the networks [2, 10].

3.2 Strength

In this article, in the experimental part, we will mainly focus on the property called vertex strength [11]. The strength is an equivalent of degree property, but it is adapted for the weighted networks. Also, if we will work with directed networks, we can distinguish between incoming and outgoing strength (same as the degree).

Vertex strength gives us information about the edges, which incoming to the selected vertex. Strength mainly focuses on how strong is the connection between the selected vertex and other vertices. We can count it as a weights edges sum of all edges joining selected vertex with other vertices. If we take in mind the direction of the edges, we sum only edges pointed in and out of the vertex. Formally, we can define the incoming vertex strength S_i^I of vertex i and outcoming vertex strength S_i^O of vertex i like this:

$$S_i^I = \sum w_{ji}, \tag{10}$$

$$S_i^O = \sum w_{ij}, \tag{11}$$

where w_{ij} is the edge weight between vertex i and vertex j.

4 CML

The CML is a dynamical system with discrete time, discrete space and a continuous state. It was introduced to study spatiotemporal chaos [7]. We are using the CML for capturing the inner dynamic of the SEA. To be able to obtain the CML from the SEA algorithm, we need a dynamical network. After that, we compute a selected property of each vertex after each algorithm iteration, and we insert the counted values of one vertex into one row of the CML. In other words, each row of the CML shows us the property of one vertex, so it captures the behavioral of one individual. In the final CML which capture the run of the DE, the time is given by the generations, space is determined by the individuals, and the state is any network property.

4.1 Control

The CML does not serve only for capturing the dynamic, but we can also attempt to control the algorithm via it. In this article, we will focus on the evolutionary static pinning control [20, 22]. Pinning means that in selected time, we use some pings (some inputs), which change the behavior of the algorithm. Static is when in selected time, we remembered the algorithm state and we try different pings, and we choose the best. Evolutionary means that the evolutionary algorithm finds the pings.

The evolutionary control of the DE composes of two evolutionary algorithms. For the convenience, we call them a master and a slave. The master algorithm will attempt to find the proper pings for the slave algorithm, and the slave algorithm will use these pings for the run, and it will optimize the cost function.

4.2 Slave Algorithm

The function of the slave algorithm stays the same. The main goal is to find an optimal solution of a cost function, but in this case, the algorithm is controlled. It should react to the pings and change the dynamic. One way, how to change the behavioral is to select some individuals for mutation more often than the others. In the controlled algorithm, we choose the individuals for the mutation according to the equation:

$$p_i = \frac{1 - \mathrm{PING}_i}{\sum (1 - \mathrm{PING}_j)}. \tag{12}$$

p_i is the probability that individual i will be selected. PING_i is the input ping on individual i.

If there are no pings, each individual is selected with the same probability, and the algorithm works like the classical version. If we use a negative ping on one individual, the probability change, and this individual will be selected less often, and vice versa, if we use a positive ping, the probability raise, and the individual will be selected more often.

As you can see from the equation, the minimal ping value is −1. With this value, the probability is 0 and individual never be selected. The maximal ping we set on value equals to 3.

4.3 Master Algorithm

The main function of the master algorithm is to find pings which cause that the slave algorithm reaches the required state. We define the cost function of the master algorithm like this:

$$f_{\text{cost}} = \sum (\text{ReqCML}_i^g - \text{RealCML}_i^g)^{2.} \tag{13}$$

ReqCML_i^g is the required state of the CML row after iteration g. For the DE algorithm, it is the required network property value for individual i after generation g. RealCML_i^g is the actual state of the CML row.

5 Experiments

For the experiments, we used CEC 2015 benchmark functions [4]. In this article, we present only the results for 2 functions, but the same experiments we can do on any other function. We picked Rotated High Conditioned Elliptic Function and Hybrid Function 1.

As a master algorithm, we used DE/Rand/1/Bin and we set up parameters like this: *NP = 10, F = 0.5, CR = 0.9, gen = 10*. Dimension is equal to the population of the slave algorithm. As a slave algorithm, we also used DE/Rand/1/Bin and we set up parameters like this: *D = 10, NP = 30, F = 0.5, CR = 0.9, gen = 500*.

5.1 Experiment I

In this part, we just want to show, how the CML looks like. In the next Figs. 1 and 2 you can see the CML which capture the property outcoming strength for selected functions.

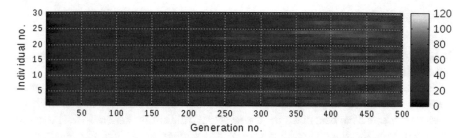

Fig. 1 CML for property outcoming strength and elliptic function

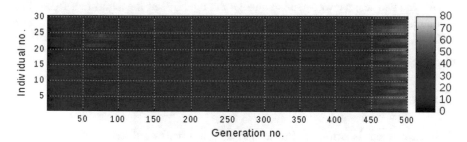

Fig. 2 CML for property outcoming strength and hybrid function

Table 1 Determined average outcoming strength

Function name	Average outcoming strength
Rotated Hogh conditioned elliptic function	54
Hybrid function 1	23

As you can see, the dynamic of the elliptic function differs from the hybrid function. For elliptic function, the property strength growths for all individuals, and at the end strength reaches the maximum value. On the other hand, for hybrid function, the strength fluctuates. At the beginning and the end of the run, it is stronger than in the middle part. Also, the maximal strength is lower according to Elliptic function.

For the next experiments, we determine the average outcoming strength for each cost function from 50 independent runs of the DE algorithm. You can see the reference values in Table 1.

5.2   Experiment II

In this experiment, we attempt to lower the outcoming strength of the individuals by 50%. Also, we want to ensure that all individuals have the same strength.

For the first 10 generations, we did not use any pings. Then, after each generation, we remembered the state of the slave algorithm, and we found, with the master algorithm, the best pings for the next slave algorithm generation. The master algorithm runs the slave algorithm several times. Each time, from the remembered state, and with different pings. The best found pings were applied, and we repeated the process again. We can see the results of the control in Figs. 3 and 4.

As you can see, we were able to change the dynamic of the algorithm. According to classical run, we reduced the outcoming strength by 50%. Also, we were able to hold the same property value for all individuals.

5.3   Experiment III

This experiment is similar to the second one, but instead of lowering the strength, we attempt to raise it by 100%. Also, we want to ensure that all individuals have the same strength. We can see results in Figs. 5 and 6.

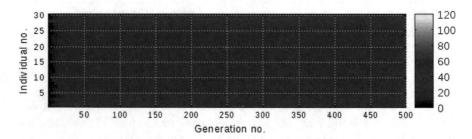

Fig. 3 CML for property outcoming strength and elliptic function—control run

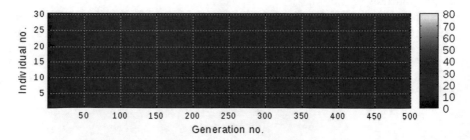

Fig. 4 CML for property outcoming strength and hybrid function—control run

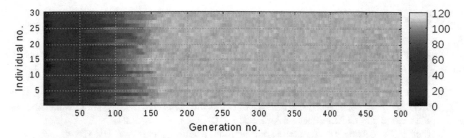

Fig. 5 CML for property outcoming strength and elliptic function—control run

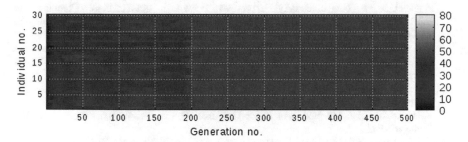

Fig. 6 CML for property outcoming strength and hybrid function—control run

As you can see, we were also able to raise the strength by 100% and hold the same property value for all individuals.

6   Conclusion

In this article, we reviewed, how we can convert a run of the SEA into a network and we show, how we can convert a network into the CML. In this article, we wanted to demonstrate that it is possible to control the DE algorithm by evolutionary static pinning control. As you can saw in the figures in the experimental part, we successfully changed the dynamic, and we were able to lower and raise the strength of the DE algorithm.

The static control showed us that it is possible to change the behavioral of the algorithm, but it brings many unnecessary evaluations of the cost function. On the other hand, now we can investigate how the pings work, and later be able to control the CML dynamically, in a real time.

Acknowledgements The following grants are acknowledged for the financial support provided for this research: Grant Agency of the Czech Republic—GACR P103/15/06700S and by Grant of SGS No. SP2017/134, VSB Technical University of Ostrava.

References

1. Albert-Laszlo Barabasi, Reka Albert, and Hawoong Jeong. Scale-free characteristics of random networks: the topology of the world wide web. Physica A: statistical mechanics and its applications, 281(1):69–77, 2000.
2. Alain Barrat, Marc Barthelemy, Romualdo Pastor-Satorras, and Alessandro Vespignani. The architecture of complex weighted networks. Proceedings of the National Academy of Sciences of the United States of America, 101(11):3747–3752, 2004.
3. Stefano Boccaletti, Vito Latora, Yamir Moreno, Martin Chavez, and DU Hwang. Complex networks: Structure and dynamics. Physics reports, 424(4):175–308, 2006.
4. Q Chen, B Liu, Q Zhang, JJ Liang, PN Suganthan, and BY Qu. Problem definition and evaluation criteria for cec 2015 special session and competition on bound constrained single-objective computationally expensive numerical optimization. Computational Intelligence Laboratory, Zhengzhou University, China and Nanyang Technological University, Singapore, Tech. Rep, 2014.
5. Swagatam Das, Sankha Subhra Mullick, and Ponnuthurai N Suganthan. Recent advances in differential evolution–an updated survey. Swarm and Evolutionary Computation, 27:1–30, 2016.
6. Marco Dorigo, Mauro Birattari, and Thomas Stutzle. Ant colony optimization. IEEE computational intelligence magazine, 1(4):28–39, 2006.
7. Kunihiko Kaneko. Coupled map lattice. InChaos, Order, and Patterns, pages 237–247. Springer, 1991.
8. Pavel Kromer, Miloš Kudělka, Roman Senkerik, and Michal Pluhacek. Differential evolution with preferential interaction network. In Evolutionary Computation (CEC), 2017 IEEE Congress on, pages 1916–1923. IEEE, 2017.
9. Mark Newman. Networks: an introduction. Oxford university press, 2010.
10. Mark EJ Newman. Analysis of weighted networks. Physical review E, 70(5):056131, 2004.
11. Tore Opsahl, Filip Agneessens, and John Skvoretz. Node centrality in weighted networks: Generalizing degree and shortest paths. Social networks, 32(3):245–251, 2010.
12. Mikail Rubinov and Olaf Sporns. Complex network measures of brain connectivity: uses and interpretations. Neuroimage, 52(3):1059–1069, 2010.
13. Eckehard Scholl and Heinz Georg Schuster. Handbook of chaos control. John Wiley & Sons, 2008.8.
14. John Scott. Social network analysis. Sage, 2017.
15. Rainer Storn and Kenneth Price. Differential evolution a simple and efficient adaptive scheme for global optimization over continuous spaces. international computer science institute, berkeley. Berkeley, CA, 1995.
16. Rainer Storn and Kenneth Price. Differential evolution a simple and efficient heuristic for global optimization over continuous spaces. Journal of global optimization, 11(4):341–359, 1997.
17. Lukas Tomaszek and Ivan Zelinka. On performance improvement of the soma swarm based algorithm and its complex network duality. In Evolutionary Computation (CEC), 2016 IEEE Congress on, pages 4494–4500. IEEE, 2016.
18. Lukáš Tomaszek and Ivan Zelinka. Conversion of soma algorithm into complex networks. In Evolutionary Algorithms, Swarm Dynamics and Complex Networks, pages 101–114. Springer, 2018.
19. Stanley Wasserman and Katherine Faust. Social network analysis: Methods and applications, volume 8. Cambridge university press, 1994.
20. Ivan Zelinka. Investigation on evolutionary deterministic chaos control–extended study. Heuristica, 1000:2, 2005.

21. Ivan Zelinka. On mutual relations amongst evolutionary algorithm dynamics and its hidden complex network structures: An overview and recent advances. Nature-Inspired Computing: Concepts, Methodologies, Tools, and Applications: Concepts, Methodologies, Tools, and Applications, page 215,2016.
22. Ivan Zelinka, Roman Senkerik, and Eduard Navratil. Investigation on evolutionary optimization of chaos control. Chaos, Solitons & Fractals,40(1):111–129, 2009.
23. Ivan Zelinka, Lukas Tomaszek, and LumirKojecky. On evolutionary dynamics modeled by ant algorithm. In Intelligent Networking and Collaborative Systems (INCoS), 2016 International Conference on, pages 193–198. IEEE, 2016.

Ensuring Security in Sharing of Information Using Cryptographic Technique

J. Sathya Priya, V. Krithikaa, S. Monika and P. Nivethini

Abstract The issue of security of information is paramount in our society. There are many hackers around us to steal our private information such as passwords. To secure such information, we intend to create ciphertext. Ciphertexts are the words that can only be obtained by authorized persons or users. Ciphertexts can be obtained by performing encryption on the plaintexts. Ciphertexts do not have any meaning. These texts can only be understood after decryption. Our work shows that the information can be retrieved by the authorized person in a few milliseconds whereas it takes much time for an unauthorized person or a third party. In order to perform encryption, we have proposed an algorithm in such a way that it combines many algorithms such as the Caesar cipher, mono-alphabetic cipher, hill algorithms. Different algorithms are applied to individual cases like a plain text that is in uppercase, lowercase, special characters, and numbers. For each case, we have used different algorithms with different keys. We hope that we can suggest an approach for protecting the private information at a higher level.

Keywords Caesar cipher · Hill cipher · Mono-alphabetic cipher
Encryption · Decryption

1 Introduction

Cryptography is a method of writing secrets. It was discovered in ancient times. Cryptography was started from a way back to 1900 B.C. There is an argument that the application of cryptography was started at the battle times plans. After the invention of computers and computer communications, many new forms of cryptography techniques had emerged. Cryptography plays a vital role in any data and telecommunications where the communications are over any untrusted medium.

J. Sathya Priya (✉) · V. Krithikaa · S. Monika · P. Nivethini
Department of IT, Velammal Engineering College, Chennai, India
e-mail: sathyapriya@velammal.edu.in

© Springer Nature Singapore Pte Ltd. 2019
M. A. Bhaskar et al. (eds.), *International Conference on Intelligent Computing and Applications*, Advances in Intelligent Systems and Computing 846,
https://doi.org/10.1007/978-981-13-2182-5_3

Cryptography can be explained as a method or technique where a message is encoded into another form which can only be understood by the receiver [1].

2   Existing Algorithms

2.1   *Caesar Cipher*

Caesar cipher was invented by Julius Caesar and it is the oldest cipher technique that has been ever known. It involves the substitution of original letters with the letters that are x places from the original letter [2].

Example:

Key: 4
Plain text: the meeting is canceled
Ciphertext: QIIXMRK MW GERGIPPIH
The above example can be explained by
Plaintext: a b c d e f g h i j k l m n o p q r s t u v w x y z
Ciphertext: E F G H I J K L M N O P Q R S T U V W X Y Z A B C D

A	B	C	D	E	F	G	H	I	J	K	L	M
E	F	G	H	I	J	K	L	M	N	O	P	Q

N	O	P	Q	R	S	T	U	V	W	X	Y	Z
R	S	T	U	V	W	X	Y	Z	A	B	C	D

So, the algorithm goes as:

$$C = E(4, p) = (p + 4) \bmod 26$$

So, the algorithm, in general, can be expressed as (Fig. 1):

$$C = E(K, P) = (P + K) \bmod 26. \tag{1}$$

Fig. 1 Caesar cipher

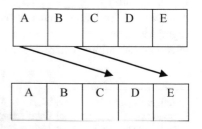

2.2 Hill Cipher

Lester S Hill invented Hill cipher in 1929. It is based on the concepts of linear algebra. In this technique of encryption, each and every letter is multiplied by an N * N matrix [3, 4]. After the multiplication, the result is then taken modulo against 26. The key to this encryption technique is N * N matrix. This key matrix is chosen in such a way that it is invertible [2].

$$n_1 = (k_{11}m_1 + k_{12}m_2 + k_{13}m_3) \bmod 26$$
$$n_2 = (k_{21}m_1 + k_{22}m_2 + k_{23}m_3) \bmod 26$$
$$n_3 = (k_{31}m_1 + k_{32}m_2 + k_{33}m_3) \bmod 26.$$

2.3 Mono-Alphabetic Cipher

Mono-alphabetic cipher is an encryption technique that depends on the fixed substitution or replacement structure [2]. In this cipher technique, each and every substitution letter is fixed and cannot be changed. Hence, the key is itself is given as an array of fixed letters that has to be substituted in original place (Fig. 2).
    Example:
    The key can be selected in any way. Hence, it is infinite in number with different combinations (Fig. 3).

3 Proposed Algorithm

3.1 Algorithm for Encrypting Numbers

Here, we have altered the Caesar cipher for the number encryption. Instead of adding a key value, we have multiplied the key with the ASCII value of the character and have done the modulo operation with 13. Thus, the equation goes as follows:

A	B	C	D	E	F	G	H	I	J	K	L	M
Q	W	E	R	T	Y	U	I	O	P	A	S	D

N	O	P	Q	R	S	T	U	V	W	X	Y	Z
F	G	H	J	K	L	Z	X	C	V	B	N	M

Fig. 2 Hill cipher

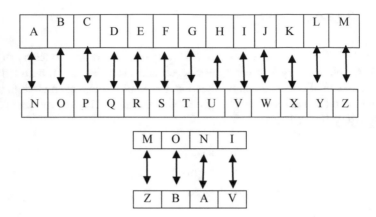

Fig. 3 Mono alphabetic cipher

$$a = (a * keynum)\%13 \tag{2}$$

where *a* is the ASCII value of the number and *keynum* is the key that we have entered for number encryption.

3.2 Algorithm for Encrypting Lowercase Alphabet

Since the situation here is to encrypt a letter rather than a block of letters we use 1 * n matrix instead of n * n matrix. Hence, it reduces the complexity and the time consumption. The formula or the cryptographic technique goes as:

$$t = t + (\text{int}) c * keyhill[i], \tag{3}$$

where *keyhill*[i] is 1 * n matrix that is 1 D array.

Here we encrypt the messages that are in lower case. Hence for converting the ASCII value back into lower case letters, we add 97 which is the starting ASCII value for lower case letters.

$$t = (t\%26) + 97. \tag{4}$$

3.3　Algorithm for Encrypting Uppercase Alphabet

Here, mono-alphabetic Cipher technique is used for encrypting higher case letters. So the encrypted letter can be obtained by using the below Eq. (5)

$$c = keymono[\text{index}] \tag{5}$$

where *keymono* [index] is the key for mono-alphabetic cipher encrypted is first has to be converted to its ASCII values. Then, 65 are to be subtracted from the value which is the ASCII value of the starting letter of uppercase letter. Then, the corresponding key is substituted for the letter that has to be encrypted.

$$\text{Index} = a - 65 \tag{6}$$

3.4　Algorithm for Encrypting Special Characters

We have used the Caesar cipher for encrypting the special characters.

The ASCII value of the special character is added with the *keyspecial* which is the key that we have entered for special character encryption. Then, the equivalent character for the resultant ASCII value is entered in the output file.

4　Advantages of Proposed System

In ancient times, the encryption was done for each and every word with only one technique. This made the encryption brutally weak. Hence these techniques are ignored. In our work, we have proposed an encryption technique for encrypting each and every letter with different algorithms. This makes the encryption stronger. It reduces the time complexity. It will be a great deal for any unauthorized user to access our data. Confidentiality, authentication, non-reputation, and data integrity can also be achieved by this algorithm.

5　Snapshots

See Fig. 4.

Fig. 4 **a** Input file, **b** data entry page, **c** output file

6   Conclusion

Nowadays, our personal and business information are vulnerable to many cyber attacks. The innovative approach presented in this paper can increase the security of such information. In future, the decryption algorithm can also be included and hence the scope of the algorithm can be increased.

References

1. R. Hwang, "Adigital Image Copyright Protection Scheme Based on Visual Cryptography", *Tamkang Journal of Science and Engineering*, vol. 2, no. 2, pp. 97–106, 2000.
2. S. Lee, J. Han, H. Bae, "The encryption method to share a secret binary image and its decryption system", *Proceedings of the 1st International Symposium on Information and Communication Technologies Dublin Ireland ACM International Conference Proceeding Series*, vol. 49, pp. 94–00, 2003.
3. M. Naor, A. Shamir, "Visual Cryptography", *Advances in Cryptography-Eurocrypt 94*, vol. 950, no. 7, pp. 1–12, 1995.
4. Andrey Bogdanov, Dmitry Khovratovich, Christian Rechberger, 2011, Biclique Cryptanalysis of the Full AES, Crypto 2011 cryptology conference, Santa Barbara, California.
5. National Bureau of Standards "Data Encryption Standard" FIPS-PUB, 46, Washington, D.C., Jan 1977. http://csrc.nist.gov/publications/fips/fips46-3/fips463.pdf.
6. Ching-Nurg Yang, "A Note on Efficient Color Visual Encryption", *Journal of Information Science and Engineering*, vol. 18, pp. 367–372, 2002.
7. M. Nakajima, Y. Yamaguchi, "Extended Visual Cryptography for Natural Images", *Journal of WSCG*, pp. 303–310, 2002.
8. Shirish Sabnis, Color cryptography using substitution method, IRJET Volume: 03 Issue: 03 | march 2016.
9. William Stallings, Cryptography and Network Security, 6th edition Pearson Education 2014.
10. https://practicalcryptography.com/ciphers/hill-cipher/.
11. Rosen, K. Elementary Number Theory and its applications. Addison Wesley Longman, ISBN 0-321-20442-5.M, 2000.

Efficient Collaborative Key Management Protocol for Secure Mobile Cloud Data Storage

L. Shakkeera and A. Saranya

Abstract Mobile Cloud Computing (MCC) is the assimilation of mobile computing, cloud computing and wireless networks to make mobile thin client devices
resource-rich in terms of storage, memory computational power, and battery power
by remotely executing the wide range of mobile application's data in a pay-per-use
cloud computing environment. In MCC, one of the primary concern is the security
and privacy of data managed in cloud. The existing techniques fail to manage key
security during key generation and key distribution processes. The primary objective
of this project work is to develop an efficient collaborative key management protocol
for secure mobile cloud data storage by implementing file encryption, key generation, key encryption, key decryption, key distribution, and file decryption methods.
In our proposed methodology, we are getting Storage as a Service (SaaS) for
accessing the secure data from DriverHQ public cloud infrastructure. The secret keys
are generated by using Pseudorandom Number Generator (PRNG) algorithm that
uses mathematical way to produce sequences of random numbers. The algorithm
achieves the security for the file data and key by eliminating key escrow and key
exposure problems. The input files are encrypted and decrypted by using Advanced
Encryption Standard (AES) algorithm. The AES algorithm is more vulnerable
against exhaustive key search attack. So, the proposed method achieves data confidentiality and data integrity in mobile cloud data. And also the work reduces
encryption and decryption computation and storage overhead in thin client mobile
devices, and minimizes the energy consumption of the mobile devices efficiently.

Keywords Mobile cloud computing · Data storage · Key management
SaaS · PRNG · Confidentiality · Integrity

L. Shakkeera (✉) · A. Saranya
Department of Information Technology, School of Computer,
Information and Mathematical Sciences, B.S. Abdur Rahman
Crescent Institute of Science and Technology, Chennai, Tamilnadu, India
e-mail: lshakkeera@crescent.education

A. Saranya
e-mail: saran.santhosh8@gmail.com

© Springer Nature Singapore Pte Ltd. 2019
M. A. Bhaskar et al. (eds.), *International Conference on Intelligent Computing
and Applications*, Advances in Intelligent Systems and Computing 846,
https://doi.org/10.1007/978-981-13-2182-5_4

1 Introduction

1.1 Mobile Cloud Computing

MCC [1] is the combination of mobile computing, cloud computing [2, 3] and wireless networks to bring resource-rich computational resources to mobile end users, network operators, as well as cloud service providers. The main purpose of MCC is to execute high-end applications like gaming, image processing, video streaming applications with rich end user experience.

The advantage of MCC brings business chances for cloud service providers and network operators. It also provides services to secure their data for small developers [4]. In future, the mobile technology combines the advantages of mobile computing and cloud computing to easy access to applications in resource-constrained cloud environment. Hence, it enhances the optimal services in the form IaaS, PaaS, SaaS, and DaaS services for mobile end business users.

1.2 Key Management

Key management [5, 6] is one of the main aspects of cryptographic system. It involves key generation, key distribution, key protection in the form key encryption and key decryption, key storage, key exchange with authorized users, key replacement, key monitoring, and key recording in log files for the use of future cryptanalysis.

Key management is the process of secure administration of public, private, secret, and master keys. The key is a string of binary zeros and ones and uses in cryptographic algorithm for encryption (ciphertext) and decryption (plaintext) processes. The algorithms provide data confidentiality, data integrity, message authentication codes (MACs), digital signatures, non-repudiation, and entity authentication. Improper management of keys leads to threats in accessing mobile application that increases business hazards.

2 Related Work

Jia and Wang [7] proposes the cloud storage auditing with key exposure resilience for the key update problem. The work discusses the verifiable outsourcing of key updates in cloud storage auditing protocol. In this protocol, key updates are outsourced to the Third-Party Authority (TPA) and are transparent for the client. TPA only sees the encrypted version of the client's secret key, while the client can further verify the validity of the encrypted secret keys when downloading them from the TPA. The work deals with key and data security with less time consumption. In meanwhile, the TPA is semi-trusted for key updating.

Pratima and Rutuja [8] discuss the authority-based data access control policy, it is only the user with particular authority can access the data stored at cloud storage, based on their authority level. The Hierarchical threshold Access Structure (HTAS) is used along with Distributed Key Generation (DKG), known as Hierarchical threshold Distributed Key Generation (HTDKG). The various levels of users are there based on their designation and some threshold is assigned for each level. At the time of data access, only specific levels users can access the specific amount of data. Because of this, all data is not revealed to any unconcern user. The third-party Key Distribution Center (KDC), which executes the HTDKG protocol and take responsibility for all key generation, distribution, and management activities. This system decreases the processing time and enhances the memory utilization by utilizing KDC. The advantages of this paper are to reduce the processing time and all levels of user can access the data in cloud, but the restriction is to use the data at all time.

Patranabis and Shrivastava [9] focuses an efficiently implementable version of key aggregate cryptosystem. The model provides as the solution for several applications like collaborative data sharing and product license distribution. It allows the users to decrypt the data using single key with constant size and broadcast efficiently to numerous users. The Key Aggregate Cryptosystem (KAC) is constructed and implemented by using elliptic curves. It is suitable to cloud-based data sharing systems. Here, the data confidentiality is achieved, but a large space is required for data storage.

Teja and Hemalatha [10] discuss a framework that utilizes Advanced Encryption Standard (AES) encryption using USB gadget. The records are getting from the cloud storage; documents are to be scrambled till the USB gadget is connected to the PC. The proposed framework recognizes the USB that contains the private key utilized for the records to be downloaded from the cloud. The paperwork is to provide complete security for the records, text files to be uploaded and downloaded at any time, file encryption and decryption. The work does not support images, audio and video files and every time the need of secret code to access the files in cloud infrastructure.

Kumar [11] proposes the cryptographic key generations, digital signatures, and authentication protocols. The proposed work contains modification to Dual Elliptic Curve Deterministic Random Bit Generator (Dual EC DRBG) and proposes a composite Cryptographically Secure Pseudorandom Number Generator (CSPRNG) using modified Dual EC DRBG. They compared the output against the standard Pseudorandom Number Generators (PRNGs) like Linear Congruential Generator (LCG), Blum Blum Shub (BBS) and modified Dual EC DRBG. The main advantages are giving more security for the data and generation and the cycle length is high. But, the model is slow in generating pseudorandom bits hence it is not suitable for application which demands greater speed.

3   Existing System

Key authority ought to be completely reliable, as it can decrypt all the ciphertext using generated secret keys without the permission of the data owner. This problem is called as key escrow problem. Mobile front-end devices like smartphones and thin clients are far more susceptible than cloud servers with reverence to privacy protection. Thus, the susceptibility in secret key protection may easily lead to the exposure of keys to unauthorized users. In accumulation, current key management techniques also require much bilinear pairing, exponentiation, and multiplication calculation especially in the decryption process. It increases the complexity of the key management system. And also the energy consumption of the mobile devices increases due to the massive amount of data processing, computation, storage, transmission, reception, encryption, and decryption processes. To overcome the problems mentioned, our proposed key management and storage system satisfies both the end users and cloud service providers.

4   Proposed System

4.1   Key Management Protocol

The proposed system focuses on efficient collaborative key management protocol focusing to improve security and efficiency of key management in mobile cloud data sharing system. The main aim and objectives of the paperwork are as follows:

- A novel collaborative key management protocol is proposed. With help of that secure key management is guaranteed which is easier to deploy and compare with previous multi-authority schemes.
- Key generation is done by Pseudorandom Number Generation (PRNG) algorithm and provides the security while transmission of both the data and key. PRNG is a mathematical formula to produce sequence of random number, so it does not predict the intruder.
- Random key is allocated to each attribute group that contains clients who share the common attribute via updating the attribute random key. The fine-grained and immediate attribute revocation is provided.
- PRNG key generation algorithm to achieve data confidentiality, backward and forward secrecy, revocation, avoid key escrow, and exposure problems.
- The proposed mechanism perfectly addresses both key escrow and key exposure problems.
- The collaborative mechanism helps distinctly reduces the client decryption overhead by including decryption server to execute most of the decryption process.

4.2 Proposed Methodology

Figure 1 depicts the block diagram of an efficient collaborative key management system. It enhances the security and efficiency of key management in mobile cloud data sharing system. For providing security for data files, the system uses Advanced Encryption Standard (AES) algorithm for encryption and decryption processes and Pseudorandom Number Generator (PRNG) algorithm for key generation. In PRNG, each time the new random key is generated, so the intruder cannot get the data files. PRNG algorithm achieves data confidentiality, backward and forward secrecy, revocation, avoids key escrow and exposure problems. Key escrow is nothing but without permission of user to access the data and key exposure is the try to access data without all the keys. Both AES and PRNG algorithms efficiently eliminate key escrow and key exposure problems.

Pseudorandom Number Generator (PRNG). PRNG algorithm uses mathematical formulas to produce sequences of random numbers. PRNG generates a sequence of random numbers. The input key length is divided into two 4 bytes of each. PRNG combines bitwise logical operation, exchange of bits position and bit manipulation in order to fulfill the confusion and diffusion mechanism. The output of key is given to the input of the second round. The main usage of random numbers to avoid replay attacks, cryptographic key generation, uses in stream ciphering. So, the intruder cannot easy to capture the secret key.

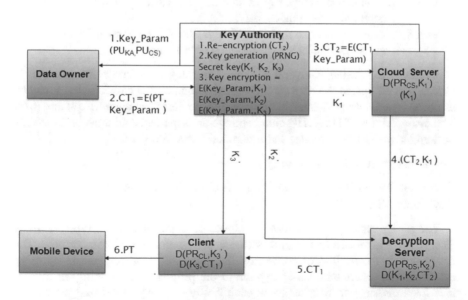

Fig. 1 Block diagram of the proposed system. *Key_Param* Key parameters, PU_{KA} Key authority's Public key, PU_{CS} Cloud server's Public key, *PT* Plaintext, CT_1 Ciphertext 1, CT_2 Ciphertext 2, K_1, K_2, K_3 Secret keys, *E* Encryption process, *D* Decryption process, K_1', K_2', K_3' Encrypted secret keys, PR_{CS} Cloud server's Private key, PR_{DS} Decryption server's Private key, PR_{CL} Client's Private key

PRNG algorithm.

Step 1   At first, 8 bytes of seed (initial) key is selected as input to PRNG generator.

Step 2   The seed key bits are converted into zeros and ones binary representation. Let, assume the given key consists of n-bits. Then the length of the key is 4*n bytes.

Step 3   The bitwise XOP operation is performed on bits of each successive bytes. 1st $\oplus$ 2nd replaces the 1st byte. 2nd $\oplus$ 3rd replaces the 2nd byte, etc. The process continues till the last byte where the last byte ('n') is XORed with 1st byte nth $\oplus$ 1st.

i.e., 1st byte—11011000
2nd byte—00101100
New 1st byte—1 1 1 1 0 1 0 0

Step 4   The successive bytes are shuffled with each other in pair. Even though 'n' is the last number, then the last byte is left unaltered.

Step 5   The final bit string of 8 * n bytes can be taken as the first pseudorandom random key 'K1'. The resulting exchanged string of bytes is divided into two halves, left and right halves of each 4 * n bytes.

Step 6   The first and last bytes are left unaltered and intermediate bytes are exchanged.

Step 7   The generated secret key in step 5 is fed back as an input to step 2 to generate the next random key.

Step 8   In order to generate more secret keys, the Steps 3 to 6 can be repeated as many as required by each mobile applications.

Cloud setup using DriverHQ. The DriveHQ is offering file storage in cloud. We can upload the files from personal computer to a cloud folder. It can be easily accessed, managed, shared, or published the files from anywhere at any time. The main features of DriverHQ cloud storage are superfast data transfer, cloud file backup DriveHQ file manager, and integrated group file sharing.

File Encryption in Key Authority

Input: Key_Param (PU$_{KA}$, PU$_{CS}$), Plaintext (PT), Output: Ciphertext (CT$_1$, CT$_2$)
Key_Param: (PU$_{KA}$, PU$_{Cs}$)

For giving security for the data in cloud, store the data in encrypted format. Encryption and re-encryption processes are done before cloud data storage. The encryption is the method of converting the data into a code/ciphertext by using cryptographic algorithms. Re-encryption is the process of encrypting an already encrypted message or information in such a way that only authorized parties can access the messages in the input files. In our work, the key authority is responsible for file encryption using Key_Param. The Key_Param is the combination of key authority's public key and cloud server's public key. It is the most effective way to achieve data security in public cloud environment. The output of the module gives encrypted form of file. Figure 2 depicts file encryption in key authority.

Key Generation Using Pseudorandom Number Generator (PRNG) Algorithm
Input: Seed Key, Output: Secret keys (K_1, K_2, and K_3)

Pseudorandom Number Generator (PRNG) algorithm generates a sequence of random numbers as secret keys. Initially, the seed key value is given as input for key generation. The seed key is considered as starting value to generate the sequence of pseudorandom integer values. By using the process of PRNG algorithm, the random keys are generated while executing the mobile application data files. In our work, we are generating K_1, K_2, and K_3 as secret keys. Figure 3 shows the key generation using PRNG algorithm.

Key Encryption and Key Distribution
Input: K_1, K_2, and K_3 and Output: K_1', K_2', K_3'

Key authority encrypts the secret keys with Key_Param (PU_{KA}, PU_{CS}) and distributes the encrypted secret keys to cloud server (K_1'), decryption server (K_2') and client (K_3'). The key encryption process achieves security for the key during the transmission. Figure 4 explains key encryption and key distribution.

Key Decryption and File Decryption
Input: K_1', K_2', K_3 and K_1, K_2, K_3' and CT_1, CT_2 and Output: K_1, K_2, K_3 and CT_1, PT

The encrypted secret keys are decrypted in cloud server, decryption server, and client using its respective private keys. PR_{CS}, PR_{DS}, and PR_{CL} are private keys of cloud server, decryption server, and client, respectively. Figure 5 shows key decryption in cloud server, decryption server, and client.

The file is decrypted in decryption server and client. In the decryption server, the re-encrypted file (CT_2) is decrypted using collaborative secret keys (K_1, K_2). The ciphertext (CT_1) is decrypted using secret keys (K_3) to retrieve plaintext (PT). Figure 6 shows file decryption in the decryption server and client.

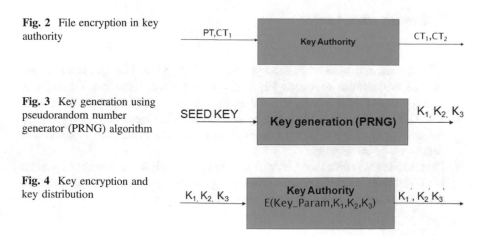

Fig. 2 File encryption in key authority

Fig. 3 Key generation using pseudorandom number generator (PRNG) algorithm

Fig. 4 Key encryption and key distribution

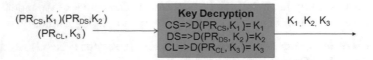

Fig. 5 Key decryption

Fig. 6 File decryption

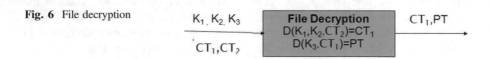

5 Experimental Results and Discussions

5.1 System Setup

The proposed key management protocol is developed using software front tools like HTML and JSP in NetBeans IDE 8.1. For backend design, JDK 1.8.0 and database connectivity with MySQL 5.0.22 are used. The connectivity between application server, database server, and user interface is connected with Apache Tomcat 5.0/6.X.

DriveHQ storage tool is used for mobile data storage in public cloud infrastructure. DriveHQ file manager makes remote storage for data files. The file manager transfer, access, share, synchronize, collaborate, and publish files remotely. The files are drag and drop to DriverHQ remote storage and also drag and drop files and folders using DriverHQ file manager, FTP and WebDAV. using web browser, The free space available for each user login is 1 GB. Nearly, 2–20 MB file is uploaded in cloud storage space. The complete system is implemented in Windows 7 64 bit operating system with high-end hardware configurations.

5.2 Results and Analysis

- AES algorithm is used for file encryption and decryption. The data and key size are 128 bits, number of rounds 10. AES is stronger and faster than Triple-DES. The key size of AES is high, so the intruder cannot access the key or cloud data.
- PRNG is used for the random key generation. PRNG algorithm has to generate the random key for every time interval. The random keys are used to secure the data in cloud environment.
- DriveHQ is the public cloud service for data storage. It is the one the important IaaS—Storage service provider. The storage is easily accessible, manageable, sharable or publishable files from anywhere and at any time.

Data Integrity Versus Number of files. Figure 7 shows the data integrity of the proposed system when the number of files is accessed from mobile devices to cloud and cloud to mobile devices. The graph clearly shows that data integrity is achieved when numbers of files are increased. Initially, the system considers 50 files with the combination of text, image, audio, and video files. The integrity of data is achieved using PRNG algorithm. PRNG algorithm generates sequence of random keys as secret keys for each data file access. This eliminates the intruder to access the files from cloud. Meanwhile, the key authority encrypts the data by using public key of cloud server and key authority. In cloud environment, the files are stored in encrypted form. This avoids the intruder cannot easily get the data, So, files cannot be altered at anytime.

Data Confidentiality Versus Number of files. Figure 8 shows the data confidentiality of proposed system when a number of files are accessed from mobile devices to cloud and cloud to mobile devices. The authorized users can access the data at anytime. Initially, the system considers 50 files are considered for data access. In our proposed work, data owner, key authority, and cloud server are assigned with unique username and password. The security access is not stored in any storage spaces. Meanwhile, the authorized users are uploading their data files into public cloud storage without the intervention of third-party authorities. The data confidentiality intern achieves data integrity in cloud.

Data Confidentiality Versus Number of users. Figure 9 shows the data confidentiality of proposed system when numbers of users are varying in mobile cloud. Initially, the system considers 50 users to access the data from DriverHQ cloud storage. Only the authorized users are accessing the cloud data by using secret key generated by key authority. The authorized users can access the cloud data with help of random key, each time a new key is generated by key generation algorithm, data confidentiality is achieved when the number of users is vary.

Fig. 7 Data integrity versus number of files

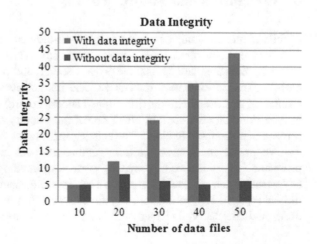

Fig. 8 Data confidentiality versus number of files

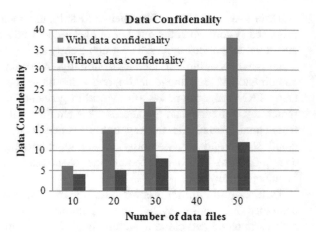

Fig. 9 Data confidentiality versus number of users

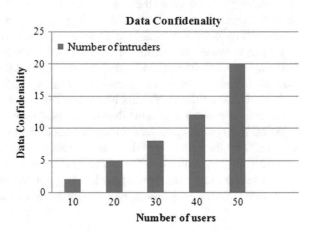

6 Conclusion and Future Work

Mobile cloud computing is cost-effective, on-demand network access, sharing of data, more convenient data access from anywhere and anytime. Therefore, enforcing the safeguard against the private, secret, and perceptive data stored in the mobile cloud is extremely imperative. The instantaneous participation of a large number of end users requires fine-grained access control policies for data sharing in mobile cloud environment. In MCC, one of the primary concerns is the security and privacy of data stored in cloud. The existing techniques fail to manage key and data. The proposed mechanism perfectly addresses both key escrow key exposure problems. The objective of proposed work is to develop an efficient collaborative key management protocol for secure mobile cloud data storage by implementing file encryption, key generation, key distribution (key encryption and key decryption), and file decryption methods. For the data storage, DriverHQ public cloud infrastructure is used. The secret keys are generated by using PRNG algorithm. It produces sequence of random numbers as secret keys. The algorithm achieves the security for the file data and key by eliminating key escrow and key exposure problems. Advanced Encryption Standard (AES) algorithm used for file encryption and decryption. Data integrity in mobile cloud storage is achieved and also reduces encryption and decryption computation and storage overhead in client mobile devices, and minimizes the energy consumption of the mobile devices. The future work focuses on variation of PRNG algorithm to increase the complexity to generate the random key number. The work will build on to extend the proposed scheme by reducing the ciphertext size, encryption and decryption cost, which are the open challenges that hinder practical application of key management scheme in mobile cloud data storage.

References

1. Tranos Zuva.: Mobile Public Cloud Computing, Merits and Open Issues. International Conference on Advances in Computing and Communication Engineering, 2016.
2. Amal Ghorbel, Mahmoud Ghorbel, Mohamed Jmaiel.: Privacy in Cloud Computing Environments: a Survey and Research Challenges. Journal of Supercomputing, (6), 2017.
3. Zheng Yan, Vasilako. S: Flexible Data Access Control Based on Trust and Reputation in Cloud Computing. IEEE 5(3), 2017.
4. Jiang Zhang, Zhenfeng Zhang, and Hui Guo.: Towards Secure Data Distribution Systems in Mobile Cloud Computing. IEEE Transactions on Mobile Computing, 16(11), 2017.
5. O Sri Nagesh, Tapas Kumar.: Secure Key Management in Mobile Cloud Computing. International Journal of Innovations & Advancement in Computer Science, 4, 2015.
6. Yan Wang, Zhi Li, Yuxia Sun.: Cloud Computing Key Management Mechanism for Cloud Storage. 3rd International Conference on Cyberspace Technology, 2015.
7. Jia, Cong Wang.: Enabling Cloud Storage Auditing With Verifiable Outsourcing of Key Updates. IEEE transactions on information forensics and security, 11(6), 2016.
8. Pratima, Rutuja, S.: Efficient Hierarchical Cloud Storage Data Access Structure with KDC, In: IEEE International Conference, 2016.
9. Sikhar Patranabis, Yash Shrivastava.: Provably Secure Key-Aggregate Cryptosystems with Broadcast Aggregate Keys for Online Data Sharing on the Cloud, 66(5), 2017.
10. TalariBhanu Teja, Vootla Hemalatha.: Encryption And Decryption – Data Security For Cloud Computing – Using Aes Algorithm, SSRG International Journal of Computer Trends and Technology, 2017.
11. Mahesh Kumar.: Hybrid Cryptographically Secure Pseudo-Random Bit Generator. In. 2nd International Conference on Contemporary Computing and Informatics, 2016.

A Comprehensive Analysis of Key Management Models in the Cloud: Design, Challenges, and Future Directions

S. Ahamed Ali, M. Ramakrishnan and N. Duraipandian

Abstract Cloud computing from the last few years has grown from a promising business concept to one of the fastest growing segments of IT Enterprise. Because of the attractive features offered by cloud computing, many enterprises are using cloud storage for storing essential information. Like any other emerging technologies cloud computing also brings many new security challenges. Thus, numerous researches were done to ameliorate this problem by introducing effective key management solutions to secure the data stored in the cloud. This paper presents a comprehensive survey on the various key management techniques that can be adopted to provide effective security to the cloud-stored data. The primary focus of this survey is to highlight the issues and challenges involved in designing an effective key management solution for the cloud. We have also done a detailed analysis of thematic taxonomy on key derivation hierarchies.

Keywords Cloud computing · Key management · Key derivation hierarchy

1 Introduction

Cloud computing is a technology which is preferred by many organizations to relocate their resources and maintain them outside their enterprise, regardless of the location of the cloud server. NIST (2009) defines cloud computing as a model for providing an enabled, convenient, on-demand, and pay-as-you-use shared

S. Ahamed Ali (✉)
Department of Information Technology, Velammal Engineering College,
Anna University, Chennai, India
e-mail: ahamedali@velammal.edu.in

M. Ramakrishnan
School of Information Technology, Madurai Kamaraj University,
Madurai, India

N. Duraipandian
Velammal Engineering College, Chennai, Tamil Nadu, India

© Springer Nature Singapore Pte Ltd. 2019
M. A. Bhaskar et al. (eds.), *International Conference on Intelligent Computing
and Applications*, Advances in Intelligent Systems and Computing 846,
https://doi.org/10.1007/978-981-13-2182-5_5

collection of computing resources that include applications and programs, storage, servers, services, and networks that can be customized easily with speed and limited management effort. The cloud supports numerous self-recovering scalable hardware and software models that allow applications to recover from all hardware and software failures. Organizations prefer choosing a provider who will offer the maximum amount of low-cost storage and bandwidth, and providing highly secure data storage platform. For example, leading cloud storage providers such as Dropbox offers 2 GB free storage space for storing data and applications and for every 1 TB they charge \$9.99 for a month and iCloud offers 5 GB free storage space for storing data and applications and for every 50 GB they charge \$0.99 for a month. Hence, cloud computing has become a hot topic for research in industries and academia due to the various advantages offered by this technology. Since the data is stored and managed from the remote site various issues like confidentiality, integrity, availability, and access control will occur. Key management is a technique used to address the problem of ensuring the storage security to the data stored in the cloud storage servers. Since key management is an important aspect in the implementation of cloud security, the purpose of this paper is to present a comprehensive survey on the various existing techniques and to delineate the various categories of noteworthy issues and challenges in the cloud security domain. The contributions of this paper are as follows.

1. Introduce the basic concepts of key management in the cloud
2. Analyze the tree-based key derivation mechanisms used to secure the data in the cloud
3. Summarize the various taxonomies used for implementing the cloud security
4. Compare and highlight the different key derivation mechanisms and their features.

The rest of this paper is structured as follows. Section 2 introduces the different categories of key management in the cloud. Section 3 discusses the taxonomy and presents a detailed survey and comparison of various access hierarchy mechanisms that were reviewed. Section 4 brief about tree-based access hierarchy mechanisms. Finally, Sect. 5 concludes the paper.

2   Basics of Key Management

Key management is the most critical component in designing a secure storage system. The most challenging problem for the cloud service provider is to convince the data owner to trust the security model adopted in the cloud server. Even though cloud service provider claims that it provides strong encryption and integrity enforcing cryptographic algorithms it is difficult for the cloud service provider to ensure that data in safe in the cloud server. Encrypting the data before storing it in the cloud is common scheme adopted by data owners to protect their data.

The algorithm used for encrypting the data is made public (known to all the users of the system) but the key used to encrypt the data is kept secret. Thus, the entire security system completely depends on the key. The data stored in the server is safe until the key gets compromised. Hence, the key used for encryption should be kept secret and should be managed efficiently. The following are the commonly used key management schemes.

Centralized Group Key Management Protocols: A single controller managing the entire group. There will be a key server which will be responsible for key distribution and management.

Decentralized Key Management Protocols: In these types of protocols, there will be no group controller. All the members of the group will generate key.

Distributed Key Management Protocols: In this type, there is no group controller as in the case of decentralized key protocols. In this scheme, all the group members or only one member are involved in group key generation.

3 Key Management Based on Access Hierarchies

Key management for access hierarchies was initially introduced by Atallah M. J et al. The hierarchy is modeled as a set of partially ordered classes; the classes are represented as a directed graph. A user who obtains the key for a certain class can also obtain an access key for all the child classes with the help of key derivation. Every class is associated with a private key. In this scheme, hash algorithms are used for a node to derive a descendant's key from its own key. In order to derive the descendant node's access key, the number of operations is linear to the distance between the nodes in the graph. Data updation are handled in the node itself and the changes made in the node will not propagate to its ancestors. The entire access hierarchy is modeled as directed graph $G = \{V, E, O\}$, where V is a set of vertices $\{v_1, v_2, ..., v_n\}$, E is set of edges $\{e_1, e_2, ..., e_n\}$ and O is a set of objects $\{o_1, o_2, ..., o_n\}$. In this model, a vertex v_i represents a class in the access hierarchy and has a set of objects associated with it. In the access hierarchy, a path from node v_i to node v_j indicates that subject has obtained the access rights of class v_i and it can also access all the objects of class v_j. The key allocation graph is represented in Fig. 1. L represents the level in the graph and H represents a one-way collision resistant hash function.

Thus for a given key space K, the key allocation will be V *union* O maps to 2^k objects and access classes. The problem with this approach is it depends on a trusted entity that can generate and distribute keys. The security of this scheme relies on the use of pseudorandom functions. Another problem in this approach is the metadata for block must be transmitted for retrieving the block from the remote server. Further, once one of the keys is compromised all the keys can be easily derived.

Security policies based confidentiality preserving mechanisms for data outsourcing was presented by S. De Capitani et al. They have presented the concept by

Fig. 1 Key access graph

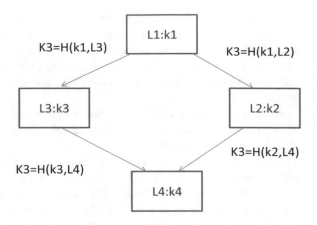

application of selective encryption by protecting the privacy of security policy. In their model, they have constructed a secure system by considering a set of users U, and set of resources R and a policy that grants access to the users U over the resources R. The system was designed using an access control matrix A that has U number of rows and R number of columns. Each entry in the access control matrices A $[u, r]$ is set to 1 if the user U has access rights over the resource R; it is set to 0 otherwise. Figure 2 denotes an access control matrix for three users U (X, Y, Z) over the three resources R (A, B, C). For example ACL $(B) = \{Y, Z\}$.

The data outsourced in the remote server is encrypted with different set of keys, and the decryption key is provided to the users which will allow the user to decrypt only the resources that he/she is authorized to access. The distribution of the decryption keys is done through key derivation scheme with the help of tokens. Each user is provided with a set of keys $K = \{k_1, k_2, \ldots, k_n\}$ and the user need to pick a key $k_i \in K$ and a set of tokens necessary to derive all the other keys. The token is defined as $t_{ij} = k_j \oplus H (k_i, l_j)$, here H is the one-way collision resistant hash algorithm and lj is a label publically available. The advantage of the approach presented by S. De Capitani et al. is the access control policy can be used to define key derivation hierarchy. The problem with this approach is if the access rights for the users are changed the service provider has to perform additional operations. The same issue will also arise if user revocation was done.

Multi-key Strategies for Data Outsourcing was proposed by Damiani et al., in their work, they have considered tuples as resources and they created access control

Access Control Matrix (A)		Resources		
		A	B	C
USERS	X	1	0	1
	Y	1	1	0
	Z	1	1	1

Fig. 2 Access control matrix

Access Control-Matrix(A)	t1	t2	t3	t4
A	1	1	0	1
B	0	1	1	0
C	0	1	1	0
D	1	1	1	1

Fig. 3 Access control matrix of Damiani's approach

mechanisms that are applicable to each row of a database. The access control mechanism is implemented in matrix form for a set of users U and for a set of resources R. Thus, authorization and access control in their scheme is applicable to record, table and database level resources. The access control list A contains with U number of rows and R number of columns. An entry in the access control matrix defines the list of actions that the user U is authorized over the resource R. In their approach, all tuples that share same access control list are encrypted with the same key. Hence the number of keys depending on the number of tuples has the same access control list. For example in Fig. 3, tuple t1 and t4 share the same key. For a user to access a tuple, he/she has to retrieve the encrypted tuple and its label (which have information regarding tuples having the same access control list) and the key.

Damiani et al. defined access control list (acl) for a resource and capability token (cap) for a tuple. From Fig. 3 the acl and cap can be written as Eqs. 1 and 2, respectively.

$$\text{acl}_{t3} = \{B, C, D\} \tag{1}$$

$$\text{cap}_B = \{t2, t3\} \tag{2}$$

To reduce the number of keys Atallah's key derivation method was used, where the elements of the set correspond to users of the system and subset containment property of the set is used to represent the relationship among the tuples. Damiani et al., approach takes acl as a input and produces key derivation hierarchy as output. The problem with this approach is this system is modeled only for users with read access rights. The number of that each user has in his/her key set depends on number of tuples have the same access control list. Hence there is a possibility of having an increased number of key set for a user.

4 Key Management Schemes Based on Tree Based Access Hierarchies

Wang et al. proposed a scheme using key derivation method as shown in Fig. 4 for secure and efficient way of accessing the data outsourced in the cloud server. In their scheme, every data block is encrypted using a private key and they have

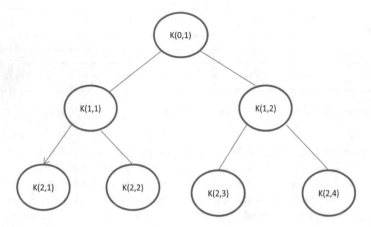

Fig. 4 Key derivation tree

designed a key derivation hierarchy that reduces the number of keys maintained by the data owner. Every key in the key derivation hierarchy is derived using parent node and with some public data. The key derivation model use the one-way collision resistant hash function. In this model, the data owner will maintain only the root node key. In order to access the encrypted data, the data owner will issue a key to users based on their access rights. The user will derive the key from the hierarchy and access the data blocks by decrypting it. The key derivation hierarchy is modeled using binary tree and hash functions.

In the hierarchy, the key $k(i, j)$, i represents the level and j represents the index of the node in the hierarchy. For any node, the left child can be calculated using Eq. 3

$$\left(d_{(i+1)(2j-1)1}, d_{(i+1)(2j-1)2}\right) = \left(H\left(d_{ij1}\|(2j-1)\right), H\left(H\left(d_{ij1}\|(2j-1)\right)\right)\right) \quad (3)$$

and its right child is calculated using Eq. 4

$$\left(d_{(i+1)(2j)1}, d_{(i+1)(2j)2}\right) = \left(H\left(d_{ij1}\|2j\right), H\left(H\left(d_{ij1}\|2j\right)\right)\right) \quad (4)$$

Thus by recursively applying this procedure, the entire key derivation can be completed.

When the leave node level of the binary tree is reached the hash value can be used as keys to encrypt the data blocks. Wang et al., also have provided mechanisms for block level data updates. The advantage of their approach is the outsourcing model is flexible and the data owner needs to maintain only few keys, the other keys are generated using key derivation hierarchy. The storage server is accessed only during updation of data. The major drawback of this method is if one of the keys in the key derivation hierarchy is compromised, then it is possible to derive all the keys enabling decryption of resources that the user is not authorized to access.

Blundo et al., proposed a key derivation hierarchy using a key derivation graph that can generate keys by combining with public tokens. They have proposed a model in which the data owner can develop a flexible policy for controlling the access to the resources. The access control matrix proposed by them is similar to the model proposed by Attalah. In order to reduce the burden of the data owner, selective encryption model was developed. Selective encryption method will relieve the data owner from managing and controlling access control. In selective encryption technique, the encryption policy is associated with the access rights over the data. Thus in this model, the data is encrypted with different keys and users are assigned with a set of keys which are sufficient for decrypting the data for which the user has granted rights. In order to avoid data owners storing and managing many keys, they designed a heuristic algorithm for developing a key derivation hierarchy that reduces the total number of keys to be distributed to users in the system. The heuristic algorithm is based on minimum spanning tree with weight minimization property. The minimum spanning tree T is defined as a user tree $T = \{V, E\}$ where V is the vertex and E is the edge with a weight function can be represented using Eq. 5

- $\quad$ Weight function $\left(W_f\right) \rightarrow: \forall\left(v_i, v_j\right) \in E, W_f\left(v_i, v_j\right) = \left[v_j \cdot \alpha / v_i \cdot \alpha\right]$ $\qquad$ (5)

$v_j \cdot \alpha$ stands for the access control list for vertex v_j and $v_j \cdot \alpha$ stands for the access control list for vertex v_i and w is the weight function. The weight of the entire tree will the sum of weight of all, $w\left(v_i, v_j\right)$. For all the internal vertex in the minimum spanning tree for any pair of vertex v_i, v_j (the children of V), the set of users who can access the V can be computed using Eq. 6

$$U = v_{i \cdot \text{acl}} \cap v_{j \cdot \text{acl}} \qquad (6)$$

This model works only for the system which has users only with read-only rights and the minimum spanning tree they have defined is an NP-hard problem.

Zhou et al. (7) presented an approach for private data management and they termed their approach as owner-write-users-read/write. This approach is based on a hierarchical key management tree that can be shared with or managed by another party without compromising the security. They have used a subtree data manager who holds the root node key, which is used to derive all the child node keys. The key derivation model used in this approach is presented in Fig. 5

The encryption key e_{ij} is associated with the decryption keys d_{ijk}, where i, j, k denotes i denotes the level of the tree, j denotes the index of the nodes in a particular level in the tree, and k denotes the index of the decryption keys. The root node (d_0) in the tree is the master key. Any node N_{ij} except the leaves can derive its child nodes of indices $i(2j - 1)$ (for the left) and $i(2j)$ (for the right). The key value K_{ij} of node N_{ij} is represented by $K_{ij} \leftarrow \left(d_{ij1}, d_{ij2}\right)$ where d_{ij1} denotes the master decryption key and d_{ij2} denotes the secondary decryption key. The one-way hash

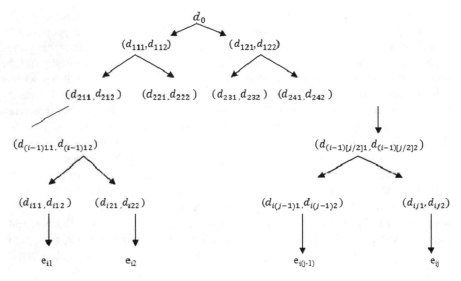

Fig. 5 key derivation model

function can be used to derive d_{ij2} from the input d_{ij1} as $d_{ij2} \leftarrow \mathrm{H}(d_{ij1})$. They have used the polynomial proposed by Mu et al. (1999) to generate key derivation tree. They have used an encryption scheme which designed based on the complexity of ElGamal encryption algorithm. This algorithm also considers "Write" applications and allows the user to re-encrypt the data. The problem in this approach is if the root node key is compromised then all the child keys can be easily derived. The security of their scheme mainly depends on the subtree data manager. The encryption key is given to user when the write operations are performed hence there is an additional overhead on the data manager to check the validity of the encryption data regularly.

5   Conclusion

This comprehensive survey paper has analyzed the various state of art methodologies for cloud-based secure key management. We have categorized the key management techniques as access control and tree-based key derivation mechanisms. We have done a survey on many interesting features of key derivation in both the mechanisms. For most of the work, we have done a detailed analysis and comparisons. We have also found that there is no specific mechanism that addresses all the key management issues and key management opens the door for many challenging security issues.

References

1. Atallah MJ, Blanton M, Fazio N, Frikken KB. Dynamic and efficient key management for access hierarchies. ACM Transactions on Information and System Security 2009; 12:18:1–43.
2. De Capitani di Vimercati S, Foresti S, Jajodia S, Paraboschi S, Pelosi G, Samarati P. Preserving confidentiality of security policies in data outsourcing. In: Proceedings of the seventh ACM workshop on privacy in the electronic society, WPES' 08. New York, NY, USA: ACM; 2008. pp. 75–84.
3. Damiani E, di Vimercati SDC, Foresti S, Jajodia S, Paraboschi S, Samarati P. Key management for multi-user encrypted databases. In: Proceedings of the 2005 ACM workshop on storage security and survivability, StorageSS' 05. New York, NY, USA: ACM; 2005. pp. 74–83.
4. Damiani E, di Vimercati SDC, Foresti S, Jajodia S, Paraboschi S, Samarati P. Key management for multi-user encrypted databases. In: Proceedings of the 2005 ACM workshop on storage security and survivability, StorageSS' 05. New York, NY, USA: ACM; 2005. pp. 74–83.
5. W. Wang, Z. Li, R. Owens, B. Bhargava, Secure and efficient access to outsourced data, in: Proceedings of the 2009 ACM workshop on Cloud computing security, CCSW' 09, ACM, New York, NY, USA, 2009, pp. 55–66.
6. Blundo C, Cimato S, De Capitani di Vimercati S, De Santis A, Foresti S, Paraboschi S, et al. Efficient key management for enforcing access control in outsourced scenarios. In: Gritzalis D, Lopez J, editors. Emerging challenges for security, privacy and trust, IFIP advances in information and communication technology, 297. Boston: Springer; 2009. pp. 364–75.
7. Miao Zhou, Yi Mu, Willy Susilo, Jun Yan Liju Dong, Privacy enhanced data outsourcing in the cloud. In Journal of Network and Computer Applications 35 (2012) 1367–1373, Journal homepage: www.elsevier.com/locate/jnca.
8. https://www.nist.gov/.
9. https://www.dropbox.com/.
10. https://www.icloud.com/.
11. https://www.nist.gov/.
12. https://www.srgresearch.com/.
13. https://www.soliantconsulting.com/.
14. https://www.idc.com/.
15. https://cloudsecurityalliance.org/.
16. https://www.gartner.com/technology/research.
17. https://www.icloud.com/.

Identification of Fronto-Temporal Dementia Using Neural Network

N. Sandhya and S. Nagarajan

Abstract Fronto-Temporal Dementia (FTD) affects the behavior, cognition, and memory retention capacity of the affected people. This study focuses on preprocessing the acquired MRI images and transforming RGB to grayscale image, detection of edge by applying Sobel algorithm, segmenting the image, extraction of features, and feeding to the classifier for classification.

Keywords FTD (Fronto-Temporal Dementia) · MRI (Magnetic resonance Imaging) · BPN (Backpropagation Network)

1 Introduction

Fronto-Temporal Dementia is a syndrome which results from gradual and progressive deterioration of the frontal and temporal lobes of the brain. These areas of the brain are responsible for linguistics capabilities, emotional functioning, personality, behavioral control, executive functioning, visuospatial skills, and making decisions and the patient shows reduced interpersonal conduct [2].

There are two different types of matter in the brain, the former being gray matter forming the outer layer which processes the information and the latter white matter which is the inner core and provides pathway (wiring) through which the information moves along. FTD impairs memory, calculations as well as copying abilities during the later stages of the disease.

N. Sandhya (✉)
Department of MCA, Hindustan Institute of Technology and Science,
Padur, Chennai 603103, Tamil Nadu, India
e-mail: Sandhya.N.Deepak@gmail.com

S. Nagarajan
Department of Mechanical Engineering, Hindustan Institute of Technology
and Science, Padur, Chennai 603103, Tamil Nadu, India
e-mail: Snagarajan1960@gmail.com

© Springer Nature Singapore Pte Ltd. 2019
M. A. Bhaskar et al. (eds.), *International Conference on Intelligent Computing
and Applications*, Advances in Intelligent Systems and Computing 846,
https://doi.org/10.1007/978-981-13-2182-5_6

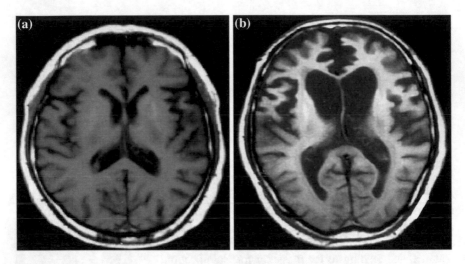

Fig. 1 FTD brain images *Source* Study of Dementia, PubMed Central

An imaging technique called MRI (Magnetic Resonance Imaging) produces good quality human body images. Bore is a tube which runs horizontally through magnet from front till back. A patient lying flat on the back is made to slide into the bore on the table. Depending on the part of the body to be examined, the patient is made to slide either from the head region or the feet region of the body. The scan begins as the region of scanning is in isocenter of the magnetic field. MRI does diagnose many injuries and deformities. MRI also reveals changes to the blood vessels in the white matter region of the brain (Fig. 1).

2 Proposed Architecture

Implementation of system is done as explained below:

1. Acquiring MRI brain images
2. Preprocessing the obtained images which means converting RGB images into grayscale images and resizing followed by detection of the edge using Sobel, Binary Dilation and Histogram equalization, and Thresholding
3. Segmenting image
4. Extracting the features
5. Classification using Artificial Neural Network (Fig. 2).

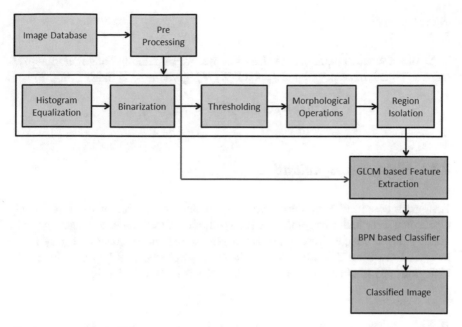

Fig. 2 Proposed architecture

2.1 RGB to Gray Conversion

In a grayscale image, the luminance of a pixel value varies from 0 to 255. To convert a color image to a grayscale image, RGB values (24 bit) are converted into grayscale values (8 bit). RGB images are converted to grayscale by a function by removing the hue and saturation information at the same time retains the luminance.

2.2 Edge Detection: Sobel Algorithm

Images should be convoluted using the Sobel filter having two kernels, X-direction kernel which detects the horizontal lines and Y-direction kernel that detects the vertical lines.

X-direction kernel (size being 3×3)

−1, 0, 1
−2, 0, 2
−1, 0, 1
Y-direction Kernel
−1, −2, −1

0, 0, 0
1, 2, 1

Define a window equaling the size of kernel for calculating convolution at pixel (X, Y). To calculate magnitude at pixel (X, Y), given magX, magY: (Fig. 3)

$$mag = sqrt(magX^2 + magY^2)$$

2.3 Histogram Equalization

Histogram equalization produces the image which has the brightness level evenly distributed over entire brightness. It is a methodology to increase image's contrast making use of image's histogram. Histogram shows the frequency of gray levels ranging from 0 to 255, which means image's pixel size is 8 bits or 1 byte. The histogram has frequency whose values varying from 0 to 255 (Fig. 4).

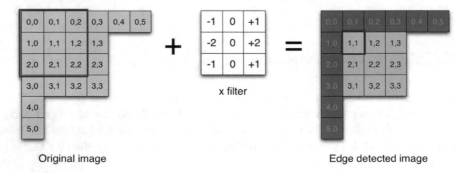

Fig. 3 Edge detection process

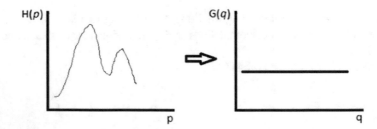

Fig. 4 Histogram equalization

2.4 Algorithm for Histogram Equalization

Step 1: Size of image: PxQ, gray level ranging from 0 to 255 forms array G with size 256 and is initialized with 0.

Step 2: Form a histogram of the image by scanning each pixel and incrementing relevant array member. G[grayval(pixel)] = G[grayval(pix)] + 1

Step 3: Produce cumulative histogram CG having size 256

$$CG[0] = G[0]CG[a] = CG[a-1] + G[a], a = 1, 2, 3, \ldots, 255.$$

Step 4: Set T[a] = Round((255*CG[a])/(PxQ))

Step 5: Rescanning the image and produce new image having grayscale value (Fig. 5).

$$New\ Image[1][m] = T[Old\ Image[l][m]]$$

2.5 Feature Extraction Using Texture Features

Feature extraction is the method of data reduction for getting the set of required parameters out of an image. Here 7 features shall be extracted using 0°, 45°, 90°, 135° angles for the extraction of Haralick features [7]. Then, spatial relationship of a pixel can be known.

The related frequencies Y(r,s,k,a) form a co-occurrence matrix. It depicts how relatively a pixel r is located at a distance s, at a distance k with an orientation a in the space.

From the MRI image input, a textural pattern is obtained.

Texture is spatial arrangement of a group of fixed pixels (also called as texture primitives) forming a simple or most basic pattern.

Texture elements are also called as Textone [4].

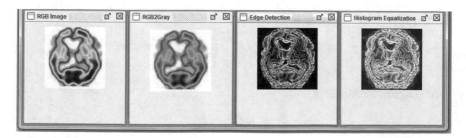

Fig. 5 Image preprocessing operations on an MRI image

Texture feature extraction methods are as follows:

- Statistical Method
- Syntactic Method
- Spectral Method

Statistical Method—these extracted features shall be depicted as vector in the multidimensional space. These are based upon the first order, second order, and higher order statistics of the input. A probabilistic decision making algorithm assigns a class to the feature vectors [5].

Syntactic Method—a full pattern of texture primitives which are arranged spatially are kept following placement rules. A homogeneity is derived between syntax of the language and structural pattern [6].

Spectral Method—spatial frequencies defines the textures and they are evaluated by autocorrelation functions.

The different texture features [3] are as follows:

Angular Second Moment

$$tf1 = \sum_{r,s=0}^{N-1} Y_{r,s}^2$$

Contrast

$$tf2 = \sum_{r,s=0}^{N-1} Y(r,s) * (r-s)^2$$

Inverse Difference Moment

$$tf3 = \sum_{r,s=0}^{N-1} \frac{Y(r,s)}{1+(r-s)^2}$$

Dissimilarity

$$tf4 = \sum_{r,s=0}^{N-1} Y(r,r) * |(r-s)|$$

Entropy

$$tf5 = \sum_{r,s=0}^{N-1} Y(r,s) * [-ln(Y(r,s))]$$

Maximum Probability

$$tf6 = \max r, sY(r,s)$$

Inverse

$$tf7 = \sum_{r,s=0}^{N-1} \frac{Y(r,s)}{(r-s)^2}$$

3 Classification Using Artificial Neural Network

Study makes use of backpropagation neural network [1, 8]. The features are extracted and fed to a neural classifier.

3.1 Back Propagation Network Algorithm

The terms used are:

x: input pattern $(x_1, \ldots, x_i, \ldots, x_n)$

t: output pattern $(t_1, \ldots t_k, \ldots, t_m)$

α: learning parameter

x_i: ith input unit

v_{0j}: jth bias layer

w_{0k}: kth bias layer

z_j: hidden unit

δ_k: weight adjusted for w_{jk} because of error in y_k being transferred back to hidden units connected to y_k

δ_j: adjusted weight for v_{ij} due to the error transferred back to unit z_j

    0: Initialize weights, learning rate.
    1: Repeat Steps 2–9 when stopping condition is not true.
    2: Repeat Steps 3–8 for the pair (x, t).
       Phase I:
    3: Input x_i (i varying from 1 to n) is obtained by input unit and transferred to hidden unit.
    4: Unit z_j (j varying from 1 to p) adding weighted input and calculating total input:

$$z_{inj} = v_{0j} + \sum_i x_i v_{ij}$$

Output is found out using activation function z_{inj} [(0,1) or (−1, +1)]

$$z_j = f\left(z_{inj}\right)$$

Output is transmitted to output layer from hidden layer.

5: Calculate the net input y_k (k varying from 1 to m)

$$y_{ink} = w_{0k} + \sum_{j=1}^{p} z_j w_{jk}$$

Output is computed by the activation function

$$y_k = f(y_{ink})$$

Phase II:

6: The calculation of error correction is δ_k (k varying from 1 to m)

$$\delta_k = (t_k - y_k)f'(y_{ink})$$

On the basis of error correction, weights, and bias are changed and updated:

$$\Delta w_{jk} = \propto \delta_k z_j$$

$$\Delta w_{0k} = \propto \delta_k$$

δk (error correction) should be sent to the previous layer in reverse direction.

7: Inputs are summated at hidden unit:

$$\delta_{inj} = \sum_{k=1}^{m} \delta_k w_{jk}$$

$$\delta_j = \delta_{inj} f'\left(z_{inj}\right)$$

Based on δ_j, weights and bias are updated:

$$\Delta v_{ij} = \propto \delta_j x_i$$

$$\Delta v_{0j} = \propto \delta_j$$

Phase III:

8: y_k shows the updated weights and bias:

$$w_{jk}(\text{new}) = w_{jk}(\text{old}) + \Delta w_{jk}$$

$$w_{0k}(\text{new}) = w_{0k}(\text{old}) + \Delta w_{0k}$$

At each hidden unit, bias and weights should be updated:

$$v_{ij}(\text{new}) = v_{ij}(\text{old}) + \Delta v_{ij}$$

$$v_{0j}(\text{new}) = v_{0j}(\text{old}) + \Delta v_{0j}$$

9: Proceed with the process until the desired outcome and the actual outcome match.

3.2 Decision

The process of Classification usually has error rate. It sometimes does not recognize the demented brain image or incorrectly classifies a healthy brain image as a demented brain image.

Error rate is depicted as follows:

TP (True Positive)—rightly classified by the algorithm as demented

FP (False Positive)—wrongly classified by the algorithm as demented but being normal

TN (True Negative)—rightly classified by the algorithm as normal

FN (False Negative)—wrongly classified by the algorithm as normal whereas the images are demented

The performance measures are listed below.

Sensitivity (True Positive Fraction): the proportion of true positive cases which are rightly classified. Sensitivity is a measure of how best the model diagnoses positive cases.

$$\text{Sensitivity} = \text{TP}/(\text{TP} + \text{FN})$$

Specificity (True Negative Fraction): the proportion of true negative cases that are classified as being normal. Specificity is a measure of how well the model diagnoses the negative cases.

$$\text{Specificity} = \text{TN}/(\text{TN} + \text{FP})$$

Accuracy: the probability of correctly diagnosing the image status (the proportion of right result either TP or TN). Accuracy is a measure of how well it identifies both groups.

$$\text{Accuracy} = \text{TP} + \text{TN}/(\text{TP} + \text{TN} + \text{FP} + \text{FN})$$

Totally, if both Specificity as well as Sensitivity are low (high), accuracy will be low (high). If sensitivity/specificity is low and the other is high, accuracy will be oriented towards either of them. Therefore, accuracy only cannot be a good performance measure.

4 Conclusion

This study converts given color image to a grayscale one, detection of edge is done by Sobel algorithm and enhances the contrast of image using histogram equalization method. Features are extracted using texture methods and the images are fed to the neural classifier. This makes use of MRI database which has both healthy brain and FTD brain images. ANN classifies the images. Implementation is done in MATLAB R2008b with Neural Network Toolbox and Image Processing Toolbox.

5 Scope for Future Study

Different supervised learning algorithms like SVM, K-means, and KNN can be used to obtain and improvise the results. The output of each supervised learning method can be tabulated and the similarities or the differences can be visualized.

References

1. N. Sandhya and Dr. S. Nagarajan, *Frontotemporal Dementia—A Supervised Learning Approach*, Springer ERCICA 2015 Vol 3, https://doi.org/10.1007/978-981-10-0287-8.
2. N. Sandhya and Dr. S. Nagarajan, *Fronto Temporal Dementia - Supervised and Unsupervised Learning Approaches*, International Journal of Control Theory and Applications, Vol 10, Number 16, 2017, ISSN: 0974-5572.
3. N. Sandhya and Dr. S. Nagarajan, *Analysis of Fronto Temporal Dementia Using Texture Features and Artificial Neural Networks, Intelligent Computing and Applications,* Springer Advances in Intelligent Systems and Computing, Vol. 632, 978-981-10-5519-5.
4. Jayashri Joshi and Mrs. A.C. Phadke, *Feature Extraction and Texture Classification in MRI*, IJCCT, Vol. 2 Issue 2, 3, 4, 2010.
5. Fukunaga K., *Introduction to statistical pattern recognition*, 2nd ed. Academic Press; 1990.
6. Pavilidis T., *Structural description and graph grammar.*, In: Chang SK, Fu KS, editors. Pictorial information systems. Springer Verlag Berlin: 1980. pp. 86–103.
7. Haralick RM., *Statistical and structural approaches to texture*, Proc IEEE. 1979;67:786–804.
8. S.N. Sivanandam and S.N. Deepa, *Principle of Soft Computing*, Wiley India Edition, Pages 74–83, 1993.

Strength and Swelling Characteristics of Expansive Soils Treated with Calcium Carbide Residue

K. Divya Krishnan, P. T. Ravichandran and P. Sreekanth Reddy

Abstract The development of infrastructure demands construction sites with soils of superior characteristics since the performance of the structures mainly depends on the soil conditions. Especially the construction over poor quality soil deposits like expansive soils needs almost care because it affects adversely the structural performance by showing high volume change with the influence of moisture content. In such kind of soil, stabilization with suitable pozzolonic binders can provide strong and durable subgrade. In practice, the use of available materials as a binder can provide an effective economical way for the improvement of poor quality soils. Based on this consideration, the uses of waste products generated from industries are being used as a stabilizer in soil improvement becomes an emerging practice in the field of Civil Engineering. In this study, Calcium Carbide Residue (CCR) generated during the production of Acetylene Gas is used to modify the properties of the clayey soil to evaluate its potential usage for the treatment. For this investigation, two soil samples of high compressible clays are treated with different percentages of Calcium Carbide residue (3, 5, 7 and 9%) and its strength and swelling characteristics were determined for the treated soils at varying curing periods of 3, 7 and 14 days. It was noticed that CCR reduced the swell potential and increases the strength characteristics (UCC) of expansive clays with increase in curing periods and percentages of binder. Thus, the use of CCR in expansive soil confirms the cost-effective chemical stabilization which provides more stable and durable soil.

Keywords Calcium carbide residue · UCC · Free swell index

K. Divya Krishnan (✉) · P. T. Ravichandran · P. Sreekanth Reddy
Department of Civil Engineering, Faculty of Engineering and Technology,
SRM Institute of Science and Technology, Kattankulathur 603203, Tamil Nadu, India
e-mail: er.divyakrishnan@gmail.com

© Springer Nature Singapore Pte Ltd. 2019
M. A. Bhaskar et al. (eds.), *International Conference on Intelligent Computing
and Applications*, Advances in Intelligent Systems and Computing 846,
https://doi.org/10.1007/978-981-13-2182-5_7

1 Introduction

Engineering field often concerned about the durability of structures especially when subgrades that exhibits instability due to poor quality of soils. Majorly the sites consisting of expansive soils, required more attention due to its poor strength and instability in volume characteristics due to the influence of environmental factors. Practically, the structures with shallow foundations on weak soils needed a ground improvement technique before the starting of construction practices. Soil modification with the addition of admixtures is an appropriate technology to improve the poor performances of the expansive soil. The effective and economic manner soil stabilization can be achieved by using waste or by-products from large-scale industries which avoids environmental pollution. Large quantities of wastes are generating from various industries, which is abundantly available and it can be utilized in a proper way to avoid its disposal problems. Improving the soil strength with waste or byproduct materials like quarry dust [1], phosphogypsum [2], fly ash [3], bagasse ash [4], and GGBS [5] are extremely cost-effective material for converting weak soil into a suitable construction platform. This study is based on the use of Calcium Carbide Residue (CCR), by-product from acetylene gas industry as an admixture to impart strength to the soil since it consists of more calcium content. The strength properties and swelling characteristics of treated soil samples are studied at different curing periods of 3, 7 and 14 days with different dosages of CCR.

2 Materials

Tests were conducted on two soil samples of expansive characteristics which were collected from two different locations, from Chennai and Andhra Pradesh. Based on the Atterberg's, Free Swell and grain size distribution, both samples are classified as Clay of high compressibility and the test results of both soil samples are tabulated in Table 1. Calcium Carbide Residue (CCR) is the admixture used to stabilize the soil samples which was collected from Andhra Pradesh. Tests on treated and untreated soils are carried out as per Indian Standards [6–10]

The grain size analysis test is performed on soil samples 1 and 2 as per IS 2720-1985 and based on the analysis the soil sample 1 contains 53% clay, 18% silt and 29% sand fraction and for the soil sample, 2 was 65% clay, 14.6% silt and 20.4% sand.

The strength characteristics are studied by conducting Unconfined Compressive Strength tests on both the samples with varying percentages of CCR (3,5,7 and 9%) at different curing periods of 3, 7, and 14 days. Swelling characteristics are also studied based on the Free Swell Index values of untreated and treated soils.

Table 1 Properties of soil samples

Description	Results	
	Sample 1	Sample 2
Free swell (%)	63	65
Liquid limit (%)	52	62
Plastic limit (%)	26	37.58
Plasticity index (%)	26	24.42
Shrinkage limit (%)	10.32	6.61
Specific gravity	2.21	2.26
Maximum dry density (g/cc)	1.67	1.507
Optimum moisture content (%)	18	23.6
CBR value (%)	3.973	4.04
UCC value (kPa)	168	154

3 Discussion of Test Results

3.1 Unconfined Compressive Strength Test

Three specimens were prepared for each dosages of CCR based on the compaction moisture contents to determine the UCS values at 3, 7, and 14 days curing periods. The UCS values obtained for treated samples of soil 1 and 2 at different curing periods are presented in Table 2.

The strength value has increased drastically with the curing period and improvement is obtained maximum of 401 and 384 kPa for both the soil samples with the 5% CCR content. Figures 1 and 2 show the stress–strain variation of both the untreated and treated samples with varying CCR percentages at 14 days of curing time.

The evident for the increase in strength for both the soil samples, where pronounced with the higher content of lime in CCR, which provided a favorable environment for the formation of pozzolanic compounds such as calcium silicate and aluminate hydrate in soil with OMC. These pozzolonic compounds favors stronger soil samples with better bonding between the soil particles. The variation of curing time also effects on the strength development of soil mixtures because it probably gives the reaction time for the soil particles and CCR for forming the bonding compounds.

Table 2 UCS values of soil 1 and 2 with varying percentages of CCR at different curing periods

Days	UCS Value, kPa							
	Soil 1				Soil 2			
	3%	5%	7%	9%	3%	5%	7%	9%
3	172	205	199	181	177	211	202	184
7	218	295	282	253	226	291	288	262
14	319	401	388	366	301	384	369	330

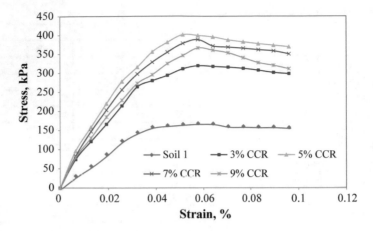

Fig. 1 Stress—Strain curve of soil 1 with CCR for different percentages at 14 days

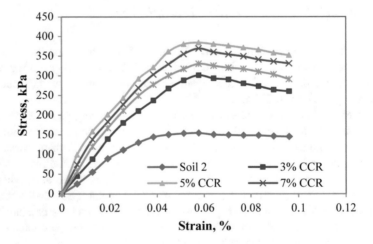

Fig. 2 Stress—Strain curve of soil 2 with CCR for different percentages at 14 days

3.2 Free Swell Index

Soil samples are used for Free Swell Tests are taken from tested UCS samples, which are dried and pulverized since it consists of various percentages of calcium carbide residue in various percentages. The swell values of treated soil samples are decreasing with the increase in content of admixture. When the CCR content reached by 9%, the Free Swell Index values are reduced to 38 and 21% from 63 to 65% for the soils 1 and 2 at the curing period of 14 days. The results obtained for the two samples are tabulated in Table 3 and the variation graph is shown in Fig. 3.

Table 3 Free swell index of soil 1 and 2 with varying percentages of CCR at 14 days

CCR (%)	Free swell index	
	Soil 1	Soil 2
0	63	65
3	54	47
5	43	38
7	40	36
9	38	21

Fig. 3 Variation of Free Swell Index value of soil with CCR for different percentages at 14 days curing period

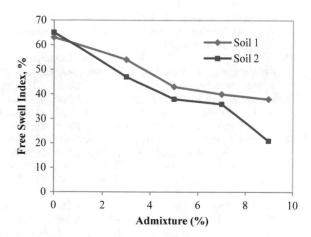

This reduction in swell values representing a suitable development in soil with CCR against the volume changing characteristics.

4 Conclusion

Based on results obtained from unconfined compressive strength test in soil treated with calcium carbide residue, results in an increase in strength value up to 5% CCR content. For this optimum content, the 14th day UCS increased to 401 and 384 kPa for both the soil samples compared with the untreated UCS values of 168 and 154 kPa.

The free swell values are decreasing gradually with the addition of CCR, which representing a suitable development in soil with CCR content against the volume changing properties of clayey soil. At an optimum content of 5% CCR, the swelling characteristic has reduced to 38 and 21% for both the samples at the curing age of 14 days.

Calcium Carbide Residue treatment is effective in improving the strength characteristics of clayey soils, suggesting more productive use of this industrial wastes as a stabilizer with more effect in a shorter time in soil with considerable environmental benefits.

References

1. R. Deepiya., N. Durga., J. Steffi., S. Ayyappan.: Stabilization of Clay Soil Using Quarry Dust and Lime. International Journal of Advanced Information Science and Technology, vol. 32 (32), pp. 111–113(2014).
2. Divya Krishnan K, P.T. Ravichandran, Investigation on Strength and Minralogical Characteristics of Phosphogypsum Stabilised Clay, International Journal of Pure and Applied Mathematics, Vol. 118, No. 20, March – 2018, pp. 2155–2160.
3. Divya Krishnan K, P.T. Ravichandran, V. Janani, R. Annadurai and ManishaGunturi, Effect of Phosphogypsum and flyash stabilization on the strength and micro structure of clay, The Indian Concrete Journal, Vol. 89, Issue 7, pp. 81–86, July 2015.
4. Neeraj kumrawat., S.K. Ahirwar.: Experimental Study & Analysis of Black Cotton Soil with CCR & BA. Intertnational Journal of Science & Management, vol. 4(3), pp. 46–53(2014).
5. A. Manimaran, Seenu Santhosh, P.T. Ravichandran, Characteristics Study on Sub Grade Soil Blended with Ground Granulated Blast Furnace Slag, Rasayan Journal of Chemistry, Vol. 11 (1), Jan - March 2018, pp. 401–404.
6. IS: 2720 (Part 1), 1983: "Methods of Test for Soil - Preparation of Dry Soil Sample for Various Tests", Bureau of Indian Standards, New Delhi.
7. IS: 2720 (Part 5-1985; 6-1972): "Methods of Tests for Soil - Determination of Atterberg's limits", Bureau of Indian Standards, New Delhi.
8. IS: 2720 (Part 10) 1991: "Methods of Tests for Soil – Determination of Unconfined Compressive Strength", Bureau of Indian Standards, New Delhi.
9. IS: 2720 (Part 16) 1987: "Methods of Tests for Soil – Laboratory Determination of CBR", Bureau of Indian Standards, New Delhi.
10. IS: 2720 (Part 40) 1977: "Methods of Tests for Soil – Determination of Free Swell Index of Soils", Bureau of Indian Standards, New Delhi.

OFDM-Based on Trellis-Coded Modulation for Optical Wireless Communication

Harshita Mathur and T. Deepa

Abstract In this paper, we propose an adaptive two-dimensional optimization for orthogonal frequency division multiplexing (OFDM) by introducing the trellis-coded modulation (TCM) and delta sigma modulator (DSM) in the optical channel. The novelty of this proposed approach is based on DSM which is able to offer an excellent robustness against noises and nonlinear distortions. Simulation result shows that this optimization for optical OFDM improves the transmission performance in terms of bit error rate (BER). The data rate of the proposed scheme has been improved without high bandwidth utilization.

Keywords Orthogonal frequency division multiplexing · Trellis-coded modulation · Delta sigma modulator

1 Introduction

With the betterment of multimedia and broadband mobile services, the need for high data rate is thriving heavily [1–3]. Due to a large number of retransmissions in frequency division multiplexing (FDM), an alteration of FDM is used called as orthogonal frequency division multiplexing, where the sub-channel frequencies are preferred in a manner such that they are orthogonal to each other. However, this cuts down the inter symbol interference (ISI) because of the channel; the loss is still very much present [4]. Considering high level of spectral efficiency, dispersion tolerance, and adequate channel equalization, OFDM has been captivated much attention [5]. System optimization can be optimized by using trellis-coded modulation (TCM) which is an advantageous approach to boost the receiver's sensitivity

H. Mathur (✉) · T. Deepa
SRM University, Chennai 603203, Tamil Nadu, India
e-mail: harshita10mathur@gmail.com

T. Deepa
e-mail: deepa.t@ktr.srmuniv.ac.in

© Springer Nature Singapore Pte Ltd. 2019
M. A. Bhaskar et al. (eds.), *International Conference on Intelligent Computing and Applications*, Advances in Intelligent Systems and Computing 846,
https://doi.org/10.1007/978-981-13-2182-5_8

dynamically. TCM is extensively applied in bandwidth limited systems which favor large coding gain without demanding for any added signal bandwidth [6].

TCM with OFDM are practiced to raise the transmission performance especially the bit error rate (BER) [7]. TCM is performed on individual subcarrier flexibly; the principle of the used method in this work is to make full usage of the low-frequency subcarriers. Along with this, high TCM formats is utilized to provide the maximum transmission capacity. In this paper, we propose OFDM based on delta sigma modulation (DSM) for M2M and C2C applications. The proposed digital multi-carrier system aims to avoid the nonlinearity of signal. The DSM is one-bit modulator that converts the continuous time OFDM signal into binary bit sequence. In this paper, the performance of optical OFDM system is analyzed in terms of BER [8]. The scheme used for optimization could not only increase the receiver sensitivity but also enhance the system capacity. However, order of DSM is not increased beyond third order as tradeoff exists between the performance and the complexity, when the order of DSM increases. In various wireless standards, adaptive schemes are that use OFDM to transmit in a wireless data network. In mobile networks, it has been used in wider extent [9].

The rest of the paper is organized as follows: Sect. 2 presents the proposed block diagram. The TCM block is introduced in Sect. 3 which shows that any increase in the distance between any sequence of TCM symbol pair results in a detection performance of active subcarriers which is much better than that of uncoded cases. Simulation results are given in Sect. 4 which gives the results of the joint optimization technique and the conclusion of the paper is illustrated in Sect. 5.

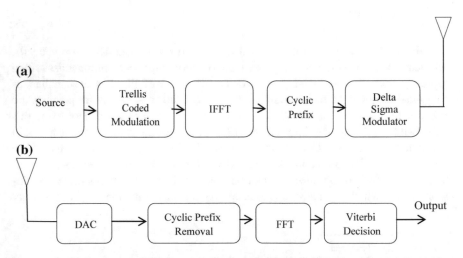

Fig. 1 Block diagram of TCM based delta sigma modulated OFDM system (**a**) Transmitter (**b**) Receiver

2 Proposed System Block Diagram

The proposed system block diagram comprises of the transmitter as well as the receiver section. Figure 1 illustrates the block diagram of TCM based delta sigma modulated OFDM system. The information bits from the source block acts like an input to the TCM block. According to the OFDM principle, multiple low-speed signals are transmitted, i.e., the high-speed signal is imparted over multiple low-speed signal. In the frequency domain, the signal is converted into the time domain after the modulation of data is performed by utilizing the inverse fast Fourier transform (IFFT). TCM starts acting on the information bits received as an input to the trellis block as soon as the channel condition of each subcarrier is known.

TCM block is designed by combining coding and modulation which is responsible for TCM being advantageous. It is performing the expansion or the increase in number of the constellation points and also produces the space for check bits. Additional bandwidth is not required for handling this approach. Because of the increase in the distance between any pair of TCM symbol sequences, growth is observed in a better detection performance of active subcarriers than that of uncoded cases. Any increase in the Hamming distance results in an increase in the diversity gain. In this section, we propose OFDM-based DSM which offers excellent robustness against noises and nonlinear distortions for M2M and C2C applications where blocks of subcarriers are filtered before transmission and reception for the purpose of eliminating ISI [10]. The proposed digital multicarrier system aims to avoid the nonlinearity of LED. The DSM is one-bit modulator which converts the continuous time signal into binary bit sequence. The receiver section retrieves the information by demodulating the sequence and hence improving the performance of the system.

3 TCM for Proposed Optical OFDM

With or without any increase in the bandwidth, data rate can be incremented by transmitting more information per symbol, i.e., more bits per channel used. By incrementing the number of possible symbol values, information content of the symbol is increased. For instance, four possible symbol levels are possible for transmitting two bits per symbol. Higher number of symbol values reduces the average power per symbol and increase the error probability for symbol if we keep constant energy per information bit. To decrease the probability of error, more code bits can be added and hence to increase the code rate [11]. In this section, 8PSK trellis is worked upon, which has 4 states selected in accordance to the state of shift register. When new symbol arrives, content of the shift register get changed.

3.1   Coding Gain

The encoding allows only some of the trajectories trough the trellis. Invalid sequences can be rejected. To achieve the positive coding gain, an increase in the error probability because of the smaller distance is maintained between the constellation points which are outweighed by the coding gain of the error correction code. Mathematically,

$$\text{Coding gain} = 10 \, \log\left[\frac{d_{\min}^2/E_b(\text{coded})}{d_{\min}^2/E_b(\text{uncoded})}\right] \tag{1}$$

where,
  $d_{\min}^2$= minimum distance between all pairs of 2 sequences

$$M_e = \min_{x_k \neq x_{k'}} \left[\sum_k d^2(x_k, x_{k'})\right] \tag{2}$$

where,
  x_k = symbol from an M-ary scheme at instant kT_s.

  To reduce the BER, forward error correction (FEC) can be used. To add redundant bits is the concept behind FEC which means more data has to be sent which mostly decreases the overall bit rate. This is not acceptable. However, bit rate is decreased means bits take longer to be sent and therefore SNR value increases and simultaneously BER value decreases. If BER lowers down more than SNR then we can receive high SNR values and get a reduction in BER for a fixed SNR; this improvement is called coding gain. To measure the coding gain, BER is specified [12].

  TCM coding gain performance depends on number of states of encoder. After a certain increase in number of states coding gain becomes constant [13]. To overcome the abovementioned problem, multidimensional constellation is one of the best solutions. BER versus signal-to-noise ratio (SNR) is improved in comparison with the uncoded modulation. TCM system also provides flexible rates and higher decoding speed at the expense of increased computational complexity for the encoding and decoding process [14].

4   Simulation Results

The simulation model follows few steps which say about the Generation of random PSK modulated symbols. Convolutional encoding is performed on them by using rate 2/3. This is passed through the additive white Gaussian noise (AWGN) channel. Demodulation is performed, i.e., soft decision demodulation on the received coded symbols. Those received coded bits are passed through the Viterbi decoder followed by the counting of the number of errors from the output of it. The same steps are repeated for the multiple SNR values.

Figure 2 shows the signal for continuous and digital TCM OFDM. The digital signal is flexible with the changes in the system unlike the analog signal and is also more secure and facilitates encryption on transmission.

The BER performance for the QPSK OFDM without TCM and comparison of it with the TCM technique is illustrated in Fig. 3. At BER of 10^{-4} level, coding gain of around 3.1 dB has been achieved in QPSK digital OFDM with TCM. It is observed that for lower SNR values, the BER with TCM is higher. The error rate with Viterbi decoding is higher uncoded BER. The reason behind this is, Viterbi decoder gives preference to the error to be divided in a random manner so that it can work efficiently. At lower values of SNR, the chances of multiple received coded bits in errors are high and then the Viterbi algorithm is unable to recover.

Received constellation diagram of digital OFDM with and without TCM has been plotted in Fig. 4. This is clearly observed that convolutional coder of rate $R = k/(k + 1) = 2/3$ and M-ary signal mapper that maps $M = 2^k = 2^2 = 4$ input points into a larger constellation of $M = 2^3 = 8$ constellation points. More subcarriers can be allocated with high modulation formats. The data rate is increased using adaptive optimization techniques.

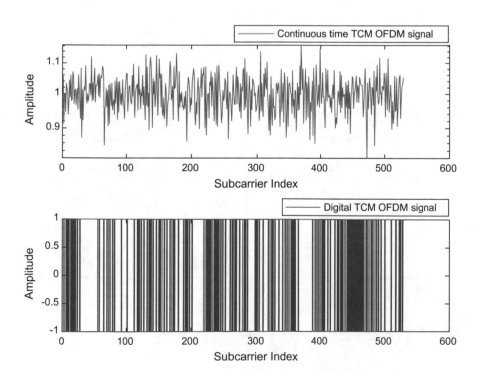

Fig. 2 Representation of continuous and digital TCM OFDM signals

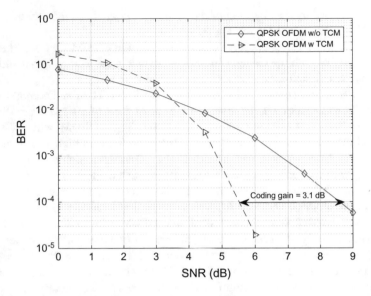

Fig. 3 BER performance of OFDM with and without TCM

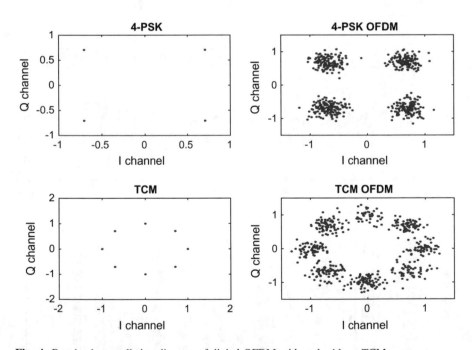

Fig. 4 Received constellation diagram of digital OFDM with and without TCM

5 Conclusion

In this paper, the joint optimization technique has been proposed for OFDM by introducing TCM and DSM which has improved the system performance and was able to offer an excellent robustness against nonlinear distortions. We considered this optimization for optical OFDM which has improved the transmission performance in terms of BER. Without high bandwidth utilization, data rate has been improved and system performance has become much better.

References

1. Z. Yu, Y. Lou, M. Chen, H. Chen, S. Yang and S. Xie, "Adaptive Three-Dimensional Optimization for Optical Direct-Detection OFDM," in *Journal of Lightwave Technology*, vol. 35, no. 9, pp. 1506–1512, May1, 1 2017.
2. D. Lavery, R. Maher, D. S. Millar, B. C. Thomsen, P. Bayvel, and S. J. Savory, "Digital coherent receivers for long-reach optical access networks," J. Lightw. Technol., vol. 31, no. 4, pp. 609–620. Feb. 2013.
3. Shahpari *et al.*, "Coherent Access: A Review," in *Journal of Lightwave Technology*, vol. 35, no. 4, pp. 1050–1058, Feb.15, 15 2017.
4. J. Zhou *et al.*, "256-QAM Interleaved Single Carrier FDM for Short-Reach Optical Interconnects," in *IEEE Photonics Technology Letters*, vol. 29, no. 21, pp. 1796–1799, Nov.1, 1 2017.
5. L. Wu, J. Cheng, Z. Zhang, J. Dang and H. Liu, "Channel Estimation for Optical-OFDM-Based Multiuser MISO Visible Light Communication," in *IEEE Photonics Technology Letters*, vol. 29, no. 20, pp. 1727–1730, Oct.15, 15 2017.
6. X. Gong, Y. Liu and L. Guo, "Joint optimization of bit and power loading for power efficient OFDM-PON," *2014 13th International Conference on Optical Communications and Networks (ICOCN)*, Suzhou, 2014, pp. 1–4.
7. L. Wei, H. Zhang and B. Yu, "Optimal bit-and-power allocation algorithm for VLC-OFDM system," in *Electronics Letters*, vol. 52, no. 12, pp. 1036–1037, 6 9 2016.
8. T. N. Vo, K. Amis, T. Chonavel and P. Siohan, "A Low-Complexity Bit-Loading Algorithm for OFDM Systems Under Spectral Mask Constraint," in *IEEE Communications Letters*, vol. 20, no. 6, pp. 1076–1079, June 2016.
9. X. Chen *et al.*, "Three-Dimensional Adaptive Modulation and Coding for DDO-OFDM Transmission System," in *IEEE Photonics Journal*, vol. 9, no. 2, pp. 1–20, April 2017.
10. E. Giacoumidis, A. Kavatzikidis, A. Tsokanos, J.M. Tang, and I. Tomkos, "Adaptive loading algorithms for IMDD optical OFDM PON systems using directly modulated lasers," J. Opt. Commun. Netw., vol. 4, no. 10, pp. 769–778, Sep. 2012.
11. Yiming Lou, Zhenming Yu, Minghua Chen, Hongwei Chen, Sigang Yang and ShizhongXie, "Experimental demonstration of 10-Gb/s direct detection optical OFDM transmission with Trellis-coded 8PSK subcarrier modulation," *2016 21st OptoElectronics and Communications Conference (OECC) held jointly with 2016 International Conference on Photonics in Switching (PS)*, Niigata, 2016, pp 1–3.
12. X. Liu, Q. Yang, S. Chandrasekhar, and W. Shieh, "Transmission of 44-Gb/s coherent optical OFDMsignal with trellis-coded 32-QAM subcarriermodulation," in Proc. Opt. Fiber Commun., 2010, Paper OMR3.

13. E. Kurniawan, A. S. Madhukumar and F. Chin,"Antenna selection technique for MIMO-OFDM systems over frequency selective Rayleigh fading channel," in IET Communications, vol. 1, no. 3, pp. 458–463, June 2007.
14. Z. Yu, H. Chen, M. Chen, S. Yang and S. Xie, "Bandwidth Improvement Using Adaptive Loading Scheme in Optical Direct-Detection OFDM," in *IEEE Journal of Quantum Electronics*, vol. 52, no. 10, pp. 1–6, Oct. 2016.

Cardiac Failure Detection Using Neural Network Model with Dual-Tree Complex Wavelet Transform

Ipsita Mohapatra, Priyabrata Pattnaik and Mihir Narayan Mohanty

Abstract Cardiac failure in the current scenario is a major issue for which many advanced health care units are increasing. In a similar way, the technology innovation related to it requires to progress adequately. Early stage/post-stage treatment for survival is highly essential. Technocrats, scientists and physicians emphasize to develop better technology as compared to earlier methods. Noninvasive method of detection is the interest of this work. It is a nonintrusive diagnostic tool and can be used in computer-aided techniques that can speed up the diagnosis process. In this paper, the approach is observed in two parts: feature extraction and classification. For detection, features are extracted using Dual-Tree Complex Wavelet Transfer (DTCWT). The features are found the best spectral features. Classification model is based on Multilayer Preceptron (MLP) Neural Network. The data are collected from MIT-BIH NSR for experiment. Classification accuracy result has been presented in the result section.

Keywords Cardiac failure · Dual-Tree Complex Wavelet Transform neural network · MLP · Hidden layer · Classification

1 Introduction

Cardiac failure is a vital circumstance owing to an irregular cardiac dysfunction, i.e., the disability of the heart in pumping blood to the body efficiently and meeting the metabolic demand of the body. Due to this disorder, enough oxygen and

I. Mohapatra (✉) · P. Pattnaik · M. N. Mohanty
Department of Electronics and Communication Engineering, ITER,
Siksha 'O' Anusandhan (Deemed to Be University), Bhubaneswar, India
e-mail: ipsitamohapatra38@gmail.com

P. Pattnaik
e-mail: priyabratapattnaik@soa.ac.in

M. N. Mohanty
e-mail: mihirmohanty@soa.ac.in

© Springer Nature Singapore Pte Ltd. 2019
M. A. Bhaskar et al. (eds.), *International Conference on Intelligent Computing
and Applications*, Advances in Intelligent Systems and Computing 846,
https://doi.org/10.1007/978-981-13-2182-5_9

nutrients are unable to reach many organs, which damage them and reduce their ability to function efficiently. Symptoms of CHF often begin slowly and occur only when the patient is very active, in the other hand, these symptoms may also begin suddenly; like heart attack or other heart-related issues. Symptoms such as slow breathing, metabolic abnormalities, LV dysfunction, fluid retention, weakness, severe anemia, and reduction in myocardial performance are commonly seen. It may lead to serious condition, i.e., stroke or sudden cardiac failure causing death.

CHF is diagnosed through physical examination, such as electrocardiography (ECG) readings, and echocardiography. ECG is non-invasive tool for detection purpose with least cost. It is a significant work to classify the heart beats appropriately for the detection of the normal and affected data. So for classification, different machine learning and data mining technique are applied to get better accuracy for the classification of different heart diseases. It is very difficult and time-consuming task for the classification of the disease by manually so we need an automated machine which can easily classify the disease with less time.

Symptoms of CHF often begin slowly and occur only when the patient is very active, on the other hand, these symptoms may also begin suddenly; like heart attack or other heart-related issues. Symptoms such as slow breathing, metabolic abnormalities, LV dysfunction, fluid retention, weakness, severe anemia, and reduction in myocardial performance are commonly seen. It may lead to serious condition, i.e., stroke or sudden cardiac failure causing death.

Here in our work, an efficient method is considered for the classification of normal and unusual ECG signal. Multilayer preceptron (MLP) which is a feed-forward neural network model with static backpropagation algorithm is implemented followed by the feature extraction technique by Dual-Tree Complex Wavelet Transform (DTCWT). This technique first cleans the informational index by supplanting missing data by nearest segment estimations of the concern class. To estimate the execution of MLP, we have used the MIT BIH NSR database which consists of number of normal and affected real-time physical ECG data.

2 Literature Review

Several methodologies for identification of cardiovascular failures have been developed in the previous couple of decades to ease the regular supervising and monitoring duty. Sudarshana et al. [1] proposed DTCWT on ECG segments of 2 s duration was considered up to six levels to obtain the coefficients. From these DTCWT coefficients, statistical features are extracted and ranked features are subjected to K-Nearest Neighbor (KNN) and Decision Tree (DT) classifiers for automated differentiation of CHF and normal ECG signals. Continuous Wavelet Transform (CWT) is introduced by Acharya et al. [2], which describes Normal, CAD, MI, and CHF ECG beat to get scaleogram. Wes et al. Presented [3] FPGA-based ECG arrhythmia detection using an Artificial Neural Network (ANN), i.e., a Multilayer Preceptor for classification resulting an accuracy of 99.82%.

Acharya et al. [4] employed HRV signals to Empirical Mode Decomposition (EMD) to attain Intrinsic Mode Functions (IMFs) and extracted the entropy-based features and later PNN and SVM were used for separating the respective features into normal and abnormal section. Autoregressive (AR) Burg method is implemented by Masetic et al. [5] and in classification phase, five different classifiers such as C4.5 decision tree, KNN, SVM, ANN, and Random forest classifier are examined. T Barman et al. used various digital filters for preprocessing then classified the ECG signal using different classification schemes such as Fuzzy Rough Nearest Neighbor, MLP, Nearest Neighbor (NN) compared to different parameters such as correctly classified samples, kappa statistics, root mean square error, TP rate, ROC, etc. Fujita et al. [6] presented the HRV signals of normal and subjects at risk with SCD techniques and features obtained are ranked using their t-value and applied to classifiers like KNN, DT, and SVM. An automated classification algorithm with three classifiers: SVM, RBFN, and MLP is presented by Hachem et al. [7] which processes small length ECG signals. Mohammad et al. [8] introduced a method for the classification of ECG which has the facility to categorize the beats into four classes as normal, ventricular, supra-ventricular, and fusion. Some researchers like Descriptor [9] implemented feature extraction method and KNN and MLP are used as classifier. Leaon et al. [10], also used the SVM, MLP, and PNN for the same data set and they got the results like 97.42, 98.24 and 97.42%, respectively Sidek et al. [11] gave a new problem detecting technique using Multilayer Perceptron (MLP). Three-dimensional reduction methods (Discrete Cosine Transform, DCT, Principal Component Analysis, PCA and re-sampling, RES) were taken for taking input vectors and Bayesian regularization was used for the training purpose.

3 Proposed Method

The ECG signals are collected from BIT-MIH NSR database. The database is of adults within age of 50–70 years old. The proposed block diagram is shown (see Fig. 1). Initially, the preprocessing is performed. Next to it, features are extracted that follows the classification process.

Spectral characteristics of a signal show the behavior related to that signal. Therefore, spectral features are considered in many cases by the researchers.

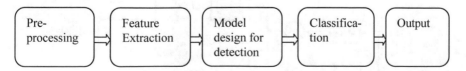

Fig. 1 Block diagram of proposed work

Fourier transform is one of them to find the spectral characteristics of signal. It is suitable for stationary type of signal and is represented by

$$X(p) = \sum_{n=0}^{N-1} x(n).e^{-j\left(\frac{2\pi}{N}\right)np}$$

$$p = 0, 1, 2 \ldots N - 1$$

(1)

where $x(n)$ = Time domain input signal, $X(p)$ = Frequency domain signal.

The nonstationary type of signal may not be evaluated with the transform. It requires the spectral characteristics in particular instant of time. The wavelet transform is one of such tool to find out the spectral components in specific time considerable works have been performed with the use of wavelet transform [12, 13].

In our purposed work Dual-Tree Complex Wavelet Transform is considered, i.e., represented by

$$F(t) = \sum_{-\infty}^{\infty} x(n)\varphi(t-n) + \sum_{i=0}^{\infty} \sum_{k=-\infty}^{\infty} y(i,n) 2^{\frac{i}{2}} \psi\left(2^i t - n\right)$$

(2)

where, $x(n)$ = Analog signal, $\varphi(t)$ = bandpass wavelet, $\Phi(t)$ = Low-pass scaling function. For some applications of discrete wavelet transform, improvements can be obtained using Dual-Tree Complex Wavelet Transform which is an expensive wavelet transform in place of critically sampled one. There are several kinds of expansive DWTs; here we consider the Dual-Tree Complex Discrete Wavelet Transform. The DTCWT of a signal x(ECG Signal) is implemented using two critically sampled DWTs in parallel on the same data, as shown (see Fig. 2).

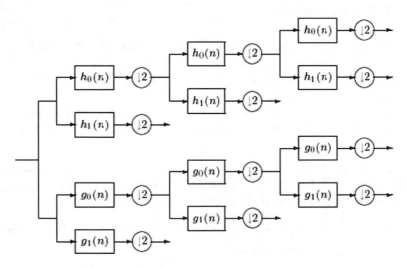

Fig. 2 Dual-Tree Complex Wavelet Transform

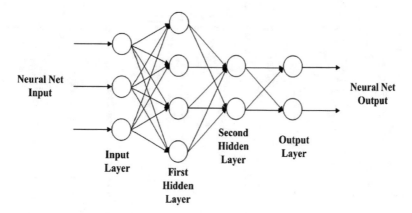

Fig. 3 Multilayer perceptron neural network

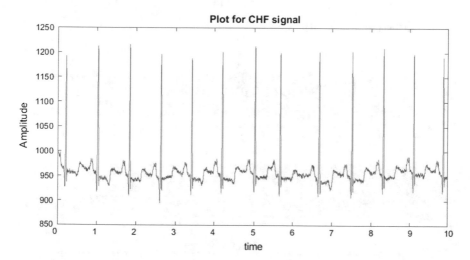

Fig. 4 ECG plot of normal arrhythmia signal

The features are free from noise. The signal can be reconstructed in these features and showed in the result section in Figs. 6 and 7. The features are applied through the Neural Network Model to classify the healthy and unhealthy signals. In this case, most popular network is considered as Multilayer Perceptron (MLP). The wavelet features are as the input to the network as shown (see Fig. 3).

$$y_k(x) = \sum_{i=1}^{d} w_{ki} x_i + w_{k0} \qquad (3)$$

Equation for the MLP has been presented in (3) and the weight updating equation has been presented in (4).

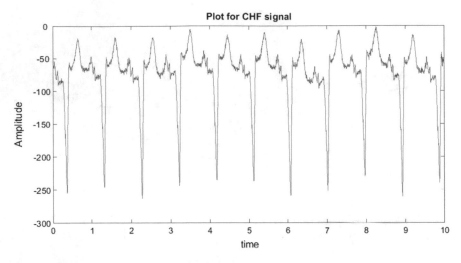

Fig. 5 ECG plot of BIDMC CHF signal

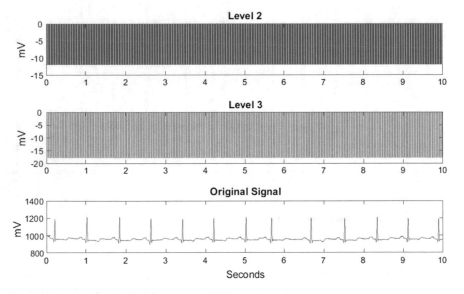

Fig. 6 After applying DTCWT in normal ECG signal

$$\Delta w_{ij} = -\eta \sum_n \delta_j x_i \tag{4}$$

where η = learning rate; δ = Error. Here, backpropagation learning algorithm is used. The feature reconstructed signal as well as classification result is explained in the result section.

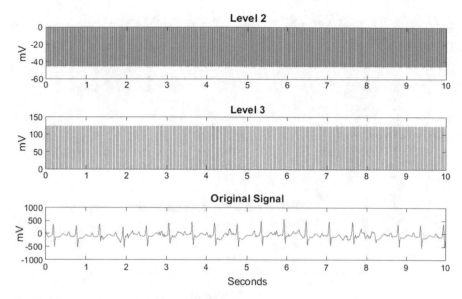

Fig. 7 After applying DTCWT in affected BIDMC signal

4 Experiment Result Discussion

4.1 Database Preparation and Training of the Network

In this study, the following Normal and CHF data were obtained from Physio-bank MIT-BIH Arrhythmia Database (MITDB) and BIDMC CHF, and open-access databases. Each data is of 10 s duration.

The signal (see Fig. 3) has the sampling frequency 250 and signal in Fig. 4 has the sampling frequency 360. The features of the above ECG signals have been extracted using Dual-Tree Complex Wavelet Transform (DTCWT) technique and are shown below (see Figs. 5 and 6).

4.2 Classification

The obtained features are applied to learning algorithm Multilayer Perceptron Neural Network. It has three layers, (1) Input layer (2) Output layer and (3) Hidden layer (Fig. 8; Table 1).

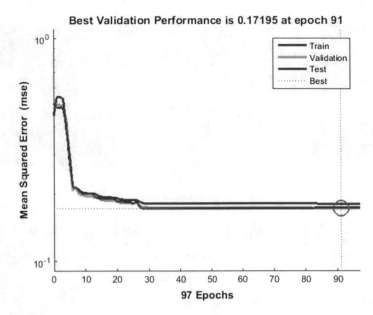

Fig. 8 Performance curve after applying MLP

Table 1 Confusion matrix

	Reference			
Patient	Patient 1	Patient 2	Patient 3	Total sum
Patient 1	44	2	0	46
Patient 2	1	35	0	36
Patient 3	1	0	35	36
Total sum	46	37	35	118

Sum of major diagonals = 114
Over all accuracy = 114/118 = 96.6%

5 Conclusion

In this work, the simple classifier s considered but with better accuracy. As it is observed from literature MLP network provides maximum 85%. Using the feature extraction technique as DTCWT, the accuracy is increased up to 96.6%. As the specific information is obtained using DTCWT and is useful with MLP classifier for detection purpose. Further, the features may be modified and also choice of classifier based on the features can be verified for better accuracy.

References

1. V. K. Sudarshan, U. R. Acharya, S. L. Oh, M. Adam, J. H. Tan, C. K. Chua, and R. San Tan.: Automated diagnosis of congestive heart failure using dual tree complex wavelet transform and statistical features extracted from 2 s of ECG signals, Computers in Biology and Medicine, vol. 83, pp. 48–58 (2017).
2. U. R. Acharya, H. Fujita, V. K. Sudarshan, S. L. Oh, M. Adam, J. H. Tan, and K. C. Chua.: Automated characterization of coronary artery disease, myocardial infarction, and congestive heart failure using contourlet and shearlet transforms of electrocardiogram signal, Knowledge-Based Systems, vol. 132, pp. 156–166 (2017).
3. M. Wess, P. S. Manoj, and A. Jantsch.: Neural network based ECG anomaly detection on FPGA and trade-off analysis, In Circuits and Systems (ISCAS), 2017 IEEE International Symposium, pp. 1–4 (2017).
4. U. R. Acharya, H. Fujita, V. K. Sudarshan, S. L. Oh, A. Muhammad, J. E. Koh, and R. San Tan.: Application of empirical mode decomposition (EMD) for automated identification of congestive heart failure using heart rate signals, Neural Computing and Applications, vol. 28, pp. 3073–3094 (2016).
5. Z. Masetic, and A. Subasi.: Congestive heart failure detection using random forest classifier, Computer methods and programs in biomedicine, vol. 130, pp. 54–64 (2016).
6. H. Fujita, U. R. Acharya, V. K. Sudarshan, D. N. Ghista, S. V. Sree, L. W. J. Eugene, and J. E. Koh.: Sudden cardiac death (SCD) prediction based on nonlinear heart rate variability features and SCD index, Applied Soft Computing, vol. 43, pp. 510–519 (2016).
7. A. Hachem, M. Ayache, L. El Khansa, and A. Jezzini.: ECG classification for Sleep Apnea detection, In Biomedical Engineering (MECBME), Middle East Conference pp. 38–41. IEEE (2016).
8. A. Mohammad, F. Azeem, M. Noman, and M. H. N. Shaikh.: A simple approach of ECG beat classification, In Signal Processing and Integrated Networks (SPIN), International Conference, pp. 641–644. IEEE (2016).
9. F. Asadi, M. J. Mollakazemi, S. A. Atyabi, I. L. I. J. A. Uzelac, and A. Ghaffari.: Cardiac arrhythmia recognition with robust discrete wavelet-based and geometrical feature extraction via classifiers of SVM and MLP-BP and PNN neural networks, In Computing in Cardiology Conference (CinC), pp. 933–93. IEEE (2015).
10. A. A. S. Leon, D. M. Molina, C. R. V. Seisdedos, G. Goovaerts, S. Vandeput, and S. Van Huffel.: Neural network approach for T-wave end detection: A comparison of architectures, In Computing in Cardiology Conference (CinC), pp. 589–592. IEEE(2015).
11. K. A. Sidek.: Cardioid graph based ECG biometric using compressed QRS complex, In BioSignal Analysis, Processing and Systems (ICBAPS) International Conference, pp. 11–15. IEEE (2015).
12. T. Barman, R. Ghongade, and A. Ratnaparkhi.: Rough set based segmentation and classification model for ECG", In Advances in Signal Processing (CASP) Conference pp. 18–23. IEEE (2016).
13. L. M. Gladence, T. Ravi, and M. Karthi.: An enhanced method for detecting congestive heart failure-automatic classifier, In Advanced Communication Control and Computing Technologies (ICACCCT), International Conference, pp. 586–590. IEEE (2014).

Design of Intelligent PID Controller for Smart Toilet of CCU/ICU Patients in Healthcare Systems

Mohan Debarchan Mohanty, Dipamjyoti Pattnaik,
Mahananda Parida, Soyam Mohanty and Mihir Narayan Mohanty

Abstract The necessity of a healthy environment is a major aspect in this modern age. Several trials have been made by multidisciplinary researchers for the development of a healthy environment. However, it is important to facilitate most of the modern healthcare resources to the unhealthy people for their betterment. Some extent of technology has an important role to support it. In this paper, a case study of water supply in healthcare system is analyzed. The paper has three aspects; first, the use of PI control for measuring the water level in a tank, second, the water distribution to several resources using PD control. The use of both the controls is further explained through an example of water distribution in a smart toilet. Lastly, the tuning of the PID controller using a fuzzy-based controller which makes the overall system robust. The simulation results confirm the advantages and demonstrate better dynamic behavior. For low energy consumption, either the input energy is decreased or the efficiency of mechanical transmission is increased. Performance of the controller is judged upon the simulation results obtained from MATLAB.

Keywords Healthcare · PI control · PD control · PID controller
Fuzzy-based controller · Robust

M. D. Mohanty (✉) · D. Pattnaik · M. Parida · S. Mohanty
Department of Electronics and Instrumentation Engineering,
College of Engineering and Technology, Bhubaneswar, Bhubaneswar, India
e-mail: mohan.debarchan97@gmail.com

M. N. Mohanty
Department of Electronics and Communication Engineering, ITER,
Siksha 'O' Anusandhan (Deemed to Be University), Bhubaneswar, India
e-mail: mihirmohanty@soa.ac.in

© Springer Nature Singapore Pte Ltd. 2019
M. A. Bhaskar et al. (eds.), *International Conference on Intelligent Computing
and Applications*, Advances in Intelligent Systems and Computing 846,
https://doi.org/10.1007/978-981-13-2182-5_10

97

1   Introduction

In the modern society, healthcare units should be improvised to provide proper care to the patients. Many e-healthcare units have already been developed but there should be minute study in each subunits for the effective use of the whole unit. Health care system is very large which consists of patient entry, patient consultation, test and diagnosis. For the system to be smart enough, there is a requirement of specialized physicians in specific fields along with their proper testing laboratories. Some of the works have been proposed earlier for the abovementioned components [1–4].

Similarly, the patients admitted in the hospitals need a healthy environment condition and that also plays an important role in the fast treatment. One such aspect in this work is the distribution of the water in the storage tanks of the healthcare units into the toilets the drinking water reservoir. The fuzzy-based tuning of the PID controller to control the inflow and outflow rate makes the system intelligent. Previously, we performed a small project where we designed a *Smart Toilet for the Bedridden People*. In this project we considered a toilet fitted in the bed itself with automatic water supply for the flush and cleaning water. Considering the combination of both the concepts, it can help in improvising the modern healthcare systems.

To control a steam engine and boiler combination, the mamdani fuzzy inference system has been explored using linguistic rules [5, 6]. In this they developed a system with if-then rules corresponding to the human language.

Several works have been done in the design of PID controller with the help of a fuzzy controller [7–9]. The authors in [7] have designed a PID controller using the root locus technique with the system transfer function and then tuned the controller with a fuzzy controller. The authors in [9] have focussed on the auto-tuning of the PID controller where they compared the tuning methods- trial and error method, ziegler nichols and the fuzzy-based tuning. They found the fuzzy-based tuning to be more effective than the other methods.

Works have been explored in the level control of coupled tanks using a PID Controller [10, 11]. The authors in [10] have done the level control using P, PI, PD, and PID controller. In [11], authors have extracted the requirements of a controller for the level measurement.

Since no mathematical model is required to design an FL controller, hence it maintains its robustness. Ease of application of FL controller makes it popular in the industry. Nevertheless, knowledge database or expert experience is desired in deciding the rules or the membership functions [4].

2 Material and Methods

The proposed scheme aims to enhance the control performance water supply in the modern health care units. As compared to the traditional controller, the FL controllers possess the following advantages; they can be developed easily, can be customized readily in natural language terms, and have a very wide range of operating conditions. Efforts are being made to provide easy and simple control algorithms by industries and researchers to accommodate increasing complexity in the processes/systems [12, 13].

Conventional control algorithms relying on linear system need to be linear zed before applied to systems, although the performance is not guaranteed, nonlinear controllers can accommodate nonlinearities however the lack of a general structure recreates in their designing [14]. Thus, linear or nonlinear control algorithm solutions are mostly developed using precise mathematical system models. Difficulties in describing these systems using traditional mathematical relations happened to provide unsatisfactory design models or solutions [10]. This is a motivating factor to be used for system design which can employ approximate reasoning to resemble the human decision process [15].

FL has been proposed by Zadeh in 1965 [15]. FL provides a specific Artificial Intelligence (AI) level to the traditional proportional, integral and derivative (PD) controllers. The use of FL in the control systems provides better dynamic response, rejects disturbance, allows low parameter sensitive variation, and removes external influences. The FL control algorithm tends to be robust, simple, and efficient. Experiments of the proposed control system have shown remarkable tracking performance and demonstrate the usefulness and validity of the hybrid fuzzy controller convincingly, with high performance under parameter variations and load uncertainties. The performance has been satisfactory for most of the reference tracks.

A conventional PID controller is widely used in industries because of its ease of operation, cost efficiency, and its range of accuracy. It constitutes of three parts—proportional, integral, and derivative control. The sum of all the controls results in the final output of the controller. Mathematically, it is given by

$$o(t) = i(t) + Kp * e(t) + Kd * \frac{de(t)}{dt} + Ki \int e(x).dx \qquad (1)$$

where

Kp Proportional gain
Ki Integral gain
Kd Derivative gain
$e(t)$ Error present in the controller
$i(t)$ Initial response
t Instantaneous time, x—Variable of integration

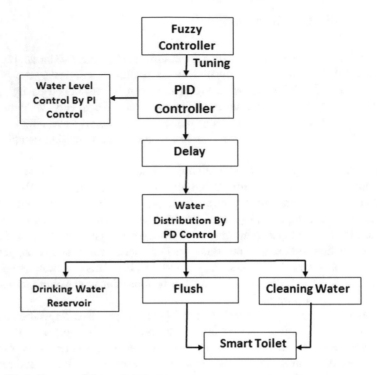

Fig. 1 Block diagram of the proposed system

The first methodology focuses upon the tuning of the PID controller achieved by the rule base of an FL controller [8]. The tuning is done in such a way that PI control of the controller will control the level of water in the tanks and the PD control of the controller will control the distribution of the water supply. The distribution of flush and cleaning water is implemented on microcontroller based Smart toilet; a toilet designed for the bedridden people which is basically works on a switching principle with some delay (Fig. 1).

Tuning of PID controller with a FL controller

The FL has the following stages of processing:

FUZZIFICATION:

Fuzzification implies the process where the preprocessing inputs are transformed into the fuzzy domain.

KNOWLEDGE BASE:

The knowledge base of FL controller is based upon database and rule base:

(i) Database: It provides major data for the functioning of the fuzzification and defuzzification modules and their respective rule base.
(ii) Rule Base: Its function is to represent the control policy in a structured manner.

FUZZY INFERENCE SYSTEM:

Fuzzy inference system has a simple I/O relationship. The input is processed from the external world by the system through a series of events called as fuzzy inference algorithm. Mamdani fuzzy is one of the widely used fuzzy inference system.

DEFUZZIFICATION:

The process in which the fuzzy sets assigned to a control output variable are transformed into a processed value. There are several methods forde fuzzification but our work is done by the mean of maximum method (MOM). In this method, the defuzzified value is taken as the element of highest membership values. As there is more than one element having maximum membership values, the mean of the maxima is taken.

Let A be a fuzzy set with membership function $\mu(x)$ defined over $x \in X$, where X is a universe of discourse. The defuzzified value is let say d(A) of a fuzzy set and is defined as,

$$d(A) = \frac{\sum_{xi \in M} (xi)}{|M|} \tag{2}$$

where

$M = \{x_i | \ \mu(x)$ is equal to the height of the fuzzy set A, $h(A)\}$ and $|M|$ is the cardinality of the set M (Table 1).

The rule base is defined as follows:

a. When |E| is larger value, Kp should be larger, Kd should be smaller for better tracking performance and the Ki should be set to zero for avoiding integral saturation and more overshoot.
b. When |E| & |δE| are medium value, neither Kp, Ki, and Kd are too large for less overshoot. Ki should be smaller but Kp and Kd should be moderate for faster response.

Table 1 Basic fuzzy rule base with 49 possible rules

Output(t)		|E|						
		NE LG	NE	MD	NE SM	ZER	PO SM	PO MD	PO LG
|δE|	NE LG	NE LG	NE	LG	NE LG	NE LG	NE SM	ZER	ZER
	NE MD	NE LG	NE L	LG	NE MD	NE MD	ZER	ZER	PO SM
	NE SM	NE LG	NE	MD	NE MD	NE SM	ZER	PO SM	PO MD
	ZER	NE LG	NE	MD	NE SM	ZER	PO SM	PO MD	PO LG
	PO SM	NE MD	NEM	SM	ZER	PO SM	PO MD	PO MD	PO LG
	PO MD	NE SM	ZER		ZER	PO MD	PO MD	PO LG	PO LG
	PO LG	ZER	ZER		PO SM	PO LG	PO LG	PO LG	PO LG

Meaning of the linguistic variables in the fuzzy interference system
NG LG—Negative Large, NG MD—Negative Middle, NG SM—Negative Small, ZER—Zero, PO LG—Positive Large, PO MD—Positive Middle, PO SM—Positive Small

c. When the |E| is smaller value, combining Kp and Ki should be increased for steady performance. Kd should be moderate for avoiding oscillation around the corresponding static value (Fig. 2).

PI Control for Tank Level Control

The PI control output consists of only the sum of the product of errors of the proportional gain and the integral gain only. The use of integral control is used for removing the steady-state error which results in an improved transient response, but also increases the system settling time. The system consists of a set point (SP) and a process variable (PV). Set Point is defined as the desired level of the tank and the Process Variable is defined as the actual level of the tank. The error is given by:

$$e(t) = \text{SP} - \text{PV} \tag{3}$$

Now for PI control, putting $Kd = 0$ in Eq. (1), we get

$$o(t) = \text{PV} + Kp * e(t) + Ki \int e(t).\mathrm{d}t \tag{4}$$

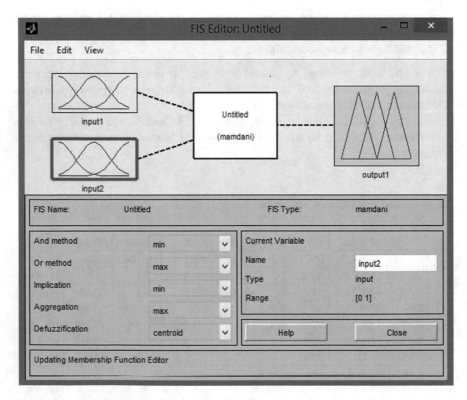

Fig. 2 Fuzzy inference system

Consider a tank with cross-sectional area A. Let the input flow rate be Q_{in} cm^3/s. The output flow Q_{out}cm^3/s in a water tank occurs through a hole in the bottom of the tank of area α [16]. The flow rate of water through the hole follows the Bernoulli equation given by

$$Qin = \alpha\sqrt{2g(SP)} \tag{5}$$

From conservation of mass property we get

$$A\frac{d(SP)}{dt} = Qin - Qout = Qin - \alpha\sqrt{2g(SP)} \tag{6}$$

PD Control for Distribution

The PD control output consists of only the sum of the product of errors of the proportional gain and the derivative gain only. The use of derivative control is to reduce both the overshoot and the settling time and has a negligible effect on rise time and the steady-state error. PD control usually gives a faster output. The flush, cleaning water, and the drinking water are the derivatives of the overall tank water. The distribution depends upon the FL tuning of the PID controller. So for the PD control Ki becomes zero which leads the overall controller output as

$$o(t) = PV + Kp * e(t) + Kd * \frac{de(t)}{dt} \tag{7}$$

The derivatives are distributed depending on the level of the tank, i.e., the process variable; if it is at a low level, the flow rate will be less, if the level is medium there will be a faster response than the low level and similarly when the process variable becomes equal to the set point flow rate becomes maximum. These IF-THEN rules of combined PI and PD controls are used for the tuning purpose.

3 Result Discussion

The experimental results and simulations indicate the following observations:

1. The fuzzy controller can regulate the output voltage of boost or buck configurations to the specified value without undammed oscillations even in case of variation in input voltage or load.
2. The FL controller does not depend on the mathematical model but on the structure's linguistic explanation.
3. The design of the present controller is based on past experience to design the MFs and various controls.

4. The response indicates that the PI control removed the steady-state error and improved the transient response, but it also increased the system settling time. Increased Ki increases overshoot and settling time making system response sluggish.
5. The PD controller decreased the system settling time significantly and to control the steady-state error, the derivative gain Kd is kept high.

For tuning, it is essential for inference rules to be modified during the tests on the equipment by reducing the number of rules or to add complementary rules. The installation tests may be replaced with PC simulation in case the control system depends on the process model.

The second-order system is considered for the better analysis of the tuning parameters. So we considered a second-order transfer function as (Figs. 3, 4, 5 and 6):

$$G(s) = \frac{1}{s^2 + 5s + 2}$$

PI control output response step values

Rise Time = 0.0885, Settling Time = 1.4796, Settling Min = 0.686, Settling Max: 1.5366, Overshoot = 55.3543, Undershoot = 0, Peak = 1.5366

PD control output response step values

Rise Time = 0.1879, Settling Time = 0.6704, Settling Min = 0.8831, Settling Max = 1.0695, Overshoot = 10.5157, Undershoot = 0, Peak = 1.0695, Peak Time = 0.3960

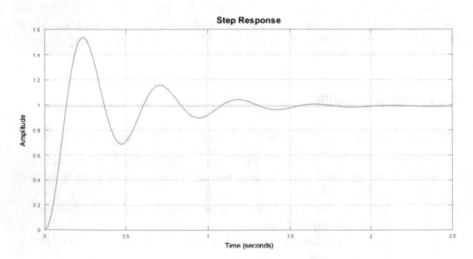

Fig. 3 Response of the tuned PI controller

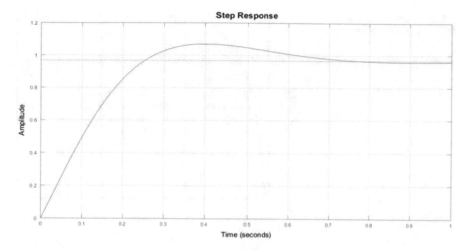

Fig. 4 Response of tuned PD controller

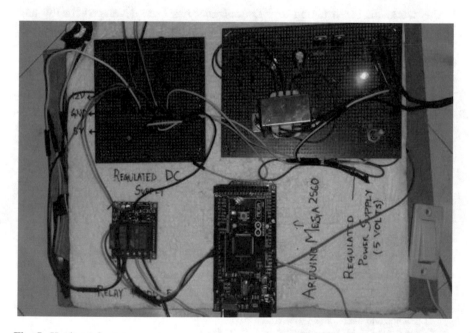

Fig. 5 Hardware layout

Fig. 6 Response of tuned
PID controller using FL

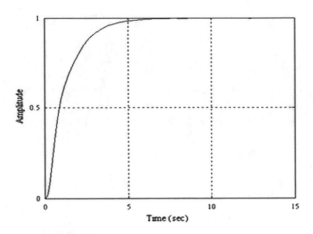

4   Conclusion

In this work, FL based tuning of PID controller has been developed for the tank
level measurement and the outflow distribution as drinking water, flush, and
cleaning water. It is implemented on a smart toilet. The smart toilet is a semiau-
tomatic toilet designed with the help of Atmega 2506 microcontroller for the
bedridden people which works on switching condition for the flush and cleaning
water. Modifications have been done to the system to make it robust and to meet the
challenges for the real-time environment. The simulation shows better performance
of the FL control as it is adaptable to the system. For the future scope, the rules may
be optimized. Also, the intelligent controller may be applied to the other environment of the healthcare.

References

1. L. Sarangi, M.N. Mohanty, and S Patnaik, "Detection of abnormal cardiac condition using
   fuzzy inference system", International Journal of Automation and Control, vol. 11(4),
   pp. 372–383,2017.
2. L. Sarangi, M.N. Mohanty, and S Patnaik, "Critical Heart condition Analysis through
   Diagnostic Agent of E-Healthcare System using spectral Domain Transform", Indian Journal
   of Science and Technology, vol. 9(38), 2016.
3. L. Sarangi, M.N. Mohanty, and S Patnaik, "Design of ANFIS Based E-Health care System for
   Cardio Vascular Disease Detection", In Recent Developments in intelligent Systems and
   Interactive Applications, Springer International Publishing pp. 445–453, 2016.
4. L. Sarangi, M.N. Mohanty, and S Patnaik,"An Intelligent Decision Support System for
   Cardiac Disease cetion", IJCTA. International Press 2015.
5. J.Yen, and R. Langari, "Fuzzy Logic. Intelligence, Control, Control, and information",
   Prentice-Hall, 1999.
6. Y.S. Zhou, and L.Y. Lai, "Optimal design for Fuzzy Controllers by Genetic Algorithms",
   IEEE Trans, On Industry Application, Vol. No.1 pp. 93–97, January February 2000.

7. Design and Performance of PID and Fuzzy Logic Controller with Smaller Rule Set for Higher Order System by S.R.Vaishnav, Z.J.Khan.
8. Design of tuning methods of PID controller using fuzzy logic by Venugopal P, Ajanta Ganguly, Priyanka Singh.
9. AUTOMATIC TUNING OF PID CONTROLLER USING FUZZY LOGIC by GaddamMallesham, AkulaRajani.
10. Analysis Of Liquid Level Control Of Coupled Tank System By Pi, Pd, Pid Controller by Surbhi Sharma, ManishaArora, KuldeepakKaushik.
11. PID Controller design for two Tanks liquid level Control System using Matlab by Mostafa. A. Fellani, Aboubaker M. Gabaj.
12. H.B.Verbruggen, and P.M. Bruiji, "Fuzzy control and conventional control. What is (And Can Be) the Real Contribution of Fuzzy Systems", Fuzzy Sets Systems, Vol. 90, 151–160,1997.
13. T.O. Kowalska, K.Szabat and K, "The Influence of Parameters and Structure of PI- Type Fuzzy- Logic Controller on DC Drive System Dyanamic", Fuzzy Sets and System, Vol. 131, pp. 251–264 2002.
14. M.S. Ahmed, U.L. Bhatti, F.M Al-Sunni, and M.El-Shafci, "Design of a Fuzzy Servo-Controller", Fuzzy Sets and Systems, Vol. 124, pp. 231–247, 2001.
15. A. Zilonchian, M. Juliano, and T. Healy, "Design of Fuzzy Logic Controller for a Jet Engine Fuel System", Control and Engineering Practices, Vol. 8, pp. 883, 2000.
16. Applied Physics and Electronics, StaffanGrundberg, Umea University.

Detection of Diabetes Using Multilayer Perceptron

Saumendra Kumar Mohapatra, Jagjit Kumar Swain
and Mihir Narayan Mohanty

Abstract Diabetes is a common disease throughout the world. It is generally in sleep mode, whenever the patient suffers from any disease, diabetes boosts them. It is a common factor of cardiac problem. Authors try to detect it using multilayer perceptron neural network in this paper. The case study is of Indian ladies with pregnancy suffer from diabetes. Data considered from PIMA database from UCI repository and are used. Eight attributes are taken as features for each subject. It has been verified for 768 patients. The common MLP classifier is utilized for attributes and the experiment is learned with R studio platform. The performance found to be better as compared to earlier methods and verified in MATLAB platform as well.

Keywords Diabetes · Neural network · MLP · Hidden layer · Classification

1 Introduction

Research in medical science improves with the aid of technology for diagnosis using modern equipment. Most people suffering from cardiac and kidney related disease due to diabetes. It causes due to the deficiency of insulin in human body and is said to be silent killer as it boosts other diseases in the human body. Diabetes is a chronic state relates with glucose in the blood. Insulin formed by pancreas and decreases the blood glucose levels. Inadequate creation of insulin causes diabetes

S. K. Mohapatra (✉) · M. N. Mohanty
Department of Electronics and Communication Engineering, ITER,
Siksha 'O' Anusandhan (Deemed to be University), Bhubaneswar, Odisha, India
e-mail: saumendramohapatra@soa.ac.in

M. N. Mohanty
e-mail: mihirmohanty@soa.ac.in

J. K. Swain
ITER, Siksha 'O' Anusandhan (Deemed to be University), Bhubaneswar,
Odisha, India
e-mail: veerjagjit@gmail.com

© Springer Nature Singapore Pte Ltd. 2019
M. A. Bhaskar et al. (eds.), *International Conference on Intelligent Computing
and Applications*, Advances in Intelligent Systems and Computing 846,
https://doi.org/10.1007/978-981-13-2182-5_11

that is categorized into two types like type 1 and type 2. Previously, these are named as insulin-dependent and non-insulin-dependent Diabetes. The Pima database is one of the most calculated population for diabetes. The samples of the calculated population concerning diabetes have been classified as positive and negative instances and is explained in [1]. The single way for the diabetes patient to live with this disease is to keep the blood sugar as normal as possible and this can be accomplished when the patient uses an exact treatment which may include proper diet and exercising. The treatment is a difficult and expensive task due to continuous modification. Physicians take the decision about the patient by verifying number of record about the patient and disease.

Machine learning is one of the technologies that allows the machines to learn. The challenge is to apply these algorithms for detection and classification of medical data and different which with or without human involvement. This approach can improve the understanding and specificity of disease recognition and also reduces the accompanied cost by passing unnecessary and costly medical tests [2]. Widespread studies about diabetes prediction has occurred for some years. Different machine learning techniques have been applied for the classification of Diabetes and which is applied on Pima Indian Diabetic database [3]. The techniques such as decision tree [4], support vector machine [5], Naive Bayes, etc., have been used for classification of the disease.

Neural network is one of the widely used machine learning technique that works on the concept of the biological neurons. It consists of an interconnected cluster of neurons as well as process information that enters into it. Neural networks have been successfully applied in an extensive mixture of both supervised and unsupervised learning applications [6]. Multilayer Perceptron (MLP) Network is one of the popular techniques. It has the capability to classify the disease by identifying the different symptoms of the patient [7]. In this work, MLP is used for classification of pregnant women. Proposed technique has been applied on the diabetes database of PIMA for Indian people and is collected from University of California (UCI).

The rest of the paper is organized as follows: The proposed Multilayer Perceptron (MLP) Network is presented in Sect. 2. The experimental results are shown in Sect. 3. Section 4 contains the conclusion and future work.

2 Proposed Method

The most common and basic method of neural network is Multilayered Perceptron (MLP) network. It uses back propagation learning method for the classification of the instances. Using this model Fig. 1 describes the proposed work.

Data is preprocessed in the first stage for the suitability of feeding to the model. Next to it, the data are divided into the graphs as training and testing data. Total 768 instances of data with eight attributes as features and a class variable are presented. The detail of the data set has been presented in Table 1. It includes no. of times

Fig. 1 Proposed method architecture

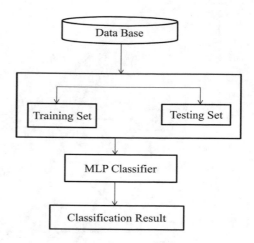

Table 1 Detailed description of the database

Feature variable	Minimum value	Maximum value	Mean	Median
Number of times pregnant	0	17	3.84	3.0
Plasma glucose concentration	0	199	120.89	117.00
Diastolic blood pressure	0	122	69.10	72.00
Triceps skinfold thickness	0	99	20.53	23.00
2 h serum insulin	0	546	79.79	30.5
Body mass index	0	67.1	31.00	32.00
Diabetes pedigree function	0.078	2.42	0.47	0.37
Age	21	81	33.24	29.00

pregnant, plasma glucose concentration, diastolic blood pressure, triceps skinfold thickness, serum insulin, body mass index (BMI), diabetes pedigree function, age, and a class variable.

The data is split into 70% training data and 30% testing data. 550 instances are taken for training and 218 instances have taken for testing data set. Both data set contains the 8 feature attributes and a class attribute.

2.1 Multilayer Perceptron (MLP)

Multilayer Perceptron is one of the feed-forward neural network classification technique. It contains number of layers. In Single-Layer Perceptron (SLP), the linearly separable problems can be solved but the nonlinear problems cannot be solved. In order to solve these complex problems MLP is used [8–10]. This technique is known as feed-forward neural network which has one or more hidden layers as described in Fig. 2. MLP is generally used for pattern reorganization and classification of input patterns and predicts the result.

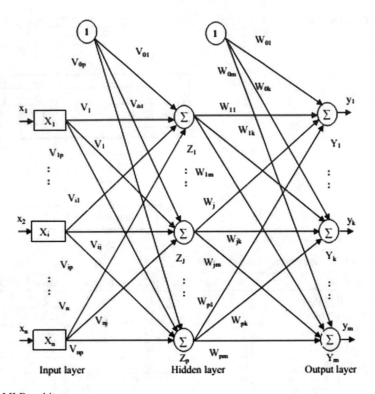

Fig. 2 MLP architecture

Before training of the network, the weights will be set randomly. After this the neurons learn from training set, which in this case consists of set of tuples (x_1, x_2, t). x_1 and x_2 are the input to the network and t is the expected output. If y is the actual output then output of the neuron depends on the weighted sum of all its neurons and can be presented as:

$$y = x_1 w_1 + x_2 w_2 \tag{1}$$

The network consists of a single hidden layer with a nonlinear activation function. The output of the network can be expressed as

$$X = f(s) = W\phi(As + p) + b \tag{2}$$

where s the input is vector and X is the output vector. The weight matrix and bias vector of the first layer are A and p, respectively [11, 12]. W is the weight matrix and b is the bias vector of the second layer. ϕ is the nonlinear element. The output is connected to the inputs of other neurons in the hidden layer and is not visible in the output [12]. The output designates as 0 and 0 represents there is nondiabetic and 1 represents for diabetes.

3 Experiment Result Discussion

Total 1768 ladies patients are considered for the experiment It is found that 268 are suffering from diabetes and rest 500 cases are in healthy. Also, it is verified for missing data and found no missing is there. In Fig. 3, the details structure of the database has been presented.

Further, the data has been divided into training set of size 550 and testing set of size 218 respectably. In the first stage, the network is trained by taking the training data from the database. In Fig. 4, the trained network has been presented with 8 input layers and 4 hidden layers.

The training phase follows the testing and validation. Finally, the classification accuracy is measured. The average accuracy is presented in Table 2.

The classification accuracy is calculated by taking the ratio of the correctly classified case to the total number of cases. The accuracy formula is presented as follows:

$$\text{Accuarcy} = \frac{TP + TN}{N} \times 100 \tag{2}$$

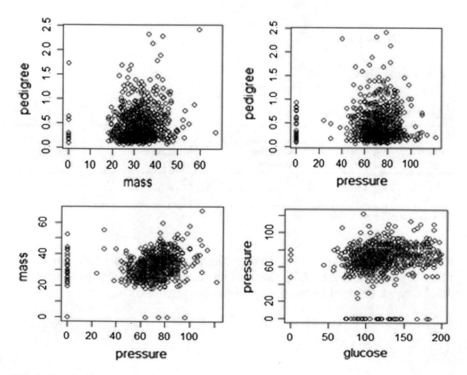

Fig. 3 The details structure of the database

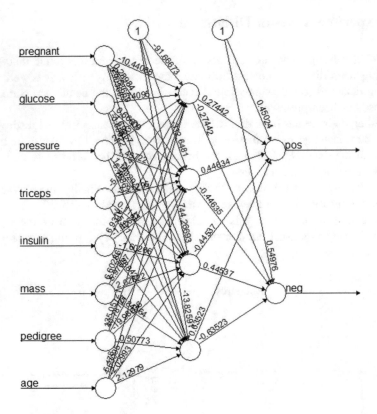

Fig. 4 Trained Neural Network

Table 2 Classification result

Reference				
Diabetes class	Total samples	Classified samples	Misclassified samples	Accuracy (%)
Positive	149	127	22	85
Negative	85	59	26	70

Average accuracy = 77.5%

where TP = Truly positive, TN = Truly Negative and N is the total number of cases. In this experiment, 77.5% classification accuracy is obtained. In Table 3, the detailed result obtained by the R studio has been presented.

Table 3 Result obtained by R studio

True positive (TP)	127
True negative (TN)	56
False positive (FP)	26
False negative (FN)	22
Sensitivity (TP/TP + FN)	0.85
Specificity (TN/FP + TN)	0.68

4 Conclusion

Though the classification result is of less amount but is efficiently classified on application of diabetic data. Further, it can be explored for better accuracy using different tools. Accurate classification can help both doctors and patients for better diagnosis of this disease. MLP has demonstrated to be an efficient tool for data analysis in case of data mining and big data. It has the capability to solve many complicated task in Artificial Intelligence research work. In this paper, MLP method has been performed for women suffering from diabetes. The PIMA Indian Diabetes data set is introduced for training the neural network. In future, other models may be used for even accurate result and also optimization techniques may be applied for comparison.

References

1. M. Nilashi, O. brahim, M. Dalvi, H. Ahmadi, and L. Shahmoradi, "Accuracy Improvement for Diabetes Disease Classification: A Case on a Public Medical Dataset", In Fuzzy Information and Engineering, vol. 9, Issue 3, pp. 345–357, 2017.
2. R. D. Canlas, "Data mining in healthcare: Current applications and issues. School of Information Systems & Management", Carnegie Mellon University, Australia, 2009.
3. http://archive.ics.uci.edu/ml/machine-learning-databases/pima-indians-diabetes/.
4. S. Perveen, M. Shahbaz, A. Guergachi, and K. Keshavjee, "Performance analysis of data mining classification techniques to predict diabetes", Procedia Computer Science, vo. 182, pp. 115–121, 2016.
5. N. Barakat, A. P. Bradley, and M. N. H. Barakat, "Intelligible support vector machines for diagnosis of diabetes mellitus", IEEE transactions on information technology in biomedicine, vol. 14, pp. 1114–1120, 2010.
6. M W. Craven, and J. W. Craven, "Using neural networks for data mining", Future generation computer systems, vol. 13.2–3, pp. 211–229, 1997.
7. L. Sarangi, M. N. Mohanty, and S. Pattanayak, "Design of MLP Based Model for Analysis of Patient Suffering from Influenza", Procedia Computer Science, Vol. 92, pp. 396–403, 2016.
8. J. S. Sonawane, and D. R. Patil, "Prediction of heart disease using multilayer perceptron neural network", In Information Communication and Embedded Systems (ICICES), pp. 1–6, IEEE, February 2014.
9. H. K. Palo, M. N. Mohanty, and M. Chandra, "Use of different features for emotion recognition using MLP network", In Computational Vision and Robotics, Springer, pp. 7–15, 2015.

10. L. Sarangi, M. N. Mohanty, and S. Pattanayak, "Design of MLP Based Model for Analysis of Patient Suffering from Influenza", Procedia Computer Science, vol. 92, pp. 396–403, 2016.
11. M. Maniruzzaman, N. Kumar, M. M. Abedin, M. S. Islam, H. S. Suri, A. S. El-Baz, and J. S. Suri,"Comparative approaches for classification of diabetes mellitus data: Machine learning paradigm", Computer methods and programs in biomedicine, vol. 152, pp. 23–34, 2017.
12. J. Tang, C. Deng, and G. B. Huang, "Extreme learning machine for multilayer perceptron", IEEE transactions on neural networks and learning systems, vol. 27, pp. 809–821, 2016.

An Efficient Design of ASIP Using Pipelining Architecture

Mood Venkanna, Rameshwar Rao and P. Chandra Sekhar

Abstract Embedded Processor are application specific and are used in various field for faster processing. ASIPs are more popular as compared to ASICs processors. Different approaches are used for the ASIP design. In this paper, ASIP is designed based on Pipelining architecture. Considering the execution time in efficient pipelining architecture is proposed. For the proposed pipelining architecture, Clock Gating technique is taken care. The architecture is implemented for Arm processor and verified with I core processor with VLIW architecture. The experiment is carried in VHDL platform. The efficacy of the proposed technology is shown in result section along with the RTL logic diagram.

Keywords Embedded processor · ASIP · Pipeline architecture
ARM-ISA · ICORE

1 Introduction

An embedded system makes our day to day life more useful and secure today. Such a modern system must take into account critical issues such as low power consumption. A high performed system may lead automatically to more power consumption and consequently heat the system that creates design issues. Among different issues of an embedded system, the reduction in power is the main design criterion that creates challenge for today's researchers. The main module of any embedded system is [1–4]:

M. Venkanna (✉) · R. Rao · P. Chandra Sekhar
Dept of Electronics and Communication Engineering, UCE,
Osmania University (OU), Hyderabad, India
e-mail: venkatmood03@gmail.com

R. Rao
e-mail: rameshwar_rao@hotmail.com

P. Chandra Sekhar
e-mail: sekharp@osmania.ac.in

© Springer Nature Singapore Pte Ltd. 2019
M. A. Bhaskar et al. (eds.), *International Conference on Intelligent Computing and Applications*, Advances in Intelligent Systems and Computing 846,
https://doi.org/10.1007/978-981-13-2182-5_12

117

General Purpose Processors (GPPs)
Application Specific Integrated Circuits (ASICs)
Application Specific Instruction set Processors (ASIPs)

An ASIP is designed for a specific application with common characteristics. It is an 'intermediate' solution between a general purpose and a single-purpose processor that provides better flexibility than a single-purpose processor. Also, it provides better performance, power, and size than a general purpose processor. During the exploration of ASIP architecture, we found that one of the important candidates of solving ASIP issues is the pipelined, multi-issue architecture having the vector processing units. The chosen core was a VLIW (Very Long Instruction Word) processor that takes into account the vector operations in which an attempt is made to modify the accommodate energy for better functionality [4–8].

Superscalar processors as well as VLIW have been designed in connection with high-performance processor architecture [7, 8]. These techniques explore the ILP or Instruction-Level Parallelism in order to provide more IPC or Instructions Per Cycle although having different instruction mechanism for scheduling. In a same cycle, the superscalar processors able to receive multiple instructions, capable of detecting ILP and provide parallel execution that leads to more power consumption and system size. However, the VLIW processors have now extra hardware for ILP detection or schedule instructions as these are software analyzed instructions thus, are occupying the instruction memories a priori. Due to the benefit of parallel execution using simple hardware, VLIW processors are more applicable to embedded systems requiring better performance.

For reduction of the flip-flops power consumption, this method intends to generate a VLIW ASIP using a clock gating system. This depends on the registers gating signals which are derived from the registers execution conditions. To minimize pipeline registers power consumption in a large-scale data path corresponding to the proposed VLIW ASIP, we extract automatically the minimum execution conditions during the generation process of ASIP.

VLIW ASIPsor scalar ASIPs must satisfy few tight constraints. These are: area requirement, power consumption, and computation performance. It requires exploration of design space so as to find an ASIP's optimum architecture parameters [8, 9]. An embedded processor evolution mechanism is required with large features capable of consuming low power in a single chip. It must be integrated in a heterogeneous system with reduced area and having low delay. ASIP design is made for the implementation of the functional unit which is either integrated with the chip or can be implemented with the concerned peripheral devices.

The traditional ASIP design gives importance to reduction in area and also the corresponding delay. However, to reduce the power consumption a more efficient ASIP with low power need to be designed. There have been different power reduction techniques such as the clock gating, Register Transfer Level (RTL), and operand isolation [10, 11].

An attempt is made to design a low power but high-performance VLWI ASIP in this work using the clock gating since the VLIW ASIPs generally have many pipeline registers. It needs better gating conditions. Further to reduce the power, it is desired to extract a minimum execution condition. However, its extraction takes more time and is prone to error in a complex data path. The design of low power VLIW ASIP solves these issues automatically. By following this method, we generate the gating signals with minimization of pipeline register functions.

In the design of ASIPs considerable progresses have been made by researchers in this area. Most of the works have focused on generation of special instruction for performance improvement and cost reduction [7–12]. It has involved parallel architectures to the design of ASIP recently. A VLIW ASIP having distributed register structure has been proposed [13]. Exploration of design space for data paths in VLIW ASIP has been proposed by Jacome et al. in [14, 15]. Kathail et al. have advocated VLIW processor design flow that allows non-Programmable hardware [16]. The customized multiprocessors have been proposed in [17] whereas the authors vary the pipeline numbers in design of ASIP in [18, 19].

2 Design Methodology

The one-pipe program of the scheduling process consists of number of sequences. The number of pipeline is selected for exploration of the design space. Then the minimal 2-pipe structure and iterative increase in number to 4-pipelines is made as the exploration progresses. Different pipelines are used to schedule the instructions in the same time slot simultaneously. Each sequence serves an instruction set to the respective pipeline in which the instruction sets are procured from a corresponding program sequence. As in every pipe we obtain a subset of the ISA target architectures the number of pipes is smaller.

The proposed ASIP architecture has been given in Fig. 1. It is a RISC-like design having with four stages. These are (a) pipelines (b) Instruction Fetch (c) Instruction Decode (d) Instruction Execute and Memory Access/Write Back.

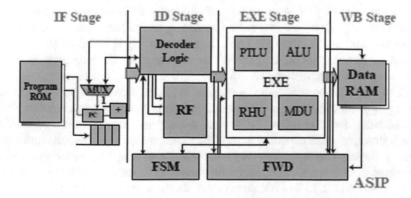

Fig. 1 Proposed pipelined architecture

A. Instruction Pipelining

ASIPs compromises between the Single-Purpose Processors (SPP) and the General Purpose Processors (GPPs). To optimize their data path for a designated operation like the embedded control or the Digital Signal Processing applications, the work designs an ASIP that relies on the CORDIC algorithm. For this purpose, the popular IDE or Xilinx Integrated Development Environment from Xilinx corporations, Inc. has been used.

Instruction pipelining segments the instruction execution into a number of single, simple cycles. In an instruction stream, this leads to successive overlapping of executed instructions. The classical pipeline architecture in RISC processors consists of five stages. These are: (a) instruction fetch (b) instruction decodes(c) instruction execution (d) memory access (e) result write back (WB). This piece of work, the overlapping of the execution of the first instruction in an instruction stream with that of the second and consequently fetching the third instruction takes place. The successive stages utilize the pipeline registers to communicate the intermediate results [2–8].

The pipeline structure of this work has seven stages. The instruction execution (EX) stage implements integer arithmetic and logic instructions with a single cycle integer multiplier. ASIP data paths are implemented using accustom functional unit that works in parallel with the base processor's EX stage. This issue and its possible remedies have been described in the following sections.

In ASIPs, multi-cycle, non-pipelined special instructions create hazards because of the conflict between the multi-cycle ISE and any other special instruction that follows immediately. Pipeline stalling with suitable number of cycles can be used to eliminate structural hazards or data hazards that use NOPs of the compiler or via hardware interlocks. However, it provides lower performance hence, most processors employ data forwarding logicin order to eliminate data hazards.

B. VLIW

In VLIWs to keep the multiple, independent functional units busy enough parallelism is needed in a code segment so as to fill the operation slots. To uncover the parallelism unrolling loops as well as scheduling code is required in a single but larger loop [17]. There are two static scheduling techniques such as local and the global toe used in any compiler. While local scheduling is based on searching the parallelism in code blocks, the global scheduling exploits it across branches and is both complex and expensive [17–20].

VLIWs have inherent technical problems due to large code size as well as having lockstep operation limitations. The size of code size increases when unrolling loops are multiplied by an unrolling factor with the original loop body. To combat this issue instruction word encoding is used that compresses and expands in the main memory during processor decoding. Further by sharing the Instructions in an immediate field corresponding to the functional units the overall instruction word can be reduced [17]. In VLIWs, there exists always a room for clever optimizations as each bit saved is translated into smaller instruction memory.

In absence of hazard detection hardware, VLIW lockstep may lead to unwanted stalls. In this, all units progress from a pipeline stage to another simultaneously in case a stall exists in any unit that creates stall in other units as well. As a result, codes perform poorly when multiple units have the same memory or register. VLIW functional units can be independent if they work asynchronously with instructions and can be accomplished using a hardware hazard detection unit as well as the static scheduling [17].

VLIW has a number of execution slots that are heterogeneous and independent to perform parallel execution if the compiler has enough capability. It is modular thus, suitable for ASIPs as it is possible to add or remove functional units for better target application. The utilization with ASIPs can be accomplished with a single dedicated scalar lane for computing. To perform intensively complex computation new instructions and functional units can be added.

C. Generation of ASIP

For generation of ASIPs automatically different scalar ASIP as well as VLIW ASIP have been proposed [7–18]. The techniques use an ADL or Micro-Operation Description (MOD) which provides the parameters of architecture such as the resources to be used, port information, data flows, etc. Use of MODs in ASIP generation helps to construct automatically the desired data path and the pipeline registers are inserted. The creation of pipeline register control logic is given as

$$enp = stallstage\, p \qquad (1)$$

where p denotes any pipeline register. The term $stagep$ is the pipeline stage number to which p belongs. $Stalln$ denotes a condition of a pipeline interlock which may occur due to multi-cycle operation or structural hazard. In case enp is true then storage of data takes place that facilitates the next pipeline stage. If enp is false, $stagep$ is stalled. As the traditional VLIW ASIP is meant for high-speed with small ASIP area Eq. (1) can be simplified further.

It is trivial to use control signal for clock gating application of ASIP pipeline registers. In this, all pipelines are supplied with clock signals in case the pipelines are no stalled. However, such operation reduces power since in an optimum VLIW case the pipelines are rarely stalled (Table 1).

Table 1 Features of the proposed ASIP

Features	Descriptions
Program ROM	4 K-Byte
Data RAM	2 K-Byte
RF (Register File)	32 × 32 Bits
PTLU	Parallel table look-up-unit
RHU	RSA and HASH unit
ALU	Arithmetic and logic unit
Pipeline	4-stage

Clock Design Mechanism for Memory Implementation

The system performance is also strongly affected by various factors besides its instruction set, the time required to move instruction and data between the CPU and memory components. The clock system is designed to implement ting memory operation execution. The average cycle is designed for implementing the clock cycle required per machine instruction is a measure of computer performances. The clock signal has various characteristics such as clock period, clock pulses, leading and trailing edge. Clock behavior depends on upon the behavior of clock elements with memory architecture with its scheduling approaches. Clock effect can be analyzed by following parameters such as Set up time, hold time and propagation delay time.

In this embedded system, the ASIP is more flexible with better performance and consumes less power. In ASIP system, the designer implements the processor as well as the memory architecture as per the desired applications. The method provides better functionality and low design complexity with retargetable compiler technology and Compiler plays a dominant role in ASIP system. The processor design is manifesting with different design metrics such as cost, power, size, and real-time developments. Commercially, two types of microprocessor RISC and CISC are used in various high-performance embedded computing. The processor unit contains pipeline unit which is controlled by DMA circuits. There are two kinds of pipeline commonly used in processor arithmetic pipeline and instruction pipeline. An instruction pipeline that has been operated using a stream of instructions follows the generalized instruction cycle overlapping fetch, decode, and execute phases. Currently, long instruction memory plays a dominant role in the pipeline mechanism. Various Pipeline mechanisms are used by various processor developer companies such as ARM, Intel and Motorola, etc. according to their performance [20, 21].

Among number of techniques, Gloria et al. [22] pointed out few important ASIP design requirements such as hardware identification and their functions for quicker application, frequently used hardware resources etc. (Fig. 2).

A theoretical analysis of VLIW instructions, SIMD instructions and operation fusions have been given in this section.

Either the data in the parallel modules is processed with a temporal offset of TCLK/N or the units DMUX and MUX must be equipped with data delay units so

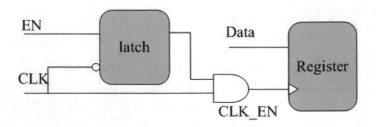

Fig. 2 The clock-gating structure

that simultaneous processing is possible. The throughput of a system with N paths is N times that of a simple system.

$$R_{T,N} = N.R_{T,1}$$

Parallel processing is possible not only for N identical but also for varying sub-circuits. The data dependency of the sub-function and the evaluation of the result must be observed for each individual case. The equation

$$Y = F(x) = \sum_{i=1}^{N} f_i(x) \tag{2}$$

The throughput is determined by the slowest sub-module f_i and the multi-operand adder. Parallel implementation of the function $Y = \sum_{i=1}^{N} f(x)_i$

$$R_{T,N} \sim \frac{1}{\max_{i} T_{D,f_i} + T_{D,ADD} + T_{D,FF}} \tag{3}$$

As an alternative to parallelization, in many cases, it is possible to describe a function F as a series circuit of sub-functions f_i. The result of one sub-function is used as the input signal of the following sub-function.

$$F = f_N(f_{N-1}..f_2(f_1(.))...) \tag{4}$$

Through the insertion of synchronously clocked, intermediate storage registers, architecture analogous to an assembly line is created. This is called pipelining. The throughput is given by

$$R_{T,N} \sim \frac{1}{\max_{I} T_{D,f_i} + T_{D,FF}} \tag{5}$$

If all sub-modules f_i are identical.

$$R_{T,1} = R_{T,1}.N \frac{T_{D,f} + T_{D,FF}}{T_{D,f} + N T_{D,FF}} \tag{6}$$

This means that in particular for large N, due to delay in the registers, the throughput no longer increases in proportion to N. Methods for extracting pipeline architectures will now be further considered. If the clock rate is chosen correctly, the delay of the logic clocks will always be smaller than one clock period minus the transfer times of the flip-flops. A similar behavior is described if one regards the logic block's computation as being instantaneous and assigns the total delay of one clock period to the data transfer flip-flop. It follows that the function of a D-FF can be treated abstractly as a delay operator [20–28].

ASIP Synthesis Steps
Few important ASIP design requirements are provided in Gloria et al. [22].

- Evaluate the possible architectural options and start the design using the application behavior.
- It is essential to Identify and introduce hardware functionalities for mostly used operation to speed up the application.

There are mainly five steps in ASIP synthesis as explained below although there have been many methods proposed for this in this field

1. *Application Analysis*: It is essential to analyze the ASIP design inputs, the test data and specific design constraints for required characteristics/requirements so that it provides the intended support for the hardware synthesis and instruction set generation. In a high-level language, each application is written and analyzed statically as well as dynamically. These analyzed information need to be stored in a suitable intermediate format for further processing.
2. *Exploration of Architectural Design Space*: Based on the design constraints, a possible set of architectures is marked for a particular application and its performance is computed for its suitability based on power consumption, hardware cost, etc.
3. *Generation of Instruction Set*: Instruction set needs to be formulated based on a particular application or selected architecture that helps the process of code and hardware synthesis.
4. *Code Synthesis*: For the intended application the synthesize code is formed using the retargetable code generator or the compiler generator.
5. *Hardware Synthesis*: The ASIP architectural template and instruction set architecture are used to synthesize the hardware is synthesized with the help of VHDL/VERLOG tools.

ICORE Design Methodology
The ICORE design consists of assembler programs as well as the VHDL hardware with respect to the synthesized core that may include the interfaces. In this, the instruction set may comprise of arithmetic and logical operation instructions, control of program flow and movement of data. It acts as a subset of a DSP instruction that may not have special instructions such as bit test, division, rounding, normalization, and loop operations. In ICORE, to suppress the propagation of unwanted values to the functional units, blocking registers or blocking gates connected to a common bus is used. To minimize the power consumption in the ROM of ICORE which happens to be very large, it is essential to minimize the word width of the ROM in ISA design [29, 30].

3 Result Discussion

The aim is to design the architecture having the separate scalar as well as the vector lanes for better performance. In order to expand the vector width or to increase vector lane parallelism, the instruction word needs to be widening or it is desired to issue more instructions per clock cycle similar to VLIW.

To describe the processor in formalism so that net list with synthesis to place and route can be extracted, an RTL hardware description language such as the VHDL or Verilog, has been used is quite the effort, and keeping design exploration iterations quick nigh impossible. The problem can be solved using machine description languages which allows to higher abstraction level processor description that increase the design exploration speed.

Pipelining increases the processor performance due to occurrence of multiple operation in subsequent clock cycles although a sing fetch/decode operation takes place during the first clock cycle.

It is observed that the pipelining is subject to different hazards. As an example, different pipelining stages may access the same memory locations when the address are generated and executed. Thus, the fetch/decode instructions in the pipelining registers are modeled within the registers.

The device utilization is shown in Tables 2 and 3 for the register and processor. The test bench of the simulation result along with the RTL logic diagram is shown in Figs. 3 and 4, respectively.

Table 2 Device utilization summary of the shift register, 3 inputs

Summary of device utilization (estimated values)			
Utilization of logic	Numbers used	Numbers available	Utilization percentage (%)
Slices size	33	3354	0
4input LUTs size	57	7168	0
Bonded IOBs size	36	221	16

Table 3 Device utilization summary of the shift register, four inputs

Summary of device utilization (estimated values)			
Utilization of logic	Numbers used	Numbers available	Utilization percentage (%)
Slices size	794	3684	22
4input LUTs size	1508	7168	21
Bonded IOBs size	55	221	24
GOLKs size	1	8	12

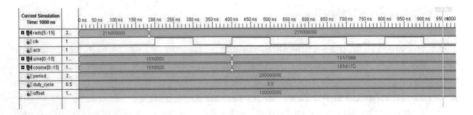

Fig. 3 Test bench of simulation

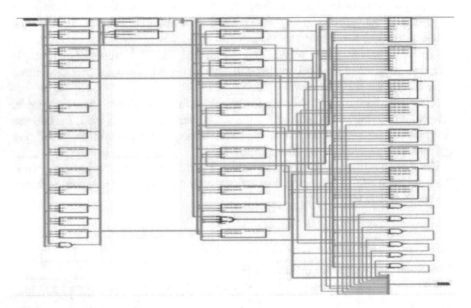

Fig. 4 Implementation of schematic diagram using RTL

4 Conclusion

In this work, we explored the pipelined architecture for ASIP. The key processing
stages are identified and the factors for hardware solution have been investigated.
We observed that the hardware solution also needs to choose suitable clock gating
technique for specific pipelined architecture. The work presents two design
approaches: (1) ASIP solution (2) modified ASIP with pipeline structure. From the
further study, we observed that the parallel nature of the modified ASIP was
incapable of rendering any additional performance which can be made possible by
changing the clocking technique to the processing pipe. Such an attempt has
improved the throughput, and flexibility. On completion of functionality variation
of the core on two levels, cycle-accurate instruction set, and RTL simulation, the
proposed solution was synthesized with two different vector widths. On comparing

the two cores to that of the previous solution, we conclude that the smaller vector width core provides great performance to area ratio while still offering a performance having room in the time budget.

References

1. K. Hwang, and Z. Xu, "Scalable parallel computing: technology, architecture", programming. Boston: WCB/McGraw-Hill, 1998.
2. K. Hwang, "Advanced parallel processing with supercomputer architectures", Proceedings of the IEEE, vol. 75, no. 10, pp. 1348–1379, 1987.
3. Frank Vahid, Tony Givargis, "Embedded System Design" pp. 9–11.
4. M. Johnson, Superscalar Microprocessor Design, Prentice-Hall, Inc., 1991.
5. K. Hwang, and Z. Xu, "Scalable parallel computers for real-time signal processing", IEEE signal processing magazine, vol. 13, no. 4, pp. 50–66, 1996.
6. K. Hwang, and Y. H. Cheng, "Partitioned matrix algorithms for VLSI arithmetic systems", IEEE Transactions on Computers, vo. 100, no. 12, pp. 1215–1224, 1982.
7. J. A. Fisher, "Very long instruction word architectures and the ELI-512", vol. 11, no. 3, pp. 140–150, ACM, 1983.
8. M. F. Jacome, G. de Veciana, and V. Lapinskii, "Exploring performance tradeoffs for clustered VLIW ASIPs", In Proceedings of the 2000 IEEE/ACM international conference on Computer-aided design, pp. 504–510, 2000.
9. B. Middha, A. Gangwar, A. Kumar, M. Balakrishnan, and P. Ienne, "A Trimaran based framework for exploring the design space of VLIW ASIPs with coarse grain functional units", In Proceedings of the 15th international symposium on System Synthesis, pp. 2–7, 2002.
10. F. Sun, S. Ravi, A. Raghunathan, and N. K. Jha, "Synthesis of custom processors based on extensible platforms", In Proceedings of the 2002 IEEE/ACM international conference on Computer-aided design, pp. 641–648, 2002.
11. P. Brisk, A. Kaplan, R. Kastner, and M. Sarrafzadeh, "Instruction generation and regularity extraction for reconfigurable processors", In Proceedings of the 2002 international conference on Compilers, architecture, and synthesis for embedded systems, pp. 262–269, 2002.
12. D. Goodwin and D. Petkov, "International conference on compilers, architecture and synthesis for embedded systems", In Proceedings of the international conference on Compilers, architecture and synthesis for embedded systems, pp 137–147, ACM Press New York, 2003.
13. R. Kastner, S. Ogrenci-Memik, E. Bozorgzadeh, and M. Sar-rafzadeh, "Instruction generation for hybrid reconfigurable systems" In ICCAD, 2001.
14. M. Jacome, G. de Veciana, and C. Akturan, "Resource constrained dataflow retiming heuristics for vliwasips", In Proceedings of the seventh CODES, pp. 12–16, ACM Press, 1999.
15. M. F. Jacome, G. de Veciana, and V. Lapinskii, "Exploring performance tradeoffs for clustered vliwasips", In Proceedings of the ICCAD, pp. 504–510, IEEE Press, 2000.
16. V. Kathail, shailAditya, R. Schreiber, B. R. Rau, D. C. Cron-quist, and M. Sivaraman, "Pico: Automatically designing custom computers", In Computer, 2002.
17. F. Sun, N. Jha, S. Ravi, and A. Raghunathan, "Synthesis of application-specific heterogeneous multiprocessor architectures using extensible processors", In Proceedings of Real Time and Embedded Technology and Applications Symposium, pp. 551–556. IEEE Computer Society, 2005.
18. S. Radhakrishnan, H. Guo, and S. Parameswara, "n-pipe: Application specific heterogeneousmult-pipeline processor design", In The Workshop on Application Specific Processors. IEEE Computer Society, 2005.

19. ARM Limited, "ARM Architecture Reference Manual", 2002.
20. M,. Venkanna, R. Rao, P. Chandra Sekhar, "Application of ASIP in Embedded Design with Optimized Clock Management", ICITKM Conference, Newdelhi, 2017.
21. T. Lang, E. Musoll, and J. Cortadella, "Individual flip-flops with gated clocks for low power datapaths," IEEE Trans. Circuits Syst. II, vol. 44, no. 6, pp. 507–516, June 1997.
22. A. D. Gloria, P. Faraboschi, "An evaluation system for application specific architectures", Proceedings of the 23rd Annual Workshop and Symposium on Microprogramming and Microarchitecture. (Micro 23), pp. 80–89, 1990.
23. http://www.xilinx.com/publications/products/cpld/logic_handbook.pdf.
24. http://en.wikipedia.org/wiki/Xilinx_ISE.
25. O. Schliebusch, A. Chattopadhyay, R. Leupers, G. Ascheid, H. Meyr, M. Steinert, and A. Nohl, "RTL processor synthesis for architecture exploration and implementation. In Design", Automation and Test in Europe Conference and Exhibition, 2004. Proceedings, vol. 3, pp. 156–160, 2004.
26. P. Mishra, A. Kejariwal, and N. Dutt, "Rapid Exploration of Pipelined Processors through Automatic Generation of Synthesizable RTL Models", Proceedings of 14th IEEE International Workshop on Rapid Systems Prototyping, pp. 226–232, June 2003.
27. M. Itoh, Y. Takeuchi, M. Imai, and A. Shiomi, "Synthesizable HDL Generation for Pipelined Processors from a Micro-Operation Description," IEICE Trans. Fundamentals of Electronics Communications and Computer Sciences, vol. E83-A, no. 3, pp. 394–400, March 2000.
28. Y. Kobayashi, S. Kobayashi, K. Okuda, K. Sakanushi, Y. Takeuchi, and M. Imai, "HDL generation method for configurable VLIW processor," Trans. Information Processing Society of Japan, vol. 5, no. 45, pp. 1311–1321, 2004. (in Japanese).
29. T, S. Bitterlich, and H. Meyr, "Increasing the Power Efficiency of Application Specific Instruction Set Processors Using Datapath Optimization", IEEE Workshop on Signal Processing, Lafayette, Louisiana, USA, Oct. 2000.
30. T. Glokler, and S. Bitterlich, "Power efficient semi-automatic instruction encoding for application specific instruction set processors", In Acoustics, Speech, and Signal Processing, 2001. Proceedings.(ICASSP'01), IEEE International Conference, vol. 2, pp. 1169–1172, 2001.

Compaction Behavior of Rubberized Soil

P. T. Ravichandran, G. Priyanga, K. Divya Krishnan
and P. R. Kannan Rajkumar

Abstract Most wastes produced around the world are from agricultural, industrial and chemical sectors in which most are biodegradable, hence they do not pose a threat to our environment while disposal. But few wastes are nonbiodegradable which stay in our environment for more than decades. Hence, proper measures should be taken to dispose these kinds of wastes by which they do not provide hindrance our environment. This study uses waste produced from scrap tires with two types of soils collected from Kancheepuram district. The rubber particles used in this study range from 75 to 425 micron at percentages of 0, 5, 10, 15 and 20%. It was observed that the maximum dry density reduces with the increment in percentage of rubber. There was reduction in optimum moisture content with the increase in rubber % due to the nonabsorbent nature of the rubber particles. From the strength tests, it was found that 10% crumb rubber was chosen as the optimum and disposal of rubber waste in clay is more recommended than in silty soil. This method can be used to effectively dispose rubber waste along with soil as filler material in various structures like highway subgrades, backfill in retaining structures and in embankments with large heights.

Keywords Crumb rubber · Standard proctor test · Compaction characteristics
Unconfined compressive strength ETP sludge · Flyash · Quarry dust

1 Introduction

Industrial growth leads to the development of a country, but it also has its own disadvantages like increase in waste production. These wastes have to be disposed properly without affecting our environment. One such waste product is scrap tires from the automobile industry. They are nonbiodegradable material which can resist

P. T. Ravichandran (✉) · G. Priyanga · K. Divya Krishnan · P. R. Kannan Rajkumar
Department of Civil Engineering, Faculty of Engineering and Technology,
SRM Institute of Science and Technology, Kattankulathur 603203, Tamil Nadu, India
e-mail: ptrsrm@gmail.com

© Springer Nature Singapore Pte Ltd. 2019
M. A. Bhaskar et al. (eds.), *International Conference on Intelligent Computing
and Applications*, Advances in Intelligent Systems and Computing 846,
https://doi.org/10.1007/978-981-13-2182-5_13

extreme wear and tear. Industrial growth leads to the development of a country, but it also has its own disadvantages like increase in waste production. These wastes have to be disposed properly without affecting our environment. Most wastes produced are from agricultural, industrial and chemical sectors in which most of the wastes are biodegradable, hence they don't pose a threat. But few wastes are nonbiodegradable which stay in our environment for more than decades. Hence proper measures should be taken to dispose these kinds of wastes by which they don't provide hindrance our environment. One such waste product is scrap tires from the automobile industry. They are nonbiodegradable material which can resist extreme wear and tear. Hence, proper measures should be adopted to dispose scrap tires in the soil effectively which would not affect our environment in any way.

Past studies show that using rubber tires in the soil usually alters the moisture content and dry density. Therefore, it is necessary for studying the compaction characteristics of the soil along with rubber to ensure that the maximum strength can be derived from the replacement. Mostly, the maximum dry density of the stabilized soil increases with the addition of admixtures like wood ash [1], flyash [2], granulated blast furnace slag [3], gypsum [4], etc. however on addition of rubber it decreases. This is due to the weight of rubber particles being less than the soil. Rubber has been incorporated in the soil in various forms such as, shreds [5], chips [6], and powdered [7] ranging from coarse to medium size. Rubber particles have been combined with clay [7], sand [8], and even with gravel [9] to find the optimum replacement ratio. Mostly 10–15% was found to be the most effective replacement ratio with all the above mentioned materials. In this study, rubber particles belonging to fine size has be adopted with two types of soil, viz., clay and silt to find out their means and effectiveness of disposal in each type.

2 Experimental Programme

The soil samples collected are named as S1 (Clay) and S2 (Silt) both collected from Kancheepuram district from 0.6 m below ground level. The sample S1 is dried for 3 days and then pulverized for further use whereas Sample S2 was collected in dry form as its laterite soil and broken into smaller pieces using hammers to pulverize them. The final form the soils used is shown in Fig. 1a, b. Basic tests [10–14] such as sieve analysis, specific gravity, Atterberg's limits are conducted and the results are tabulated in Table 1. From the results, according to BIS specifications [15] Sample S1 is classified as CH, i.e., clay with high compressibility and Sample S2 is MH, i.e., Silt with high compressibility. The rubber particles used range from 60 to 80 mesh (0.075 to 0.425 mm) and shown in Fig. 1c. The various materials that were used in this investigation and their properties are explained below. The investigations were carried out in three series, each involving six different mixes of materials, thus having 18 different mixes.

    (a) Sample S1   (b) Sample S2   (c) Crumb Rubber

Fig. 1 Pictorial representation of materials

Table 1 Properties of soil samples S1 and S2

Soil	Specific gravity	Sieve analysis		Atterberg's limit				Swell (%)
		Sand (%)	Silt and clay (%)	LL (%)	PL (%)	SL (%)	PI (%)	
S1	2.43	38	62	52	21	5	31	70
S2	2.31	46	54	60	22	6	38	40

3   Results and Discussions

3.1 *Compaction Characteristics of Soil Samples with and Without Rubber*

Standard proctor test [16] is conducted according to BIS to determine the density and moisture content necessary for the soil to give its highest strength. The soil samples S1 and S2 is mixed with 0, 5, 10, 15, and 20% rubber and the moisture content and density is calculated for each combination with different moisture contents and graph is plotted with the density at y-axis and moisture content at x-axis to find the maximum density and optimum water content corresponding to the highest point. The compaction characteristic curve for soil S1 and S2 is shown in Fig. 2 with the values tabulated in Table 2. It was observed that the compaction characteristics reduces with the gradual increment of rubber particles with the soil. This is because the weight of the rubber particles is less than that of the soil particles and the density becomes less. Also, the moisture content decreases due to the nonabsorbent nature of the rubber particles, so as the percentage of rubber added increases the water needed for the soil becomes less.

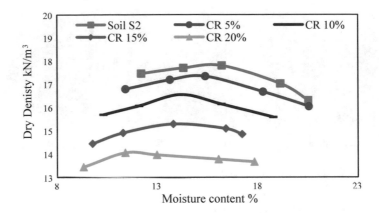

Fig. 2 Compaction characteristic curve of Soil (Sample S2)

Table 2 Proctor results of sample S1 and S2

CR %	OMC (%)		MDD (kN/m³)	
	Sample S1	Sample S2	Sample S1	Sample S2
0	16	16.2	18.1	17.88
5	14.5	15.38	17.92	17.33
10	13.8	14.17	17.12	16.54
15	12.9	13.8	16.39	15.28
20	12.5	11.41	15.9	14.05

3.2 Strength Characteristics of Soil Samples with and Without Rubber

Unconfined Compressive tests [17] were conducted according to BIS to access the strength of these rubberized soil. The maximum dry density and optimum moisture content determined for each soil rubber combination is used to prepare the respective UCS samples. The change in strength for both soil samples S1 and S2 with and without rubber is shown in Fig. 3. It is observed that 10% addition of rubber gives the maximum strength for both samples after which it decreases. Even though the strength increases, for soil sample S2, there was difficulty in compacting and preparing the samples as there was less binding. Hence rubber was more workable with clay than silty soil.

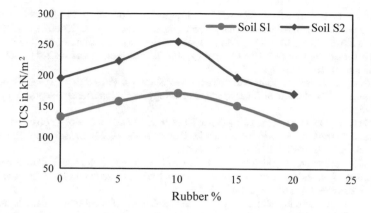

Fig. 3 UCS results of soil sample S1 and S2

4 Conclusions

From the experiments conducted on two different types of soil sample with and without rubber along with the results obtained the following conclusions can be arrived

(1) From proctor results, it is observed that the compaction characteristics decrease with percentage addition of fine rubber.
(2) Unconfined compressive test results indicate that 10% rubber replacement is the optimum percent to derive the maximum strength from the soil rubber combinations.
(3) It is recommended to dispose the rubber waste in clay soil more than in sand or silty soils to avoid less binding nature of the rubber particles.
(4) By this method, rubber wastes can be disposed in the soil safely but it is necessary to study the compaction characteristics of the soil with rubber before applying in the field.

Compressive Strength and Water Absorption tests were performed on the sludge prepared bricks and the obtained results were used in order to draw conclusions. The procedures and the results are discussed.

References

1. M. Usha Rani., J. Martina Jenifer.: Analysis of strength characteristics of black cotton soil using wood ash as stabilizer. International Journal of Research in Science and Technology, vol. 6 (1), pp. 171–179 (2016).
2. S. Bhuvaneshwari., R. G. Robinson., S. R. Gandhi.: Stabilization of expansive soils using Flyash. Fly ash utilization program (FAUP), TIFAC, DST, Delhi, Vol. 8, pp. 5.1–5.10 (2005).

3. Anil Kumar Sharma., P. V. Sivapullaiah.: Improvement of Strength of Expansive soil with waste Granulated Blast Furnace Slag. Geo Congress, ASCE, pp. 3920–3928 (2012).
4. Divya Krishnan. K., Mani Deepika., P. T. Ravichandran., C. Sudha., Ajesh K. Kottuppillil.: Study on Behavior of Soil with Phosphogypsum as Stabiliser. Indian Journal of Science and Technology, vol. 9(23), pp. 1–5 (2016).
5. S. Dukare., P. Barad.: Effect of Rubber Tyre Shred on Properties of Black Cotton Soil. International Conference on Science and Technology for Sustainable Development, pp. 80–87 (2016).
6. Priyanga G, Divya Krishnan K, P. T. Ravichandran, Characteristics of Rubberized Soil with Ground Granulated Balst-Furnace Slag as Binder Material, Materials Today: Proceedings, Vol. 5, Issue 2, April—2018, pp. 8655–8661.
7. P. T. Ravichandran., A. Shiva Prasad., K. Divya Krishnan, P. R. Kannan Rajkumar.: Effect of Addition of Waste Tyre Crumb Rubber on Weak Soil Stabilization. Indian Journal of Science and Technology, vol 9(5), pp. 1–6 (2014).
8. Changho lee., Hosung shin., Jong-sub lee.: Behavior of sand-rubber particle mixtures: experimental observations and numerical simulations. International journal for numerical and analytical methods in geomechanics, vol. 38, pp. 1651–1663 (2014).
9. Ishtiaq Alam., Umer Ammar Mahmood., Nouman Khattak., Use of Rubber as Aggregate in Concrete: A Review. International Journal of Advanced Structures and Geotechnical, 2015, vol. 04(02), pp. 92–96 (2015).
10. IS 2720 (Part 3) 1980, "Method of tests for soils— Determination of specific gravity", Bureau of Indian Standard, New Delhi, Reaffirmed 2002.
11. IS 2720 (Part 4) – 1985, "Methods of test for soils- Grain size analysis", Bureau of Indian Standard, New Delhi, Reaffirmed 2006.
12. IS 2720 (Part 5) 1985, "Determination of Liquid and Plastic Limit", Bureau of Indian Standard, New Delhi, Reaffirmed 2006.
13. IS 2720 (Part 6) 1972, "Determination of Shrinkage Factors", Bureau of Indian Standard, New Delhi, Reaffirmed 2001.
14. IS 2720 (Part 40) 1977, "Determination of Free Swell Index of Soils", Bureau of Indian Standard, New Delhi, Reaffirmed 2002.
15. IS 1498 1970, "Classification and Identification of Soils for General Engineering Purposes", Bureau of Indian Standard, New Delhi, Reaffirmed 2007.
16. IS 2720 (Part 7) 1980, "Determination of Water Content-Dry Density Relation Using Light Compaction", Bureau of Indian Standard, New Delhi, Reaffirmed 2011.
17. IS 2720 (Part 10) 1991, "Determination of Unconfined Compressive Strength", Bureau of Indian Standard, New Delhi, Reaffirmed 2006.

Fractional Segmental Transform for Speech Enhancement

**Rashmirekha Ram, Hemanta Kumar Palo
and Mihir Narayan Mohanty**

Abstract The Fractional Fourier Transform (FrFT) can be interpreted as a rotation in the time-frequency plane with an angle α. It describes the speech signal characteristics as the signal changes from time to frequency domain. However, to locate the fractional Fourier domain frequency contents and multicomponent analysis of nonlinear chirp like signals such as speech the Short-Time FrFT (SFrFT) can provide an improved time-frequency resolution. By representing the time and fractional frequency domain information simultaneously, the SFrFT can filter out cross terms and distortion in a signal adequately for better signal enhancement. The method has experienced with better Signal to Noise Ratio (SNR) and Perceptual Evaluation of Speech Quality (PESQ) under different noisy conditions as compared to the conventional FrFT in our results.

Keywords Speech enhancement · Fractional Fourier transform
Short-time fractional Fourier transform · Signal to noise ratio · Perceptual
evaluation of speech quality

1 Introduction

In spite of advanced recording and communication environment, additive noise is inherent with a signal and corrupts it. Many methods have been adopted over the last few decades particular to make the speech signal free from environmental noise to make it perceptually better and intelligible. However, the performance of these

R. Ram (✉) · H. K. Palo · M. N. Mohanty
Department of Electronics and Communication Engineering, ITER, Siksha 'O' Anusandhan
(Deemed to be University), Bhubaneswar, Odisha, India
e-mail: ram.rashmirekha14@gmail.com

H. K. Palo
e-mail: hemantapalo@soa.ac.in

M. N. Mohanty
e-mail: mihir.n.mohanty@gmail.com

© Springer Nature Singapore Pte Ltd. 2019
M. A. Bhaskar et al. (eds.), *International Conference on Intelligent Computing
and Applications*, Advances in Intelligent Systems and Computing 846,
https://doi.org/10.1007/978-981-13-2182-5_14

enhancement methods is not always satisfactory that motivates the present study. The application domain includes the design of hearing aids, the mobile voice communication system such as the Skype, Internet Protocol (IP) telephony, robust speech, emotion recognition systems, etc. [1–5]. Nevertheless, the task of research becomes complex due to many environmental noise factors such as traffic, vehicular movements, electrical transmission and distribution, airports, road works, etc., which makes the speech communication noisy.

There have been many speech enhancement algorithms proposed by the research community with some remarkable successes. These front-end enhancement techniques are broadly classified into either single or multi-channeled. The single channel algorithms are applied in case there is only a single microphone available for the user and have been popular in many real-time applications [1, 2]. The spectral subtraction, statistical models, etc., are most popular single channel approaches used to enhance the speech signals. While the spectral subtraction algorithm subtracts the short-term noise spectrum estimates to derive the clean speech spectrum, the iterative Wiener filtering approach uses an all-pole model for speech enhancement. A major drawback of these methods is that the enhanced speech is generally subjected to a musical noise or the annoying artifact that makes the Minimum Mean Squared Error (MMSE) the state-of-the-art methods in this field. On the other hand, the conventional unsupervised MMSE methods are based on either the speech and noise statistics or the additive nature of the environmental noise, hence unable to follow the out of scene acoustic noisy conditions adequately [7]. Another point of interest with regards to these enhancement approaches is the use of Short Time Fourier Transform (STFT) of the analyzed signal for better speech quality in the presence of low additive noise although, its performance degrades drastically in the presence of high-additive noise [6].

Recent years witnessed the application of Fourier Transform (FT) in the fractional domain for filtering the noisy speech signal with a better result as compared to the traditional FT. It is the generalized form of the traditional FT but unlike FT, it decomposes the signal as linear chirps. In this, the signal under consideration rotates at a certain angle $\propto$ along the axis u with respect to the time axis instead of multiples of $\pi/2$ in FT. Thus, it is possible to remove the noisy parts and cross terms in the nonstationary signal that makes FrFT more appealing for speech processing [8, 9]. In their approach, the noisy speech signal has been investigated in the presence of Babble, car and White Gaussian Noise (WGN). The results prove to be better both in terms of Signal-to-Noise Ratio (SNR) and Perceptual Evaluation of Speech Quality (PESQ) when FrFT has been compared with the traditional FT filtering approach. Briefness in realization, simplicity, ease of implementation and low computation complexity makes FrFT more robust than other methods [10, 11].

However, FrFT approach is unable to describe the spectral transitions in a nonstationary environment. A better representation of a nonstationary signal such as speech can be made in case the signal is viewed on a stationary platform with the application signal segmentation. This motivates the authors to explore STFT in a time-localized short window in the fractional domain so as to spectrally represent the enhanced signal in a better way. The resultant Short-Time Fractional FT is denoted as SFrFT has been compared with the FrFT method of filtering in terms of improved SNR and PESQ in the result section.

The work considers the detailed methodology proposed for enhancement of speech signal in Sect. 2. Section 3 tabulates the comparison results of both FrFT and SFrFT approach for speech enhancement with few graphical representations. Section 4 concludes the work, with the possible future directions.

2 Methodology

The block diagram of the proposed approach is shown in Fig. 1. The detailed algorithm is explained below.

The FrFT or Fr of a signal $s(t)$ is represented as

$$S_p(u) = Fr[s](u) = \sum_{-\infty}^{\infty} s(t) K_p(t, u) dt \tag{1}$$

Fig. 1 The proposed enhancement technique

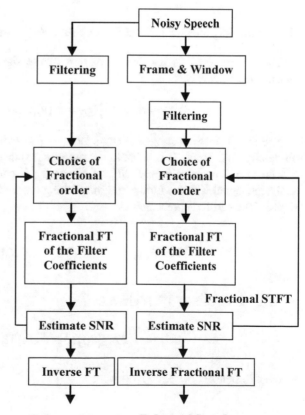

Enhanced Speech   Enhanced Speech

where $K_p(t, u)$ is the kernel function of the FrFT and is given as

$$
K_p(t, u) = \begin{cases} \sqrt{\frac{1 - j\cot\alpha}{2\pi}} \exp\left(j\left(\frac{t^2 + u^2}{2}\right)\cot\alpha - jtu\csc\alpha\right), & \alpha \neq n\pi \\ \delta(t - u), & \alpha = 2n\pi \\ \delta(t + u), & \alpha = (2n \pm 1)\pi, \end{cases} \tag{2}
$$

where Fr denotes the FrFT operator, $\alpha = p\pi/2$ is the angle of rotation of the transformed signal with transform order p, and n is an integer.

The technique consists of mainly three steps as follows:

i. A product of the signal by a chirp as given by
$$
\varphi(t) = (\sqrt{1 - j\cot\alpha})\exp\left(j\left(\frac{t^2}{2}\right)\cot\alpha s(t)\right)
$$

ii. Obtaining the FT with its argument scaled by $\csc\alpha$, i.e.

$$
\tilde{S}_p(u) = \varphi'(\csc\alpha u) \text{ with}
$$

$$
\varphi'(u) = \frac{1}{\sqrt{2\pi}} \sum_{-\infty}^{\infty} \varphi(t) \exp(-j\omega t) dt;
$$

iii. Performing another product by a chirp, $S_p(u) = \exp\left(j\left(\frac{u^2}{2}\right)\cot\alpha\right)\tilde{S}_p(u)$

The inverse FrFT is the decomposition of the signal under consideration by chirps and is given by

$$
s(t) = Fr_{-p}[S_p](t) = S_p(u)K_{-p}(t, u)du \tag{3}
$$

The SFrFT does not subject to cross terms or distort the time-frequency structure on de-chirping. Hence it is suitable for time-frequency analysis of speech signal.

The SFrFT or the pth order SF_p of a signal s is computed by multiplying the analyzing signal with a window and then taking the fractional FT of the windowed signal. This can be represented as

$$
ST_p[s](t, u) = \int_{-\infty}^{\infty} s(t')\varphi(t' - t)K_p(t, u)dt' \tag{4}
$$

And, the inverse SFrFT is given by

$$
s(t') = \int Fr_{-p}[S(t, u)\varphi(t' - t)](t, t')dt
$$

where ST_p denotes the SFrFT operator

3 Results and Discussion

The database used in this work consists of different speech signals recorded in a noise-free environment as far as possible. Each signal sample is passed through a low order front-end digital system comprising of preemphasis filtering, normalization and mean subtraction stages. This is done to spectrally flatten the signal and to make it less susceptible to finite precision effects later in the signal processing. To obtain the framed signal a frame size of 30 ms with 10 ms overlapping between frames has been used. The popular and effective Hamming window suitable for speech signal processing application has been chosen before application of the filtering algorithms adapted in this work. The clean speech signal 'The birds can fly on the smooth point' has been obtained from a male speaker with a sampling frequency of 8 kHz with 8-bit precision as shown in Fig. 2. The clean speech signal is added to different noises such as babble, car, airport, and street of different dB for the task of analyzing the signal. The noisy signal in the presence of babble noise of 5 dB is shown in Fig. 3.

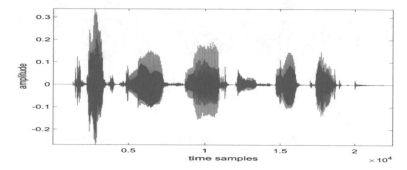

Fig. 2 Clean signal 'The birds can fly on the smooth point'

Fig. 3 Noisy signal (Babble noise with SNR of 5 dB)

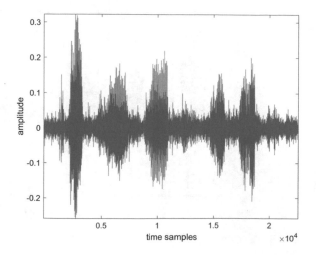

Figures 4 and 5 provide the effect of fractional FT and fractional STFT in enhancing the speech signal under consideration, respectively. The fractional SFrFT has indeed reduced the noise and cross terms in the signal in comparison to the fractional FT as observed from these Figures. For better analysis in terms of SNR and PESQ, fractional orders of different magnitude that ranges from −2 to 2 have been verified for the said task. An order of 1.2 has provided the optimum performance and hence retained for simulation of the signals. The enhanced signal outputs have been extracted using the inverse of the fractional FT and fractional STFT outputs.

Table 1 provides the SNR with different filtering approaches adopted in this work in the presence of babble, car, airport, and street noises. A noise signal of 0, 5,

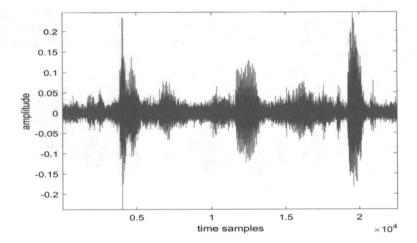

Fig. 4 The enhanced signal using fractional FT

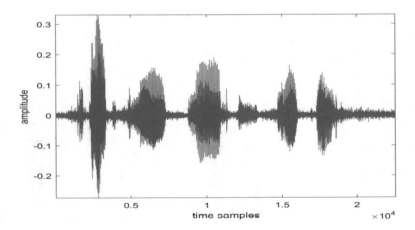

Fig. 5 The enhanced signal using fractional STFT

Table 1 Comparison of enhancement methods in terms of SNR

Noise type	Input SNR (dB)	Fractional FT	Improvement in SNR (dB)	Fractional STFT	Improvement in SNR (dB)
Babble noise	0	1.21	1.21	1.32	1.32
	5	5.89	0.89	6.12	1.12
	10	10.97	0.97	11.11	1.11
	15	16.21	1.21	16.78	1.78
Street noise	0	1.11	1.11	1.87	1.87
	5	6.21	1.21	6.83	1.83
	10	10.97	0.97	11.53	1.53
	15	15.64	0.64	16.68	1.68
Car noise	0	1.23	1.23	1.76	1.76
	5	6.14	1.14	6.78	1.78
	10	10.89	0.89	11.12	1.12
	15	15.98	0.98	16.17	1.17
Airport noise	0	1.31	1.31	1.89	1.89
	5	6.26	1.26	6.77	1.77
	10	11.04	1.04	12.02	**2.02**
	15	16.18	1.18	16.96	1.96

Table 2 Comparison of enhancement methods based on PESQ score

	Input SNR (dB)	Fractional FT	Fractional STFT
Babble noise	0	1.32	2.21
	5	1.69	1.87
	10	1.84	1.95
	15	2.21	2.32
Street noise	0	1.87	2.11
	5	2.01	2.32
	10	2.11	2.43
	15	2.32	2.53
Car noise	0	1.69	2.52
	5	2.11	2.43
	10	2.63	2.73
	15	2.98	3.01
Airport noise	0	2.01	2.34
	5	2.31	2.54
	10	2.42	2.55
	15	2.53	2.76

10, and 15 dB corresponding to different types has been tested for validation of our speech enhancement results. It is shown in Table 1 for comparison purpose. The fractional short-time FT method of filtering has shown to outperform the fractional FT method as our results reveal. This has been validated with an increase in PESQ as observed in Table 2.

4   Conclusion

It is observed that the filtering approach used to enhance the speech signals in the short-time fractional domain has provided a clearer signal as compared to the conventional fractional FT approach. The improvement has been experienced both in terms of SNR and PESQ as observed in the result section. The findings have been validated in presence of different types of noisy environments such as Babble, airport, car, and street noise. In future, the work may be extended to the application of other time-frequency analysis and efficient feature extraction methods as a comparison platform for efficient enhancement.

References

1. Loizou, P.C.: 'Speech Enhancement: Theory and Practice' (CRC Press, Boca Raton, FL, USA, 2007).
2. P. Mowlaee, J. Stahl, & J. Kulmer, "Iterative joint MAP single-channel speech enhancement given non-uniform phase prior". *Speech Communication*, 86, pp. 85–96, 2017.
3. H. Barfuss, C. Huemmer, A. Schwarz, & W. Kellermann, "Robust coherence-based spectral enhancement for speech recognition in adverse real-world environments", *Computer Speech & Language,* 46, pp. 388–400, 2017.
4. R. Ram, M. N. Mohanty, "Comparative Analysis of EMD and VMD Algorithm in Speech Enhancement", *Journal of Natural Computing Research (IJNCR)*, 6(1), pp. 17–35, 2017.
5. A. Bhowmick, M. Chandra, "Speech enhancement using voiced speech probability based wavelet decomposition", *Computers & Electrical Engineering*, pp. 1–13, 2017 (in press).
6. R. Ram, H. K. Palo, M. N. Mohanty, "An Adaptive Method for Emotional Speech Enhancement and Recognition Using PNN", *International Journal of Control Theory and Applications*, 8(5), pp. 2395–2403, 2015.
7. Y. Xu, J. Du, L. R. Dai, C. H. Lee, "A regression approach to speech enhancement based on deep neural networks", *IEEE/ACM Transactions on Audio, Speech and Language Processing (TASLP)*, 23(1), pp. 7–19, 2015.
8. R. Ram, M. N. Mohanty, "Use of Fractional Domain for Speech Enhancement", *Int. J. Of Imaging and Robotics*, 18, pp. 85–93, 2018.
9. R. Ram, M. N. Mohanty, "Design of Fractional Fourier Transform based Filter for Speech Enhancement", *International Journal of Control Theory and Applications*, 10(7), pp. 235–243, 2017.
10. J. Wang, "Speech Enhancement based on Fractional Fourier transform", WSEAS *Transactions on Signal Processing, 10*, pp. 576–581, 2014.
11. R. Tao, Y. Lei, Y. Wang, "Short-time fractional Fourier transform and its applications", *IEEE Transaction on Signal Processing, 58*(5), pp. 2568–2580, 2010.

Throughput Maximization in WPCN: Assisted by Backscatter Communication with Initial Energy

Richu Mary Thomas and Subramani Malarvizhi

Abstract This paper proposes a backscatter-assisted wireless powered communication network (WPCN), which includes a Hybrid Access Point (HAP) and multiple users. The communication protocol is that the users first harvest energy in the downlink (DL) and use this tapped energy to transmit their information signals in the uplink (UL). Furthermore, in conventional WPCN protocol, with only harvest-then-transmit (HTT) mode, urgent data transmission is not possible since users need to harvest sufficient energy before transmitting information. Here each user is equipped with some initial energy, which generalizes the previous works on energy transfer and initial energy. Our goal is to maximize the system energy efficiency while guaranteeing the user's quality of service via joint time allocation and power control in both DL and UL. Both HTT and backscatter modes are employed at the users in the proposed model to improve the system performance. An optimization problem is formulated to maximize the sum-throughput by finding the optimal transmission policy, including the optimal users' working mode permutation and time allocation. Simulation results demonstrate the superiority of the proposed model.

Keywords Wireless powered communication network · Harvest-then-transmit mode · Backscatter communication · Resource allocation

R. M. Thomas (✉) · S. Malarvizhi
Department of ECE, SRM Institute of Science and Technology, Kattankulathur, Kancheepuram, India
e-mail: richumarythomas_thomas@srmuniv.edu.in

S. Malarvizhi
e-mail: malarvizhi.g@ktr.srmuniv.ac.in

© Springer Nature Singapore Pte Ltd. 2019
M. A. Bhaskar et al. (eds.), *International Conference on Intelligent Computing and Applications*, Advances in Intelligent Systems and Computing 846,
https://doi.org/10.1007/978-981-13-2182-5_15

143

1 Introduction

The trade-off between the information transmission and the energy transfer in SWIPT (Simultaneous Wireless Information and Power Transfer) in a MIMO (Multiple Input Multiple Output) wireless systems was studied in [1]. The authors in [2] have proposed HTT (Harvest Then Transmit) protocol for WPCN (Wireless Powered Communication Network) where users first harvest energy in DL (Down Link) WET (Wireless Energy Transfer) stage and then transmit the information signals in UL (Up Link) WIT (Wireless Information Transfer). OFDM (Orthogonal Frequency Division Multiplexing) was employed by [3] to divide the broadband channel into many sub-channels to transmit information and energy. WPCN started to be studied in [2, 4–9]. A WPCN consists of a HAP (Hybrid Access Point) and multiple numbers of users with rechargeable batteries or supercapacitors inside them, providing energy. In a WPCN, the users work in TDMA (Time Division Multiple Accessing) schedules and optimal time allocation policy is maintained. In [8], multiple numbers of antennas were employed at the HAP to increase the system performance, thus transmitting more energy to the users. Energy beamforming was introduced in HAP in [9]. Later, HAP in FD (Full Duplex) mode and HD (Half Duplex) modes had been studied in [10, 11]. In [12], ambient backscattering was put forward, where the users harvested energy from the surrounding sources such as a TV tower. But this model was limited by the transmission range (Fig. 1).

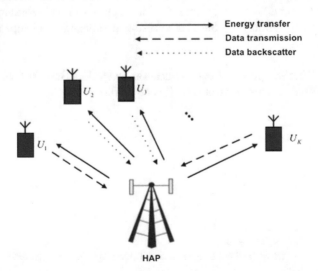

Fig. 1 Backscatter-assisted wireless powered communication network

2 System Model

A WPCN which consists of one HAP with two antennas and N wireless powered users are considered. The HAP is assumed to work with perfect Interference Cancellation techniques. One antenna of HAP is exclusively used for transferring energy to the users in the downlink mode. The second antenna of the HAP is used for receiving information signals from users in the uplink mode, working simultaneously. The users are provided with supercapacitors for storing an initial energy Q, prior to the transmission process. This stored energy can be the leftover of the previous transmissions or from ambient backscattering. Users can adaptively switch between HTT and backscatter modes and transmit or backscatter information to the HAP in TDMA mode. The block structure for the BAWPCN system is described in Fig. 2. In the proposed system, the HAP continuously broadcasts energy with constant power, denoted as P, to all users during one block. The block is divided into $K + 1$ main time slots, denoted as t_i, $i = 0, ..., K$. Ui is allocated with t_i, $i = 1, ..., K$. The main time slot ti is further divided into three parts, denoted as α_i, β_i, and $t_i - \alpha_i - \beta_i$, $i = 1, ..., K$. α_i and β_i are utilized for the HTT mode, and the remaining part is used for the backscatter mode. During DL WET, the power station transmits with power Po and time τ_0. The energy obtained from noise signals and UL WIT signals are assumed to be very small. Thus the amount of energy harvested at the user n can be expressed as

$$E_n^h = \eta \tau_0 P_0 Q_n h_n$$

where efficiency $\eta \in (0,1]$ is the energy conversion efficiency of the receiver and h_n is the forward channel gain.

A. *During HTT mode*

1. Uplink WIT

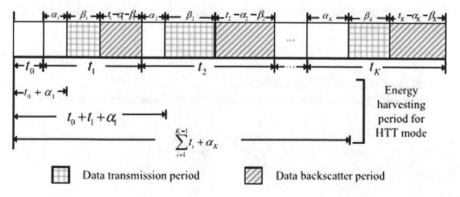

Fig. 2 Block structure for backscatter-assisted wireless powered communication network

The achievable throughput of user N is

$$B_N = \tau_N W \, \log_2 \left(1 + \frac{P_n Q_n g_n}{\sigma^2} \right),$$

where P_n is the transmit power.

2. Downlink WET

Achievable throughput of user N is

$$E_N = \eta_n P Q_n h_n \left(\sum_{j=0}^{N-1} t_n + \alpha_n \right); \quad n = 1, \ldots, N$$

3. Achievable throughput of N for HTT mode is given by

$$\begin{aligned}
R_n^h &= \beta_n W \log_2 \left[1 + \frac{E_n g_n}{\beta_n \sigma^2} \right] \\
&= \beta_n W \log_2 \left[1 + \frac{\sum_{j=0}^{N-1} t_n + \alpha_n}{\beta_n} \gamma_n \right], \quad n = 1, \ldots, N
\end{aligned}$$

where W is the bandwidth and $\gamma_n = \frac{\eta_n P h_n g_n}{\sigma^2}$

B. *During Backscatter mode*

The impedances of the HAP receiver antenna and the transmission antenna of the users are deliberately mismatched so that there will be a backscatter of data. The carrier wave from the HAP (to transfer energy to Ni) itself is used by the Ni to encode data (0 or 1) onto the uplink signal; then this information is decoded by the HAP.

4. The throughput of the user, which is dependent on time is obtained by the following equation:

$$R_n^h = (t_n - \alpha_n - \beta_n) B_n^b$$

Sum throughput of all users = R
= Harvest-Then-Transmit mode throughput + Backscatter mode throughput

$$R = \sum_{n=1}^{N} R_n^h + \sum_{n=1}^{N} R_n^b$$

3 Sum-Throughput Maximization

The sum-throughput maximization of a BAWPCN system can be done by optimizing the transmission policy. An optimization problem is first formulated. We first consider a special case, i.e., $N = 2$, to obtain some understanding. Based on these insights, the general case ($N \geq 3$) is further solved. Denote $t = [t_0, ..., t_N]$, $\alpha = [\alpha_1, ..., \alpha_N]$, and $\beta = [\beta_1, ..., \beta_N]$. The optimization problem is formulated as follows:

$$\max_{t,\alpha,Q,\beta} R(t, Q, \alpha, \beta)$$

$$\text{s.t.} \sum_{i=0}^{N} t_n \leq T,$$

$$\alpha_i + \beta_i \leq t_i, \quad i = 1, \ldots, N,$$

$$\alpha_i \geq 0, \quad i = 1, \ldots, N,$$

$$\beta_i \geq 0, \quad i = 1, \ldots, N,$$

$$Q_i \geq 0, \quad i = 1, \ldots, N,$$

$$ti \geq 0, \quad i = 0, 1, \ldots, N.$$

4 The Optimal Solution for Single User Case

The optimization problem for single user case could be expressed as

$$\max_{t_0 t_1 \beta_1} \beta_1 W \log_2 \left(1 + \frac{t_0}{\beta_1}\gamma_1\right) + (t_1 - \beta_1)B_1^b$$

$$\text{s.t.} \ t_0 + t_1 = \bar{T} \tag{P3}$$

$$0 \leq \beta_1 \leq t_1$$

$$t_0 \geq 0$$

The relationship between t_0 and β_1 in the following has to be derived to solve the equation P3:

$$\max \beta_1 W \log_2 \left(1 + \frac{t_0}{\beta_1}\gamma_1\right)$$

$$\text{s.t.,} \ t_0 + \beta_1 = \bar{T}$$

$$t_0, \beta_1 \geq 0$$

On solving it,

$$\sum_{i=1}^{N} \frac{\gamma_i}{1 + \gamma_i \frac{T_0}{\beta_1}} = v \ln 2$$

$$t\left(\gamma_i \frac{T_0}{\beta_1}\right) = v \ln 2; \quad 1 \leq i \leq N$$

$$t(x) = \ln(1 + x) - \frac{x}{1 + \dot{x}}$$

$$t\left(\gamma_i \frac{T_0}{\beta_1}\right) = t\left(\gamma_j \frac{T_0}{\beta_j}\right), \quad i \neq j$$

$$\frac{\gamma_1}{T_1} = \frac{\gamma_2}{T_2} = \ldots \frac{\gamma_N}{T_N} = \text{ constant } C$$

$$\ln(1 + C_{T_0}) - \frac{C_{T_0}}{1 + C_{T_0}} = \frac{\gamma}{1 + C_{T_0}}$$

This equation can be modified as

$$z \ln z - z - \gamma + 1 = 0, \quad \text{where } z = 1 + \frac{\gamma T_0}{1 - T_0}$$

The optimal time allocation to DL WET is

$$T_0 = \frac{Z_1^* - 1}{\gamma_1 + Z_1^* - 1} \overline{\overline{T}} \text{and}$$

the optimal time allocation to UL WIT is

$$\beta_1 = \frac{\gamma_1}{\gamma_1 + Z_1^* - 1} \overline{\overline{T}}$$

$$\text{where } \hat{T} = \begin{cases} \overline{T}; & B_1^t \geq B_1^b & \ldots(a) \\ 0; & B_1^t < B_1^b & \ldots(b) \end{cases} \tag{1}$$

5   Optimal Solution for Two-User Case

The optimal time allocation to DL WET is

$$\overline{T} = \frac{Z_2^* - 1}{\gamma_2 + Z_2^* - 1} \hat{T} \text{ and}$$

the optimal time allocation to UL WIT is

$$\beta_2 = \frac{\gamma_2}{\gamma_2 + Z_2^* - 1} \overline{\overline{\hat{T}}}$$

$$\text{where} \quad \hat{T} = \begin{cases} T; & B_2^t \geq B_2^b \quad ...(c) \\ 0; & B_2^t < B_2^b \quad ...(d) \end{cases} \tag{2}$$

Combining (1) and (2), permutations of 2 users' working mode can be summarized as

If conditions (a) and (c) are satisfied, then users N_1 and N_2 work in HH mode. If conditions (a) and (d) are satisfied, then users N_1 doesn't work, and N_2 works in backscatter mode (0B). If N_1 and N_2 work in HH mode, t_0 can be rewritten as follows

$$t_0 = \begin{cases} 0, \\ \frac{Z_1^*-1}{\gamma_1+Z_1^*-1} \frac{Z_2^*-1}{\gamma_2+Z_2^*-1} \end{cases} \text{T and}$$

$$\beta_1 = \begin{cases} 0, \\ \frac{\gamma_1}{\gamma_1+Z_1^*-1} \frac{Z_2^*-1}{\gamma_2+Z_2^*-1} T \end{cases}$$

$$t_1 = \begin{cases} 0, \\ \frac{Z_3^* \cdot 1}{\gamma_2+Z_3^*-1} T, \quad \text{and} \\ \frac{\gamma_1}{\gamma_1+Z_1^*-1} \frac{Z_2^*-1}{\gamma_2+Z_2^*-1} T \end{cases}$$

$$\beta_2 = \begin{cases} 0, \\ \frac{\gamma_2}{\gamma_2+Z_3^*-1} T, \\ \frac{\gamma_2}{\gamma_2+Z_2^*-1} T \end{cases}$$

$$t_2 = \begin{cases} T, \\ \frac{\gamma_2}{\gamma_2+Z_3^*-1} T, \\ \frac{\gamma_2}{\gamma_2+Z_2^*-1} T \end{cases}$$

For two-user case, it is proved that the last user could be scheduled and work in only one mode in optimal condition. Reverse schedule order scheme gives more sum throughput. As the number of users increases, normalized throughput also increases.

6 Multi-user Case

For two-user case, the three permutations possible are "0B", "BH" and "HH". For three user case, the six permutations possible are "0BB", "0BH", "BHB", "BHH", "HHB" and "HHH" which can be reduced to "00B", "0BH", "BHH" and "HHH".

For four user case, the five permutations possible are "000B", "00BH", "0BHH", "BHHH", and "HHHH". So for N users, $N + 1$ permutations are possible. With the increase in the number of users, sum-throughput also increases.

7 Simulation

The simulation result is shown here to evaluate the proposed transmission scheme with initial energy for the users. The bandwidth is set as $W = 1$ MHz. The initial energy that is given to each user before transmission is set to 1.5 units. Assume that the forward and backward channel power gains are reciprocal and modeled as $h_i = g_i = 10^{-3} d_i^{\kappa i} \rho_i^2$, where di is the distance between the HAP and N_i, κi is the path loss exponent, ρ_i is the short-term channel fading, and ρ_i^2 is an exponentially distributed random variable with unit mean (Fig. 3).

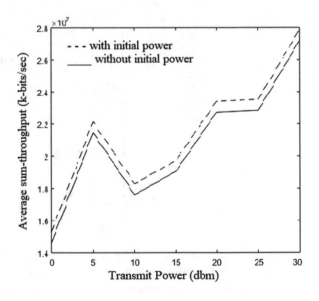

Fig. 3 Average sum-throughput versus transmit power for two-user case

Acknowledgements I take this opportunity to thank my guide Professor Dr. S. Malarvizhi of Dept. of ECE for her constant support and enlightenment throughout this work. I heartily thank, Mr. Vijayakumar, Associate Professor of Dept. of ECE for his guidance and suggestions during this paper. I would not forget to remember Miss. Arumbu, research scholar, Dept. of ECE for her encouragement and moreover for the timely support till the completion of this work.

References

1. R. Zhang and C. K. Ho, "MIMO broadcasting for simultaneous wireless information and power transfer", IEEE Trans. Wireless Commun., vol. 12, no. 5, pp. 1989–2001, May 2013.
2. H. Ju and R. Zhang, "Throughput maximization in wireless powered communication networks", IEEE Trans. Wireless Commun., vol. 13, no. 1, pp. 418–428, Jan. 2014.
3. K. Huang and E. Larsson, "Simultaneous information and power transfer for broadband wireless systems," IEEE Trans. Signal Process., vol. 61, no. 23, pp. 5972–5986, Dec. 2013.
4. L. Liu, R. Zhang, and K.-C. Chua, "Multi-antenna wireless powered communication with energy beam-forming," IEEE Trans. Commun., vol. 62, no. 12, pp. 4349–4361, Dec. 2014.
5. G. Yang, C. K. Ho, R. Zhang, and Y. L. Guan, "Throughput optimization for massive MIMO systems powered by wireless energy transfer", IEEE J. Sel. Areas Commun., vol. 33, no. 8, pp. 1640–1650, Aug. 2015.
6. J. Li, H. Zhang, D. Li, and H. Chen, "On the performance of wireless energy-transfer-enabled massive MIMO systems with superimposed pilot-aided channel estimation," IEEE Access, vol. 3, pp. 2014–2027, Oct. 2015.
7. Y. L. Che, L. Duan, and R. Zhang, "Spatial throughput maximization of wireless powered communication networks", IEEE J. Sel. Areas Commun., vol. 33, no. 8, pp. 1534–1548, Aug. 2015.
8. S. H. Kim and D. I. Kim, "Hybrid backscatter communication for wireless powered communication networks," in Proc. ISWCS, Poznań, Poland, Sep. 2016, pp. 265–269.
9. H. Ju and R. Zhang, "Optimal resource allocation in a full-duplex wireless powered communication network", IEEE Trans. Commun., vol. 62, no. 10, pp. 3528–3540, Oct. 2014.
10. X. Kang, C. K. Ho, and S. Sun, "Full-duplex wireless-powered communication network with energy causality," IEEE Trans. Wireless Commun., vol. 14, no. 10, pp. 5539–5551, Oct. 2015.
11. M. Kaus, J. Kowal, and D. U. Sauer, "Modelling the effects of charger distribution during self-discharge of supercapacitors," Electrochim. Acta, vol. 55, no. 25, pp. 7516–7523, Oct. 2010.
12. D. T. Hoang, D. Niyato, P. Wang, D. I. Kim and Z. Han. (Aug. 2016). "The trade-off analysis in RF-powered backscatter cognitive radio networks." [Online]. Available: https://arxiv.org/abs/1608.0178.

Feature Extraction Through Segmentation of Retinal Layers in SDOCT Images for the Assessment of Diabetic Retinopathy

N. Padmasini and R. Umamaheswari

Abstract Diabetic mellitus causes microvasculature changes in the retina which leads to diabetic retinopathy and may cause blindness if left unchecked. Spectral-Domain Optical Coherence Tomography (SDOCT) is a noninvasive imaging modality which could give precise information about the retinal layers. SDOCT retinal images of 75 subjects with uncontrolled diabetic mellitus for more than 2 years duration and images of 30 subjects with controlled diabetes or in normal condition are considered. The speckle noise in the images is smoothened using anisotropic diffusion filtering technique, and segmentation of Retinal Nerve Fiber layer (RNFL) along with Ganglion Cell Layer (GCL) and Inner Plexiform Layer (IPL) complex is performed using the axial gradient canny edge detection combined with a level set method. Textural features are obtained from the segmented layers, and classification of abnormality is done using SVM. The results showed that the retinal nerve fiber layer along with GCL+IPL complex thickness was reduced in subjects with even minimal diabetic retinopathy.

Keywords SDOCT · Inner plexiform layer · Ganglion cell layer
Support vector machine

This work is supported by UGC Minor Research Grants—MRP-6119/15 (SERO/UGC).

N. Padmasini (✉)
Department of BME, Rajalakshmi Engineering College, Chennai, India
e-mail: padmasini.n@rajalakshmi.edu.in

R. Umamaheswari
Department of EEE, Velammal Engineering College, Chennai, India
e-mail: umamaheswari@velammal.edu.in

© Springer Nature Singapore Pte Ltd. 2019
M. A. Bhaskar et al. (eds.), *International Conference on Intelligent Computing and Applications*, Advances in Intelligent Systems and Computing 846,
https://doi.org/10.1007/978-981-13-2182-5_16

153

1   Introduction

Globally among working-age adults, Diabetic Retinopathy (DR) is the leading cause of blindness [1]. Due to population, life style, obesity and aging   the number of people affected due to DR is expected to rise and it is estimated to be 300 million cases by the year 2025 [2]. Therefore, even though the early-stage diagnosis of diabetic retinopathy is very critical but it becomes essential. Optical Coherence Tomography (OCT) is a noncontact medical imaging modality to measure different aspects of biological tissues especially in retinal analysis. As shown in Fig. 1, a low coherence interferometer is used in OCT scanner to produce a two- or three-dimensional image of the internal tissues. In macular edema, the cystoid fluid has less optical scattering than the retinal tissues which surround it, and hence OCT is highly effective in this case [3].

In the Spectral-Domain Optical Coherence Tomography (SDOCT), the eye is illuminated with the light from superluminescent source. The backscattered light from the region of interest is combined with light from the reference, in which an interference signal is produced. This interferometric signal has lesser frequency with lesser depth of origin and vice versa. By making using of a dispersive detector, the information from the interferometric fringes is detected. A grating in dispersive detector is used to spread out the various frequencies onto a CCD detector. Using Fourier transform of the interferogram, it is able to obtain the scattering amplitude. The deeper objects in the image give rise to a higher frequency in the interferogram [4]. SDOCT is also known as Fourier-domain OCT as the distances are encoded in the Fourier transformation of the reflected light frequencies as shown in Fig. 1. As the retinal structure is accurately quantified in SDOCT, it is mainly used in retinal imaging which helps to expand research in retinal diseases and improve the diagnosis. A typical SDOCT image with all the retinal layers labeled is shown in Fig. 2. Extracting structural information becomes   vital for the quantification of retinal layers [5]. OCT is the only retinal imaging modality to provide information about diabetic macular edema. The morphologies of macular edema can be subdivided

Fig. 1 SDOCT imaging system. LS—Low coherent source, BS—beam splitter, REF—reference mirror, DG—diffraction grating, CAM—detector as a spectrometer, SMP—sample arm

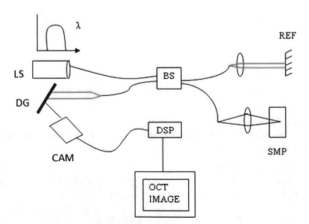

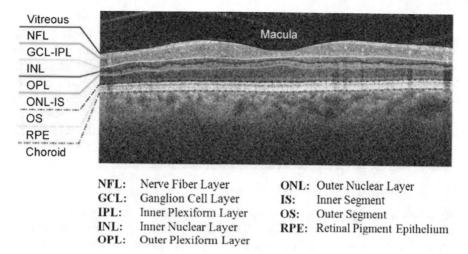

NFL:	Nerve Fiber Layer	ONL:	Outer Nuclear Layer
GCL:	Ganglion Cell Layer	IS:	Inner Segment
IPL:	Inner Plexiform Layer	OS:	Outer Segment
INL:	Inner Nuclear Layer	RPE:	Retinal Pigment Epithelium
OPL:	Outer Plexiform Layer		

Fig. 2 A typical macula scan showing the retinal layers

and classified based on the OCT findings. As a result, the physicians were able to give variability in treatment for various conditions. The prognosis of visual acuity varies depending on the macular edema morphology [6]. It is seen that in patients with very much minimal amount of DR, the pericentral area of the mean ganglion cell layer (GCL) was about 5.1 μm thinner with 95% confidence interval. Also, it has been reported that the mean Retinal Nerve Fiber Layer (RNFL) thickness in the peripheral macula is thinner when compared to the subjects with controlled diabetes. Moreover, there exists linear correlation ($R = 0.53$) between the thickness of GCL layer and diabetes duration [7]. In this present work, segmentation of RNFL along with GCL and IPL complex is carried out using axial gradient canny edge detection combined with a level set method. The statistical as well as run-length features are extracted from the segmented image, and these features are fed to an SVM with RBF for classification of abnormality.

2 Materials and Methods

The 2-D SDOCT retinal images were obtained from Aravind Eye Hospital, Puducherry, India. Seventy-five abnormal images and thirty normal images are used in this study. The scanning speed of the SDOCT (Cirrus HD-OCT) machine is 27,000 A-scans/s. This scanner can acquire a retinal volume of $200 \times 200 \times 1024$ voxels covering $6 \times 6 \times 2$ mm^3 in 1.48 s. The voxel size is $30 \times 30 \times 2$ μm, and the voxel depth is 8 bits in grayscale. True 3D views of the retinal structure could be obtained using this SDOCT scanner. Two types of scans—macula scan (fovea centered scan) and ONH—optic nerve head-centered scan are mainly acquired to identify the retinal diseases. The nine layers of the retina are very well seen in a macula scan. At the early stage of DR, the Retinal Nerve Fiber Layer (RNFL) thickness may be reduced because of the retinal ganglion cell death and axonal degeneration. It is found that thickness of RNFL is reduced in the peripheral and pericentral regions (42 ± 3 vs. 33 ± 9 μm and 27 ± 2 vs. 18 ± 5 μm, respectively), in the healthy versus diabetic retinopathy patients [8–13]. In the work by [14], they have stated that in the peripheral and pericentral areas of macula, the Inner Plexiform Layer (IPL), the GCL and the RNFL were thinner in patients with minimal DR compared to controls. From the multiple linear regression analysis, it is evident that DR is the only most significant reason for this.

A. *Preprocessing of SDOCT retinal images*

As coherent wave pattern is used in the SDOCT scanner to acquire images, speckle noise is prominent in these images. This speckle noise masks the low contrast lesions, which leads to misinterpretation by the physician while diagnosis [15]. Speckle is a multiplicative, randomly distributed granular noise of size 1–2 μm. In our previous work [16], we have used anisotropic diffusion filtering for the effective removal of the speckle noise and the same is used here.

B. *Proposed Methodology*

Figure 3 shows the block diagram of the proposed methodology. The segmentation is carried out using an axial gradient canny edge detection along with region-based level set method, the RNFL along with GCL, and IPL complex is segmented.

In order to extract structural information from different layers of this SDOCT image, it is necessary to obtain the gradient in the axial direction. This is performed using a canny edge detection operator. The edge points detected from this operator could exactly localize the center of the edge. Then, the vertical intensity difference between the layers is utilized to generate edge weights that create a pathway [17].

$$w_{xy} = 2(g_x + g_y) + w_{min} \tag{1}$$

where

w_{xy} is the weight assigned to the edge connecting the nodes x and y.
g_x is the vertical gradient of the image at node x.
g_y is the vertical gradient of the image at node y.
w_{min} is the minimum weight (1×10^{-5}).

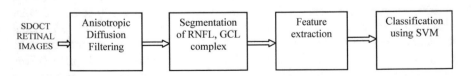

Fig. 3 Diagram of proposed methodology

The level set method is most commonly used in image processing, in which contours or boundaries are represented as the zero level set of a higher dimensional function, which is known as a level set function. This method has the capability of representing contours/surfaces with complex topology. In level set methods, the function takes positive and negative values, to represent a separation of the region into two disjoint regions [18]. These regions are defined with their membership functions as given below:

$$M_1(\emptyset) = H(\emptyset) \tag{2}$$

$$M_2(\emptyset) = 1 - H(\emptyset) \tag{3}$$

where H is a Heaviside function, and similar to unit step function, its value is 0 and 1 for negative and positive arguments, respectively.

$$H(n) = 0, \quad n < 0 \text{ and}$$
$$H(n) = 1, \quad n \geq 0$$

The energy $\in$ can be represented as a function of $\emptyset$, bias b, and vector c. By minimizing energy, we have obtained the results of the image plotted by separating each region and the boundary is traced. This minimization is done by an iterative process; the energy is minimized in each iteration with respect to each of its variables $\emptyset$, b, and c.

(1) *Energy minimization with respect to* $\emptyset$

The minimization of f($\emptyset$,b, c) with respect to $\emptyset$ for fixed b and c is achieved using standard gradient descent method, by solving the gradient flow equation.

$$\frac{\partial \emptyset}{\partial t} = -\frac{\partial F}{\partial \emptyset} \tag{4}$$

(2) *Energy minimization with respect to c*

For fixed $\emptyset$ and b, the optimal c minimizes the energy $\in$, which is given by the equation, which is given by the equation,

$$c = \frac{\int (b * k) Iudy}{\int (b^2 * k) udy} \tag{5}$$

where k is the kernel function, I represents the image, and u represents the membership function.

(3) *Energy minimization with respect to b*

For fixed $\emptyset$ and c, the optimal b minimizes the energy ϵ, given by the equation

$$b = \frac{IJ^{(1)} * k}{J^{(2)} * k} \tag{6}$$

where $J^{(1)} = \sum cU$ and $J^{(2)} = \sum c^2 U$. The convolutions with a kernel function k are the slowly varying property of the derived optimal estimator b of the bias field.

The retinal layers are segmented from the contour region plotted image. All the operations are performed using MATLAB installed in an Intel (R) Core (TM) i7 CPU 860, 64-bit OS, and 8 GB RAM processor at 2.80 GHz.

C. *Feature Extraction and Classification*

Statistical as well as run-length features arc computed for the single largest connected component in the processed image, from which the textural features are obtained. The statistical features include entropy, skew, kurtosis, and energy. Entropy gives the amount of information present in the image [19]. Skewness is a measure of asymmetry or the lack of degree of symmetry. When viewed from the center point of the dataset, if it looks same on both the right and left sides, then the dataset is said to be symmetric [20]. Energy describes the measure of information while an operation is formulated for estimation of Maximum A Priori (MAP) along with Markov random fields. It can be a negative measure for minimization and a positive measure for maximization. The run-length features include Run Percentage (RP), Long-Run Emphasis (LRE), Short-Run Emphasis (SRE), High Gray-Level Emphasis (HGRE), Low Gray-Level Emphasis (LGRE), Run-Length Nonuniformity (RLN), and Gray-Level Nonuniformity (GLN). A run is a chain of pixels with similar image values in a straight scan direction. The horizontal run length is the length of the run in the horizontal scan direction, and the vertical run length is the length of the run in the vertical scan direction. For the diagonal direction, the total number of pixels is multiplied by $\sqrt{2}$. Short-run emphasis dominates if the texture is fine-grained, similarly long-run emphasis dominates if the texture is coarse-grained in nature. LGRE and SRE are orthogonal to each other, and the former dominates when there are many runs of low gray value. Similarly, HGRE and LRE are orthogonal to each other and the former dominates when there are many runs of high gray value. GLNU increases when the histogram is dominated by few gray-level outliers. Similarly, RLNU increases when the histogram is dominated by few run-length outliers. RP gives information on the overall homogeneity [21]. All the abovesaid four statistical and seven run-length features are inputted to an SVM with RBF classifier.

D. *SVM with RBF*

Support Vector Machine (SVM) performs well in high-dimensional spaces, and hence used for solving pattern recognition problem. Given a training set of M data points $\{y_j, x_j\}$ $j = 1$ to M, where $x_j \in R$ is the jth input pattern and $y_j \in R$ is the jth output pattern, the support vector machine approach aims at constructing a classifier of the form:

$$y(x) = \text{sign}\left[\sum_{x=1}^{N} \alpha_k y_k \varphi(x, x_k) + b\right] \quad (7)$$

where α_k, y_k are positive real constants and b is a real constant

In the SVM, an optimum hyperplane is estimated by minimization of Eq. (8) subjected to the condition, $y_i(w^T x + w_0) \geq 1 - \xi_i$;

$$J(w) = \frac{1}{2}\|w^2\| + c\sum_{i=1}^{N} \xi_i \quad (8)$$

where ξ_i is a slack variable [22] and the constant $c > 0$ is user-defined. The radial basis function kernel, also called the RBF kernel, or Gaussian kernel, is a kernel that is in the form of a radial basis function (more specifically, a Gaussian function) [23]. The RBF kernel is defined as

$$k_{\text{RBF}}(x, x') = \exp\left(-\gamma x - x'^2\right) \quad (9)$$

In this work, we have classified using SVM with "'Linear", "Polynomial", "Quadratic", and "Radial basis function (RBF)" kernels, among which the RBF kernel outperformed all other kernels. The various parameters of the classifier such as accuracy rate, specificity, sensitivity, positive predictive accuracy, and negative predictive accuracy are estimated as given in Table 1.

Table 1 Extracted statistical and run-length features

Features	Normal	Abnormal—DR cases
Entropy	4.349 ± 0.456	4.481 ± 1.229
Kurtosis	2.373 ± 1.030	5.860 ± 2.891
Skew	0.997 ± 0.501	2.013 ± 0.684
Energy	2.238e13 ± 1.021e13	2.904e13 ± 2.443e13
SRE	0.726 ± 0.001	0.789 ± 0.002
LRE	4.007 ± 0.008	3.242 ± 0.058
GLN	64.206 ± 1.005	69.389 ± 2.109
RP	0.620 ± 0.001	0.682 ± 0.002
RLN	189.407 ± 0.076	246.790 ± 1.089
LGRE	0.3103 ± 0.00001	0.304 ± 0.0008
HGRE	23.535 ± 0.051	19.095 ± 0.152

3   Results and Discussion

SDOCT images of 75 subjects with uncontrolled diabetes and 30 subjects with controlled or normal diabetes are considered. The images are first filtered using anisotropic diffusion filtering for the removal of speckle noise. Using axial gradient canny edge detection along with region-based level set method, the RNFL layer with GCL and IPL complex is segmented. From this processed image, the features are obtained from the single largest connected component of the image.

From Fig. 4, it is evident that the thickness of RNFL complex is less in patients with diabetic retinopathy. The extracted features from all the normal as well as abnormal images were tabulated as shown in Table 1.

All the above feature parameters are fed to an SVM classifier which uses RBF as its kernel function. The accuracy of the classifier is found to be 98.1%. The classifier statistics is shown in Table 2.

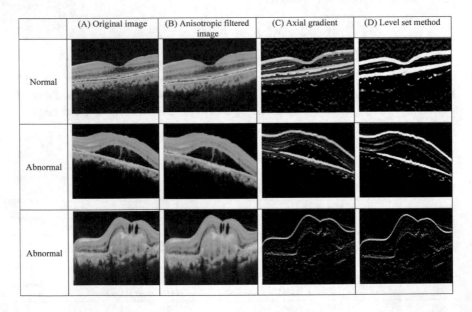

Fig. 4 Processed SDOCT normal and abnormal images

Table 2 Statistics of the classifier

Statistic parameter	Value	95% CI
Sensitivity	98.67%	92.79–99.97%
Specificity	96.67%	82.78–99.92%
Positive likelihood ratio	29.6	4.31–203.37
Negative likelihood ratio	0.01	0.00–0.10
Disease prevalence	71.43%	61.79–79.82%
Positive predictive value	98.67%	91.50–99.80%
Negative predictive value	96.67%	80.52–99.51%
Accuracy	98.10%	93.29–99.77%

CI—confidence interval

4 Conclusion

The retinal layer analysis of SDOCT images of diabetic subjects reveals that the RNFL complex is thinner than that of the normal subjects. The automated diagnosis on these images could be obtained by a simple and specific method of segmentation which involves axial gradient and a level set technique. All the 105 images are segmented, and the features from the single largest component in the processed image are estimated and fed to an SVM classifier. The results of the classifier proved the efficacy of this method is close to that of the expert manual grading results. This computationally factual method could assist the physician in accurate diagnosis of Diabetic Retinopathy. In the future, this work could be extended to accurately estimate the individual retinal layer thickness automatically so that the disease could be treated at the earlier stage itself.

Acknowledgements All the images used in this work are obtained from Aravind Eye Hospital, Puducherry, India as per the guidelines of their ethical committee in the hospital. The authors would like to thank Dr. Manavi D. Sindal; M.S., F.M.R.F., Senior consultant, Vitreo-Retina services, and her team of Aravind eye hospital, for their guidance and generous support in providing the SDOCT data.

References

1. American Academy of ophthalmology, Diabetic Retinopathy PPP - Updated 2017.
2. King, H., Aubert, R. E., & Herman, W. H. (1998), "Global burden of diabetes, 1995–2025: prevalence", numerical estimates, and projections. Diabetes care, 21(9), 1414–1431.
3. James G. Fujimoto, David Haung, Eric A. Swanson, Charles P. Lin, Joel S. Schuman, William G. Stinson, Michael R. Hee, Thomas Flotte, Kenton Gregory and Carmen A. Puliafito, "Optical Coherence Tomography", in Science New Series, Vol. 254 No. 5035, 1178–1181, November 1991.
4. Schuman, Joel S. "Spectral domain optical coherence tomography for glaucoma (an AOS thesis)." Transactions of the American Ophthalmological Society 106 (2008): pp 106–426).
5. Niu S, Chen Q, de Sisternes L, Rubin DL, Zhang W, Liu Q, "Automated retinal layers segmentation in SD-OCT images using dual-gradient and spatial correlation smoothness constraint", Computers in biology and medicine. 2014 Nov 1; 54:116–128.

6. Forooghian, Farzin, et al. "Evaluation of time domain and spectral domain optical coherence tomography in the measurement of diabetic macular edema." Investigative ophthalmology & visual science 49.10 (2008): 4290–4296.

7. van Dijk, Hille W., Frank D. Verbraak, Pauline HB Kok, Mona K. Garvin, Milan Sonka, Kyungmoo Lee, J. Hans De Vries et al. "Decreased retinal ganglion cell layer thickness in patients with type 1 diabetes." Investigative ophthalmology & visual science 51, no. 7 (2010): 3660–3665.

8. Oshitari, T., Hanawa, K., & Adachi-Usami, E. (2009), "Changes of macular and RNFL thicknesses measured by Stratus OCT in patients with early stage diabetes", Eye, 23(4), 884–889.

9. Oshitari, T. (2006), "Non-viral gene therapy for diabetic retinopathy", Drug development research, 67(11), 835–841.

10. Oshitari, T., & Roy, S. (2007), "Common therapeutic strategies for diabetic retinopathy and glaucoma", Current Drug Therapy, 2(3), 224–232.

11. Barber, A. J., Lieth, E., Khin, S. A., Antonetti, D. A., Buchanan, A. G., & Gardner, T. W. (1998), "Neural apoptosis in the retina during experimental and human diabetes", Early onset and effect of insulin, Journal of Clinical Investigation, 102(4), 783.

12. Sugimoto, M., Sasoh, M., Ido, M., Wakitani, Y., Takahashi, C., & Uji, Y. (2005), "Detection of early diabetic change with optical coherence tomography in type 2 diabetes mellitus patients without retinopathy", Ophthalmologica, 219(6), 379–385.

13. DeBuc, D. C., & Somfai, G. M. (2010), "Early detection of retinal thickness changes in diabetes using optical coherence tomography", Medical Science Monitor, 16(3), MT15–MT21.

14. Van Dijk, H. W., Verbraak, F. D., Kok, P. H., Stehouwer, M., Garvin, M. K., Sonka, M., & Abramoff, M. D. (2012), "Early Neurodegeneration in the Retina of Type 2 Diabetic Patients Retinal Neurodegeneration in Type 2 Diabetes", Investigative ophthalmology & visual science, 53(6), 2715–2719.

15. Tsantis, S., Dimitropoulos, N., Ioannidou, M., Cavouras, D., & Nikiforidis, G. (2007). "Inter-scale wavelet analysis for speckle reduction in thyroid ultrasound images", Computerized Medical Imaging and Graphics, 31(3), 117–127.

16. Padmasini, N., K. S. Abbirame, R. Umamaheswari, and S. Mohamed Yacin. (2015), "Reduction of Speckle Noise in SDOCT Retinal Images by Fuzzification and Anisotropic Diffusion Filtering" , Journal of Biosciences Biotechnology Research Asia (BBRA journal), 15(1).

17. LaRocca, F., Chiu, S. J., McNabb, R. P., Kuo, A. N., Izatt, J. A., & Farsiu, S. (2011), "Robust automatic segmentation of corneal layer boundaries in SDOCT images using graph theory and dynamic programming", Biomedical optics express, 2(6), 1524–1538.

18. Li, C., Huang, R., Ding, Z., Gatenby, J., Metaxas, D. N., & Gore, J. C. (2011), "A level set method for image segmentation in the presence of intensity inhomogeneities with application to MRI", IEEE Transactions on Image Processing, 20(7), 2007–2016.

19. Haralick, R. M., & Shanmugam, K. (1973), "Textural features for image classification", IEEE Transactions on systems, man, and cybernetics, (6), 610–621.

20. MacGillivray, H. L. (1986), 'Skewness and asymmetry: measures and orderings", The Annals of Statistics, 994–1011.

21. Tang, X. (1998), "Texture information in run-length matrices. IEEE transactions on image processing", 7(11), 1602–1609.

22. Ricci, E., & Perfetti, R. (2007), "Retinal blood vessel segmentation using line operators and support vector classification." IEEE transactions on medical imaging, 26(10), 1357–1365.

23. Chen, Sheng, Colin FN Cowan, and Peter M. Grant. "Orthogonal least squares learning algorithm for radial basis function networks." IEEE Transactions on neural networks 2.2 (1991): 302–309.

Internet of Things (IoT) Enabled Wireless Sensor Network for Physiological Data Acquisition

R. Prakash and A. Balaji Ganesh

Abstract Wearable Health Monitoring System (WHMS) plays a major role in telemedicine research. The study explores Internet of Things (IoT) enabled wearable medical device that handles physiological signals, such as pulse rate and core body temperature along with activity monitoring. The indigenously developed nodes, such as sensor node, destination and relay nodes are successfully deployed in indoor environments. The cooperative wireless network is established and eventually empowered with IoT architecture. The hybrid wireless protocols, such as Radio-Frequency (RF), Wi-Fi and Bluetooth are made to collaborate with each other efficiently and performance characteristics are analyzed. It is strongly envisaged that efficiency of existing wireless sensor network can be improved by integrating IoT architecture which is subsequently proved with the results obtained.

Keywords Cooperative IOT · Wireless sensor networks

1 Introduction

The advancements in Wireless Sensor Network (WSN) empower various applications, such as structural monitoring [1], military surveillance, industrial measurements [2] and most importantly in healthcare applications [3, 4]. Several smart biomedical sensors are built as wearable units that also enhance the smart healthcare monitoring and also help professional to monitor vital physiological information from remote locations. IoT (Internet of Things) is a proven concept and along with WSN and it envisages to enhance all services and devices that significantly reduce human intervention and assure better quality human life [5, 6]. The data collected from wearable sensors are made available and accessible anywhere and anytime

R. Prakash · A. Balaji Ganesh (✉)
Electronic System Design Laboratory, Velammal Engineering College,
Chennai, India
e-mail: abganesh@velammal.edu.in

© Springer Nature Singapore Pte Ltd. 2019
M. A. Bhaskar et al. (eds.), *International Conference on Intelligent Computing and Applications*, Advances in Intelligent Systems and Computing 846,
https://doi.org/10.1007/978-981-13-2182-5_17

with the advent of IoT enabled wireless sensor network. The continuous monitoring of human vital parameters should be supported with huge data repository. According to GSMA, the number of wearable devices connected might be 15 billion and the number would increase to 24 billion during 2015 and 2020, respectively [7, 8]. An integrated computing device with internet connection in a hospital as well as in home environment guarantees better healthcare services to the patients. The major challenging tasks of WSN in hospitals are privacy, sensor placement, intrusiveness, safety and data handling [9]. The successful delivery of collected information with minimum end-to-end delay at destination is a primary need for all WSN applications [10]. This could be achieved by incorporating cooperative and network coded communication in the existing WSN [11, 12]. Recently, a cooperative communication enabled implant node for home care is described the WSN mechanism in off-body relay nodes [13]. The network coding has optimum usage of available resources with minimum network traffic and other network characteristics. The destination node collects patient information from multiple sources and reconstructs bio-signal by using XOR-based decoding. This has been widely appreciated because it reduces network traffic, provides better broadcast communication between wireless networks and also it enhances throughput [14]. Every successful delivery is acknowledged by using various schemes, including Forward Error Correcting (FEC) Hybrid Automatic Retransmission Request (HARQ) and Automatic Retransmission reQuest (ARQ) [14]. The relay selection in a cooperative communication is performed on the basis of sink availability, link quality, residual power, and distance. The relay node selection differs based on *how* and *when* a sensor node link is required to establish a connection with a relay node [15]. Researchers investigate various security as well as network level issues [16]. The Ultrawide Band (UWB) depends on a WSN in a hospital network can also offer major benefits, such as extended communication range, higher data rate (until 27 Mbps) and low power consumption [9, 17]. Zhang et al. have proposed a XOR-based Hybrid Automatic Retransmission Request (XOR^2-HARQ) that used double XOR operation to retrieve the lost number of packets [18]. Lin et al. explore an activity recognition system by using a smart insole which is integrated with pressure sensors [19]. Recent researchers find the possibility of using the commercially sensor nodes, such as Zolertia Z1, Cricket, MicaZ, IRIS TelosB, and LOTUS or configured themselves to validate the efficiency of various algorithms [20, 21]. Sensing Health with Intelligence, Modularity, Mobility and Experimental Reusability (SHIMMER) is considered as one of the major providers of customizable sensor nodes for activity monitoring and healthcare application [22, 11]. The current trends in IoT for healthcare applications is classified in many ways depends on the functionality, benefits, and perspective. Moreover, the technique focuses on data prevention, early critical pathology detection and home care monitoring that prevents expensive clinical admissions.

2 System Architecture

The experiment presents cooperative communication enabled IoT architecture to monitor physiological as well as postural activity of a user in home environment which is illustrated in Fig. 1 In this study, the source transmits the packets to destination by utilizing services of intermediate relay nodes; thus reduces the probability of packet loss. The smart band or smart node (S) is developed as a wearable device. The relay node (R_n) may be placed in adjacent with air conditioning machine in each room of the user. The reason for this deployment is to switch on/off the air conditioning machine. Each relay node is interfaced with a temperature, humidity, and carbon dioxide sensor and also it acts as an actuation unit. The destination server (D) is configured to receive data both from smart band and relay nodes. The battery operated smart band or source node is fixed firmly by using upper arm and soft band in the user which sends the information directly to destination server. Smart band or source node also transmits its data to destination server (D) through nearby relay node (R_n). The relay node is also configured to receive control signal from destination server.

As illustrated in Fig. 1, the destination server (D) is designed as a rack that consists of two controllers, namely CC430 and CC3200 from Texas Instruments. The microcontroller, CC430 is configured to acquire and process physiological as well as activity data collected from the user. The microcontroller, CC3200 acts as transceiver that transmits and receives the packets to and from web server and also

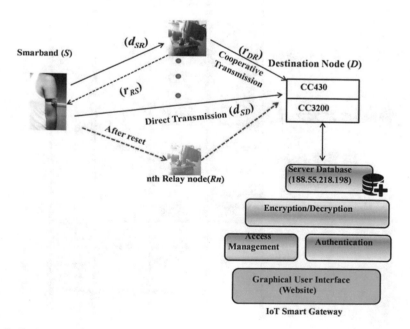

Fig. 1 System overview

it efficiently handles the services, such as data handling, decision-making, and abnormal detection and eventually responds with alert messages. The route discovery mechanism in adaptive relaying scheme is initiated by route request (*RREQ*) to all the individual relay nodes in the network. The relay nodes respond with an acknowledgement frame (*RACK*) to smart band. The nearby relay node is chosen by calculating RSSI and Link Quality Index (LQI) among the replied acknowledgement frame and eventually, the link is established. The smart-band multicasts its message (*S_SNT*) to the established links. The functionality of link establishment in the network is performed in Fig. 2.

The relay node calculates the distance of smart band by analyzing RSSI during each transmission. When RSSI threshold limit exceeds a reset signal is generated from relay node and reestablishes a new link. Eventually, a new *RREQ* is broadcasted that is initiated by source node. The sequential data initiated from smart band to destination server and respective transmission is represented as

$$d_{SD} = \{d(1), d(2)\ldots d(n)\} \tag{1}$$

where n denotes the sequential data is directly transmitted from smart band. The relay node forwards the composed message from smart band along with its individual message and forms a weighted data,

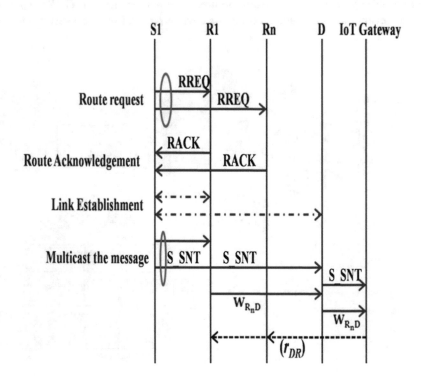

Fig. 2 Link establishment

$$W_{R_mD} = d_{SD} + j_d \qquad (2)$$

where m represents the mth relay node that is chosen for relayed transmission. J_d is the actual message of the relay node. The received data, Z at the destination node is expressed as

$$Z = \left\{ \frac{d_{SD} + y_k}{W_{R_mD} + y_k} \right\} \qquad (3)$$

where y_k represents the additive noise in the corresponding channel. The selected relay node estimates the distance of smart band by calculating the RSSI of transmission. During each transmission, $RSSI_{SR}$ gets reorganized and best RSSI is stored as J_{RSSI}. A reset signal r_{RS} is forwarded to smart band with maximum transmission power when $RSSI_{SR}$ becomes smaller than $(J_{RSSI} - P)$

$$RSSI_{SR} < (J_{RSSI} - P) \qquad (4)$$

where P is constant and it is based on the dimensions of each room in a home environment where relay node is deployed. The relatively best RSSI value has been calculated from received data Z and corresponding physiological data updated to the server database by using CC3200 network processor.

3   Results

It is presumed that incorporation of IoT architecture into existing wireless sensor network would promote the functional efficiency of healthcare application. The presumption is eventually confirmed by obtaining performance characteristics, such as end-to-end delay and execution time. Figure 3 shows end-to-end delay associated with cooperative communication based wireless sensor network is further reduced by 60.1% with the advent of IoT architecture. The reduced end to delay proves that the system experiences merely very limited data computational complexity with minimum data transmission load and consumes lesser memory.

The execution time is associated with data collection and handling, abnormal detection, alert processing, actuator control and encoding/decoding mechanisms which is mostly dependent on services of central processing unit (CPU). Since data acquisition and processing are to be shared by CC430 the IoT architecture finds flexibility in handling the services. The instruction processing includes fetch, decode, operand fetch, and execute. The execution time of the system processor has been calculated and represented in Eq. (5).

$$\text{Execution Time} = \text{Instructin Count} * \text{CPI} * \text{Clock Cycle} \qquad (5)$$

Table 1 depicts the comparison results on execution time between existing cooperative enabled wireless sensor network and IoT architecture. It is observed

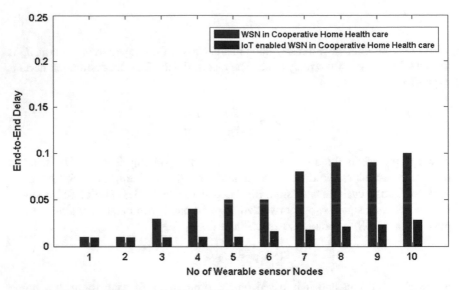

Fig. 3 End-to-End delay of cooperative IoT in home health care

that the execution time in IoT enabled destination node yields more than 35.1% better results than typical WSN.

Three types of modes, such as *acc* mode, *pulse* mode, and *temperature* mode have been programmed. The smart-band firmware is programmed in such a way that either user can manually change their power mode or caregiver can access the smart-band power modes through web server in order to reduce power consumption. In *acc* mode, the accelerometer sensor alone to be powered and all sensors go to sleep mode. The *pulse* mode powers up oximetry sensor and smart band sends the pulse rate data to the server which helps to monitor a patient sleep pattern during deep sleep. The *temp* mode provides core body temperature alone. Table 2 shows average current consumption values in various power modes. It should be noted that smart band is configured to sense and collect all interfaced sensors by default. Further, the transmission power for the smart band is fixed 18 dBm and found it consumes 194 mA for each transmission.

Table 1 Shows the comparison results on execution time between typical wireless sensor network (WSN) and IoT enabled WSN for home healthcare monitoring

Number of wearable sensor nodes	Execution time in typical WSN (ms)	Execution time in IoT enabled WSN (ms)
1	20	8
2	30	19
3	50	31
4	65	48
5	72	61

Table 2 Shows the average current consumption in respective power modes

S. No	Type of mode	Average current consumption (mA)
1	Acc	0.332
2	Pulse	70
3	Temp	0.125

4 Conclusion

The work presents an experimental evaluation of cooperative IoT architecture in home environments for monitoring physiological and postural activity signals. The study presents a successful implementation of sequential level of security in data transmission and processing. Also, the work validates the functional efficiency of IoT in existing wireless sensor network by evaluating its characteristics, such as execution time and end-to-end delay in a real-time hospital sensor network.

Acknowledgements The authors gratefully acknowledge the financial support from Science for Equity Empowerment and Development Division under Department of Science and Technology, New Delhi, India by sanctioning a project - File No.: SSD/TISN/047/2011-TIE (G) to Velammal Engineering College, Chennai.

References

1. Xu, Ning, Sumit Rangwala, Krishna Kant Chintalapudi, Deepak Ganesan, Alan Broad, Ramesh Govindan, and Deborah Estrin. "A wireless sensor network for structural monitoring." In *Proceedings of the 2nd international conference on Embedded networked sensor systems*, pp. 13–24. ACM, 2004.
2. He, Tian, Sudha Krishnamurthy, John A. Stankovic, Tarek Abdelzaher, Liqian Luo, Radu Stoleru, Ting Yan, Lin Gu, Jonathan Hui, and Bruce Krogh. "Energy-efficient surveillance system using wireless sensor networks." In *Proceedings of the 2nd international conference on Mobile systems, applications, and services*, pp. 270–283. ACM, 2004.
3. Akyildiz, Ian F., Weilian Su, Yogesh Sankarasubramaniam, and Erdal Cayirci. "Wireless sensor networks: a survey." *Computer networks* 38, no. 4 (2002): 393–422.
4. Nadeem, Adnan, Muhammad Azhar Hussain, Obaidullah Owais, Abdul Salam, Sarwat Iqbal, and Kamran Ahsan. "Application specific study, analysis and classification of body area wireless sensor network applications." *Computer Networks* vol. 83, pp. 363–380.2015.
5. Sarita A., Manik L. D.: Internet of Things –A Paradigm Shift of Future Internet Applications. Institute of technology Nirma University, 1–7 (2011).
6. Al-Fuqaha, Ala, Mohsen Guizani, Mehdi Mohammadi, Mohammed Aledhari, and Moussa Ayyash. "Internet of things: A survey on enabling technologies, protocols, and applications."*Communications Surveys & Tutorials, IEEE* 17, no. 4 (2015): 2347–2376.
7. http://gigaom.com/2011/10/13/internet-of-things-will-have-24-billiondevices-by-2020/.
8. The Mobile Economy, GSMA, 2013.
9. Sliman, Jamila Ben, Ye-Qiong Song, Anis Koubâa, and Mounir Frikha. "A three-tiered architecture for large-scale wireless hospital sensor networks." In *Workshop Mobi Health Infin conjunction with BIOSTEC*, p. 64. 2009.

10. Kirby, Karen K. "Hours per patient day: not the problem, nor the solution."*Nursing Economics* 33, no. 1 (2015): 64.
11. Prakash, R., Ganesh, A. B. and Girish, S. V., 2016. Cooperative wireless network control based health and activity monitoring system. *Journal of medical systems*, *40*(10), p. 216.
12. Prakash, R., Ganesh, A. B. and Sivabalan, S., 2017. Network Coded Cooperative Communication in a Real-Time Wireless Hospital Sensor Network. *Journal of medical systems*, *41*(5), p. 72.
13. Ntouni, Georgia D., Athanasios S. Lioumpas, and Konstantina S. Nikita. "Reliable and Energy-Efficient Communications for Wireless Biomedical Implant Systems." *IEEE Journal of Biomedical and Health Informatics,* 18, no. 6 (2014): 1848–1856.
14. Ostovari, Pouya, Jie Wu, and Abdallah Khreishah. "Network coding techniques for wireless and sensor networks." In *The Art of Wireless Sensor Networks*, pp. 129–162. Springer Berlin Heidelberg, 2014.
15. Ibrahim, Ahmed S., Ahmed K. Sadek, Weifeng Su, and KJ Ray Liu. "SPC12-5: relay selection in multi-node cooperative communications: when to cooperate and whom to cooperate with?." In *Global Telecommunications Conference, 2006. GLOBECOM'06. IEEE*, pp. 1-5. IEEE, 2006.
16. Alcaraz, Cristina, Pablo Najera, Javier Lopez, and Rodrigo Roman. "Wireless sensor networks and the internet of things: Do we need a complete integration?." In *1st International Workshop on the Security of the Internet of Things (SecIoT'10)*. 2010.
17. Cao, Huasong, Victor Leung, Cupid Chow, and Henry Chan. "Enabling technologies for wireless body area networks: A survey and outlook."Communications Magazine, IEEE 47, no. 12 (2009): 84–93.
18. Zhang, Zhang, Tiejun Lv, Xin Su, and Hui Gao. "Dual xor in the air: A network coding based retransmission scheme for wireless broadcast broadcasting." In *Communications (ICC), 2011 IEEE International Conference on*, pp. 1–6. IEEE, 2011.
19. Lin, Feng, Aosen Wang, Lora Cavuoto, and Wenyao Xu. "Towards Unobtrusive Patient Handling Activity Recognition for Reducing Injury Risk among Caregivers." (2016).
20. Ammari, Habib M. (Ed.) (2013). The Art of Wireless Sensor Networks, Volume 1: Fundamentals. Springer Science & Business Media.
21. John G. Webster, Halit Eren (Ed.) 2014. Measurement, Instrumentation, and Sensors Handbook, Second Edition. CRC Press.
22. Burns, Adrian, Barry R. Greene, Michael J. McGrath, Terrance J. O'Shea, Benjamin Kuris, Steven M. Ayer, Florin Stroiescu, and Victor Cionca. "SHIMMER™–A wireless sensor platform for noninvasive biomedical research." *Sensors Journal, IEEE* 10, no. 9 (2010): 1527–1534.

QR Code-Based Highly Secure ECG Steganography

P. Mathivanan, S. Edward Jero and A. Balaji Ganesh

Abstract In concepts such as connected health, patient data needs to be protected when they are transmitted over the internet to the caregiver. The current letter proposes embedding the QR code of patient data into discrete wavelet transform coefficients of the cover ECG Signal. The imperceptibility of hidden data is estimated by performance metrics such as peak signal-to-noise ratio, percentage residual difference, and Kullback–Leibler distance. The proposed approach is demonstrated on MIT-BIH database. Watermarking in the low coefficient values deteriorate the signal less and results in lower bit errors. It is shown that larger watermarks can be accommodated with a larger version of QR code for more or less the same imperceptibility. It is found that QR code version 40 allows embedding of maximum 2632 bytes of patient data with zero bit errors. Since QR code has an inbuilt error correction code, it provides additional layer of security to patient data and the proposed approach can be used for secure transfer of patient data.

Keywords ECG steganography · QR code · Adaptive watermarking and threshold selection

P. Mathivanan (✉)
Department of Electronics and Communication Engineering,
Velammal Engineering College, Chennai 600066, India
e-mail: mathivanan@velammal.edu.in

S. Edward Jero
Karuvee Innovations Pvt. Ltd, Chennai, India
e-mail: edwardjero@gmail.com

A. Balaji Ganesh
Department of Electrical and Electronics Engineering,
Velammal Engineering College, Chennai 600066, India
e-mail: abganesh@velammal.edu.in

© Springer Nature Singapore Pte Ltd. 2019
M. A. Bhaskar et al. (eds.), *International Conference on Intelligent Computing and Applications*, Advances in Intelligent Systems and Computing 846,
https://doi.org/10.1007/978-981-13-2182-5_18

171

1   Introduction

Developments in communication technology have made concepts such as connected health or remote health monitoring possible. In such efforts, while patient's medical information is transmitted to the caregiver over the internet, policies such as Health Insurance Portability and Accountability (HIPPA) [1] exist to ensure patient data protection. ECG signals are one of the widely transmitted medical information and techniques like steganography, [2] which allows hiding patient data into it. However, embedding patient data alters the ECG signal and care needs to be taken to avoid modifying characteristic points such as QRS complex, P, T, and U waves which otherwise will affect diagnosability. Usually, steganography is performed in transform domain where the cover signal (ECG) is decomposed into frequency subbands. Subsequently, patient data is embedded into a chosen subband using one of the watermarking methods such as least significant bits, quantization, and singular value decomposition methods [2].

This letter proposes an approach where patient data is converted into Quick Response (QR) code and embedded into the ECG signal using a swap approach. The site at which the data is embedded is chosen through a threshold selection approach such that the characteristic points are not modified. The QR code provides additional security to patient data compared to other approaches owing to an inbuilt error correction coding mechanism [3]. Since the data is embedded in the 1D signal directly, data loss in conversion of 1D signal to 2D images [2] is eliminated completely. The performance of ECG steganography using QR code is evaluated in this letter.

2   Quick Response Code

QR code implements error correction feature to the data using Reed–Solomon coding algorithm [4]. Hence, the resultant binary matrix of QR code has the capability to restore its data even if the code is damaged. The error correction level and the data capacity of QR code versions are predefined. For instance, data capacity of version 40 is 4296 alphanumeric characters, 7089 numeric data, 2953 binary data, or 1817 kanji data. In this letter, the patient data is converted to QR code. The QR code is referred to as watermark since it will be embedded in the cover ECG signal. The readers are referred to [7, 8] for further details on QR code.

3   Discrete Wavelet Transform (DWT)

DWT decomposes an input signal by convolving it with one chosen mother wavelet function as in Eq. 1.

$$x(t) = \frac{1}{\sqrt{M}} \sum_k W_\varphi(j_0, k)\varphi_{j_0,k}(t) + \frac{1}{\sqrt{M}} \sum_{j=j_0}^{\infty} W_\psi(j_0, k)\psi_{j_0,k}(t) \qquad (1)$$

W_φ and W_ψ are the approximate and detailed coefficients of a signal $x(t)$ in order. $1/\sqrt{M}$ is a normalizing factor. $\psi_{j_0,k}(t)$ is the mother wavelet function; $\varphi_{j_0,k}(t)$ is the scaling function; where j_0 is an arbitrary starting scale and $j \geq j_0$. Wavelet is a mathematical expression of an oscillating wave that exists only in a finite interval and zero elsewhere [2]. As a result, the input signal is decomposed into several frequency bands and each band is represented by a coefficient matrix. In the ECG signal, the characteristic points lie in the low-frequency band. Hence, high-frequency band is a logical choice to embed the watermark. Readers are referred to [2, 5] for further details on DWT.

4 Methodology

The proposed work consists of four major steps as given in Fig. 1, including (i) signal decomposition and patient data conversion, (ii) threshold selection and watermarking, (iii) reconstruction of cover ECG signal, and (iv) performance evaluation. We use the ECG signals from MIT-BIH database [6] to demonstrate the proposed approach. A one-level DWT using Daubechies 4 (db4) wavelet is applied to decompose the cover ECG signal. The patient data which is the watermark is converted into a QR code. The coefficients to be modified are chosen using a threshold scheme and watermarking is achieved through the swap approach. Upon inverse DWT, the watermarked ECG is obtained which can be transmitted over the internet. At the receiver's end, the watermarked ECG can be subjected to DWT and by performing the reverse of swap, the watermark can be retrieved. The proposed approach also allows reconstruction of cover ECG signal.

4.1 Threshold Selection

The number of coefficients in high-frequency subband to be modified depends on the size of the watermark. For a watermark size of N, in the ordered coefficients of the high-frequency band, N coefficients on both sides of a near-zero coefficient are chosen to be modified. This, we call the threshold boundary and denote by $(\mu_{\min}, \mu_{\max})$ such that it accommodates a watermark of size N in $2N$ coefficients. The readers are referred to [7] for the detailed study of threshold selection.

Fig. 1 QR code-based ECG steganography architecture

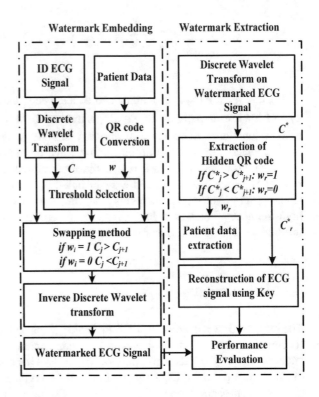

Watermark embedding

We propose using the swap scheme for watermarking. Unlike quantization schemes, in the swap scheme, we do not embed the watermark. Instead, we swap or modify the coefficients depending on the watermark. The swap scheme uses two coefficients for watermarking (w) one bit. Let C_i and C_j be two subsequent coefficients in the ordered set of coefficients in the high-frequency band. If $w = 0$, and $C_i > C_j$ or $w = 1$ and $C_i < C_j$, then the two coefficients will be swapped. The swap rules for all other possibilities are provided in Table 1. Factor α is chosen to be 0.1. The flag k denotes no change, addition of α and swapping operation as 1, -1, and 0, respectively. Inverse DWT on the coefficients modified through the swap scheme and the coefficient of the remaining subbands provide the watermarked ECG signal.

Table 1 Details of the data embedding process

Condition	Watermark bit	Operation	Flag
$C_j > C_j + 1$	1	No swap	1
$C_j < C_j + 1$	1	Swap C_j and $C_j + 1$	0
$C_j = C_j + 1$	1	Add α to C_j	-1
$C_j < C_j + 1$	0	No Swap	1
$C_j > C_j + 1$	0	Swap C_j and $C_j + 1$	0
$C_j = C_j + 1$	0	Subtract α from C_j	-1

4.2 Watermark Extraction

Upon receipt of the watermarked ECG at the care giver's end, the watermark can be extracted by subjecting the watermarked ECG to DWT and performing the reverse of the swap scheme on the coefficients of the high-frequency band. The hidden watermark (QR code) is retrieved during watermark extraction and QR decoder is used to convert the QR code into patient data. In order to achieve this, a watermarked ECG signal is decomposed using DWT and the hidden QR code is retrieved as given in Eq. 2. Here, the watermarked coefficients are identified using flags which are stored during watermarking.

$$w = \begin{cases} 1 & \text{if}(C_i > C_j) \\ 0 & \text{if}(C_i < C_j) \end{cases} \tag{2}$$

4.3 Cover Signal Reconstruction

Usually, cover signals cannot be restored in ECG steganography. However, the proposed approach allows the reconstruction of cover ECG signal using reverse watermarking method along with the stored indices, flag k, and α.

5 Performance Metrics

The performance of the proposed method is evaluated using the metrics such as Peak Signal-to-Noise Ratio (PSNR), Percentage Residual Difference (PRD), and Kullback–Leibler distance (KL) [2]. Imperceptibility of the watermark is estimated using PSNR where higher PSNR denotes better imperceptibility and vice versa. The residual error in the watermarked ECG is measured using PRD. Estimation of histogram difference between cover and watermarked ECG signals are obtained using KL distance. Percentage of errors in the retrieved patient data is calculated using Bit Error Rate (BER).

6 Experimental Results

The proposed watermarking method by swapping two coefficients and adding or subtracting a constant (α) with the first coefficient modifies the cover signal. The imperceptible changes between cover ECG, C and stego ECG signal, C^* is shown in Fig. 2. The difference between reconstructed ECG C^*r and cover ECG signal C is shown in Fig. 3. It can be observed that modified coefficients distort the signal little and it is not visible to the naked eye

Fig. 2 Cover ECG versus
watermarked ECG

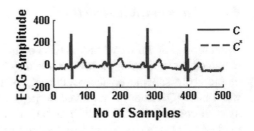

Fig. 3 Cover ECG versus
reconstructed ECG

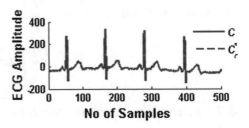

Table 2 Performance metrics for different threshold values

QR code (Version 40)							
		16,272			19,090		
μ_{max}	μ_{min}	PSNR	PRD	KL	PSNR	PRD	KL
5	−5	44.91	0.09	3.8e-05	46.35	0.08	1.4e–05
15	−15	38.75	0.09	3.8e-05	41.31	0.08	3.5e–04
25	−25	36.78	0.09	0.0013	38.82	0.08	7.1e–05
35	−35	34.73	0.09	6.7e-05	37.08	0.08	7.2e–05
55	−55	32.21	0.09	0.0122	33.33	0.08	1.9e–04

Metrics such as PSNR, PRD, and KL distance are used to analyze robustness and imperceptibility of hidden patient data. The proposed approach with a QR code version of 40 resulted in PSNR of 44 dB for signal 16,272. As the threshold bound became wider in order to accommodate larger watermark sizes, the PSNR decreased as given in Table 2.

The same observations are made for another signal 19,090 also. In order to validate the approach, it was demonstrated on different watermark sizes and versions and the results are presented in Table 3. This study is performed using the QR code versions 6–40 that supports up to 2632 bytes of patient data. In addition, PSNR decreases to the increasing watermark size. However, the proposed approach allowed retrieval of the entire hidden data with zero BER.

Table 3 Performance of the proposed approach for different watermark sizes, QR code versions, and signals

Version	(41 × 41) − 6		(109 × 109) − 23		(177 × 177) − 40	
Size (bytes)	64		878		2632	
ECG signal	PSNR (dB)	PRD %	PSNR (dB)	PRD %	PSNR (dB)	PRD %
16,420	42.89	0.08	41.70	0.05	41.02	0.06
16,272	41.20	0.09	36.82	0.09	36.65	0.09
16,483	39.18	0.06	37.10	0.05	38.17	0.06
16,539	43.21	0.09	40.03	0.09	38.49	0.09
19,090	41.66	0.06	37.99	0.08	39.34	0.08
19,088	45.71	0.05	42.41	0.09	42.57	0.11
19,140	45.45	0.05	39.95	0.08	39.28	0.09
19,830	43.69	0.06	39.78	0.10	39.64	0.12

7 Conclusion

In this letter, we have investigated the use of QR code and DWT with watermarking using a swapping approach, for ECG steganography. When the threshold bound is small, higher imperceptibility is observed but watermark size is limited. As the bounds got wide, the watermarked signals deteriorated more. With larger versions of QR Code, a larger watermark can be embedded with relatively lesser deterioration to the signal. A 5% decrease in PSNR is observed for a 41 times increase in watermark size. One of the major benefits of using QR code is that it provides additional security to the hidden patient data using the inbuilt error correction coding mechanism.

References

1. Law, P.: Health Insurance Portability and Accountability Act of 1996. Public Law 104–191. *US. Statut. Large.* 110, 1936–2103.
2. Jero, S. E., Ramu, P., Ramakrishnan, S.: 'Discrete Wavelet Transform and Singular Value Decomposition Based ECG Steganography for Secured Patient Information Transmission'. *J. Med. Syst.*, 2014, Oct; 38(10):132. https://doi.org/10.1007/s10916-0140132-z.
3. Uma, M. S., Jude, H. D.: 'Frequency domain QR code based image steganography using Fresnelet transform. *AEU - Int. J. Electron. Commun.*, 69 (2): 539–544, 2015.
4. Lin, Y. S., Luo, S. J., Chen, B. Y.: 'Artistic QR Code Embellishment. *Comput. Graph'. Forum*, 32(7): pp 137–146, 2013.
5. Ibaida, A., Khalil, I.: 'Wavelet-based ECG steganography for protecting patient confidential information in point-of-care systems', *IEEE Trans. Biomed. Eng.*, 2013, 60, (12), pp. 3322–3330, https://doi.org/10.1109/tbme.2013.2264539.

6. Moody, G. B., Mark, R. G.: 'The MIT-BIH Arrhythmia Database on CD-ROM and software for use with it' *Proc. Comput. Cardiol.*, pp. 45–50, 1990.
7. Jero, S. E., Ramu, P.:'Curvelets-based ECG steganography for data security,' in *Electronics Letters*, vol. 52, no. 4, pp. 283–285, 2 18 2016. https://doi.org/10.1049/el.2015.3218.
8. Lin, P. Y., Chen, Y. H.: "High payload secret hiding technology for QR codes." EURASIP Journal on Image and Video Processing 2017.1 (2017): 14.

Real-Time Human Detection and Tracking Using PEI Representation in a Dynamic 3D Environment

M. Mahalakshmi, R. Kanthavel and N. Hemavathy

Abstract In this paper, we present an improved methodology for detecting and tracking the various posture and movement of people in a crowded and dynamic environment with the help of a single RGB-D camera. The RGB-D cameras are also called as low depth cameras or ranging cameras. The depth camera provides depth information for each pixel. The indigenous RGB-D pixels are transformed into a new point ensemble image (PEI) and human detection and tracking in a 3D space can be accomplished in a more effective and accurate manner. PEI representation, unlike height map representation, projects all the points in the cell into the grid. First, the detector locates the human physically from the probable candidates who are then carefully filtered in a supervised learning and classification second stage. The statistics of color and height are then computed for associating data to generate the 3D orientation of the tracked individuals. We use classifiers such as JHCH and HOHD. The statistics of color and height are then computed for associating data to generate the 3D orientation of the tracked individuals. In tracking, we try to estimate the similarity criteria in order to compare the current frame and the detected response. We have used RANSAC matching algorithm in the tracking stage. The qualitative and quantitative experiments are performed using the different datasets that show a promising improvement by improving the accuracy of the system to 97%. We have concentrated on the false positives and miss rates in the detecting stage and track lost error and ID switch error in the tracking stage. We have produced significant improvements by reducing these errors and even works well in a highly occluded environment. As we concentrate only on the upper part of the

M. Mahalakshmi (✉) · N. Hemavathy
Department of Electronics Communication Engineering, Velammal
Institute of Technology, Anna University, Chennai, India
e-mail: strimaha@gmail.com

N. Hemavathy
e-mail: hemavathy@velammal.edu.in

R. Kanthavel
Department of Electronics Communication Engineering,
V V College of Engineering and Technology, Tirunelveli, India
e-mail: kanthavel2005@gmail.com

© Springer Nature Singapore Pte Ltd. 2019
M. A. Bhaskar et al. (eds.), *International Conference on Intelligent Computing and Applications*, Advances in Intelligent Systems and Computing 846,
https://doi.org/10.1007/978-981-13-2182-5_19

body, occlusion is not going to affect our system anywhere. The results of both qualitative and quantitative experiments obtained show a promising improvement by reducing track lost error thereby improving accuracy.

Keywords RGB-D camera · Point ensemble image · Supervised learning 3D orientation

1 Introduction

Detection and tracking of moving objects [1] are challenging research problems, faced in the unconstrained surveillance condition. In the recent years, there had been advancements in the detection and tracking system by processing video sequences [2] instead of processing static images. Compared to still image, video sequences [1, 3] provide more information about the object as well as the environment over a period of time. Tracking and detecting of humans [4] in real time is used in a number of applications including surveillance, activity monitoring, and gate analysis.

The increase in the necessity of video surveillance used in web-based monitoring application or smartphone-based application, it is required that the surveillance video application should be capable of producing various video data's with regards to the computational capabilities of the terminal devices. Surveillance systems must be able to detect and track objects moving in its field of view, classify these objects, in particular, humans and detect their activities. Even though tracking [5] is considered as an important process, it is the part where most of the errors occur due to change in illumination, shadow, occlusion, and reflections in natural environment. The video sequences are processed in order to find the main events of interest, which depends greatly on the high-level description of the video stream, which in turn relies on the accurate detection and tracking of moving objects [6] and the relationship between their trajectories and the scene.

In our proposed modelm we use a RGB-D camera to detect and track the plausible postures and movements of the target individual in a dynamic 3D environment. Our proposed method consists of three stages. In the first stage, the information captured by the RGB-D camera is transformed into point ensemble image representation. The stage involves an unsupervised detector to obtain the probable positions of the human body which are then refined with the help of a classifier which employs two features which are histogram of height distance (HOHD) and joint histogram of color and height (JHCH) and finally, the data is carried out to detection responses to generate three-dimensional trajectories.

Our contributions in this project are (1) An enhanced PEI representation which helps in reducing the overlapping in the original image at the same time preserves all the information of the pixels. (2) A human plausible location technique is employed to prevent the need for a number of procedures to scan the entire point cloud, thereby facilitating quick reduction in search space. (3) HOHD and JHCH are two features which are employed to characterize the shape and appearance of the human in 3D space. (4) A progressive framework which is a single-pass framework helps the system to achieve real-time performance as it reduces the number of positions of the candidate to be examined by the latter stages.

There exist a number of papers on detecting and tracking objects. Papers related to our proposed model are discussed as follows. Bhaskar et al. [7] provided a system to detect and track object in an environment that is influenced by illumination changes, occlusions, and improper placing of camera. This system used DRA (Dynamic Reverse Analysis) approach for detection and E-RBPF (Enhanced Rao-Blackwellized Particle filter) for tracking. This model has increased accuracy and robustness but they do not track large number of targets in crowded and dynamic environment. Xi et al. [3] considered multiple objects tracking as an integer programming of flow network model and then converted to standard linear programming to achieve optimal solution. It has the merit of lower calculation complexity, better tracking accuracy, and robustness. The speed of this tracking algorithm on video sequence is of 25 fps. The global optimum solution of linear programming can be found more quickly. He et al. [8] proposed a system to automatically track the variable number of objects in the monocular and un-calibrated camera. They extended the previous work which used maximum a posteriori (MAP) data association problem for tracking object. They used an improved greedy algorithm to reduce the amount of ID switches. A linear hypothesis method is proposed to fill up the gaps in the trajectories. The time gap allowed in this system is only for less than 30 frames. And, this system is based on the assumption that object has a constant speed. So, these ideas failed in the case of occlusion and speed changes of the target object.

Kwalek and Kepski [9] had tried to eliminate too much false alarms generated by inertial sensors in the previous works. This paper proposed to use tri-axial accelerometer to detect the fall and if the target is in motion. Threshold has been set to differentiate false alarms and actual alarms. This proposed model is suitable only for indoor use because of the limitation of the Kinect sensor. Zoidi et al. [10] proposed an appearance-based approach for tracking object using Kalman filtering and sampling technique. They are used for object position prediction and for selecting the required regions, respectively. The changes in positions of the object are simultaneously handled in left and right channel video frames. The advantage of this system is that it is suitable for application in 3D movies, TV, and video content. But this model does not gather any information on depth data that are collected using cameras such as Kinect. Maziran [11] has proposed a method which is a development of probabilistic estimation theory via particle filter. The chosen new features of mobile objects are completely analyzed through a neural network. The captured frames are of two-dimensional data matrices, the extracted new features may be utilized as the third dimension. The main advantage is its accuracy on detecting the objects under random movement. The main problem of this system is finding appropriate neural network inputs.

Sabirin and Kim [12] presented a method for detecting and tracking moving objects for video surveillance application. Spatial–temporal graph has been used and graph matching is performed to identify the real moving object even under occlusion. Spatial base layer is mapped to spatial enhancement layer. This model has achieved higher spatial resolution and processing time down to 27%. But they lack handling occlusion in compressed domain because of the limitations of available features to be extracted. Choi et al. [13] presented a paper on tracking, interacting with people from a mobile vision platform. Tracking is done by finding the MAP solution of a posterior probability and is solved using RJ-MCMC particle filtering method. The characteristics are joint formulation of all variables, the combination of multiple observation cues, allowing people to interact, automatically detecting people, and automatic detection of static features for camera estimation.

Kim et al. [14] proposed a solution to detect 3D objects and they produced output more same as reality. They allow the user to add new objects quickly. This model uses multiple keyframes for single object. This model achieves fast keyframe generations and also supports polygonal or circular-based models, whereas the previous works supports only box type objects. It has good scalability in terms of number of objects. But this model has the drawback of computational complexity and the memory grows linearly with the number of objects in the database. Dan et al. [15] proposed a counting system based on the combination of depth and vision data. Here, a video-plus-depth camera is used which is mounted on the ceiling and bidirectional matching algorithm is used for detecting the trajectory of the object. This proposed system consists of the lost depth data recovery algorithm, people detection based on the human modeling, and people tracking using both depth [16] and color data.

Levi and Weiss [17] had proposed a system to detect objects from a small training database. Instead of using standard linear features, they have used local EOH as features to improve performance. These histograms significantly improve the ability of the system to learn from small learning databases. It detects only profile face and chair. To detect many types of object, color, and texture features have to be combined with EOH. Ess [18] proposed a system which jointly estimates camera position, stereo depth, object detection, and tracking. The different modules were integrated in a graphical model. But in this model, very close pedestrians for which only part of the torso is visible are often missed by the pedestrian detector. Enzweiler et al. [19] proposed a paper for handling partial occlusion. The degree of visibility is computed by examining occlusion boundaries. Multi-cue classification improves performance by the factor of two. But this only handles partially occluded and non-occluded pedestrians. Luber et al. [20] presented a paper on 3D people detection and tracking using RGB-D data. Multi-cue person detector and online detector are used. Data are collected in a populated indoor environment using a setup of three Microsoft Kinect sensors. Online classifier contributes to find the correct observations in cases when the priori detector fails. Oron [21] have proposed a locally order less tracking. This algorithm automatically estimates the quantity of disorder in the object. It is implemented using the Earth Mover's Distance (EMD). In order to account for the quantity of disorders in the object, they adjust costs online while tracking. This system correctly estimates the noise parameters. This method provides a general framework for all the noise models. The different noise models can be used for the appropriate applications.

Enzweiler et al. [19] proposed a paper for handling partial occlusion. The degree of visibility is computed by examining occlusion boundaries. Multi-cue classification improves performance by the factor of two. But this only handles partially occluded and non-occluded pedestrians. Ess [18] proposed a system which jointly estimates camera position, stereo depth, object detection, and tracking. The different modules were integrated in a graphical model. But in this model very close pedestrians for which only part of the torso is visible are often missed by the pedestrian detector. Ess et al. [18] has proposed a system to encounter issues in multi-person tracking in a heavily rushed pedestrian region. They have placed a stereo rig in that platform and they estimate camera position, stereo depth, object detection, and tracking. They provide robust tracking performance. In this paper, the speed and of the performance of the system must be improved. There are failures in this system such as false positives on trees, missing detections, and obstacles in detecting path.

Dalal and Triggs [21] have proposed a system for robust visual object recognition. They use Histogram of Oriented Gradient (HOG) that outperforms the existing features. They introduce a very highly challenging dataset containing about 1800 human images with a variety of poses and with cluttered background. The disadvantage of this system is that it still needed to be optimized and speeded up for better detecting performance. Fleuret et al. [22] have proposed a multi-camera people tracking system. This model can handle occlusions in all the frames more effectively. They process individual trajectories separately thereby avoiding confusions that happen among one another. They have used map estimation that can be error-prone. This algorithm will fail to detect correctly and locate all target humans if there are too many people in the scene.

Geraldo Silveira and Ezio Malis have developed a visual tracking system under light changes. They align images of Lambertian, non-Lambertian objects under shadows, interreflections, and specular reflections. The second-order optimization technique is used that directly reduces the intensity variations. The system needs to be still extended to improve robustness and reliability. Zhu [23] has given a human detection system using a cascade of histograms of oriented gradients. They combined the cascade-of-rejecters approach to the histograms of oriented gradients (HOG) features. They are AdaBoost feature selection. They speed up computation by using the integral image representation. This system can process 5–30 frames/s. Levi and Weiss [17] had proposed a system to detect objects from a small training database. Instead of using standard linear features, they have used local EOH as features to improve performance. These histograms significantly improve the ability of the system to learn from small learning databases. It detects only profile face and chair. To detect many types of object, color, and texture features have to be combined with EOH. Felzenszwalb [24] has proposed a learning model to recognize objects. They trained their system to detect people in images using a single shape model. They suggest that learning techniques can be extended to represent objects using multiple model shapes. This system is applicable only for

a limited set of objects. There is also a problem of learning a set of canonical views
to represent an object.

2 Methodology

Our aim is to develop a human detection and tracking system using a single RGB-D
camera as shown in the Fig. 1, thereby making it suitable for real-time application
and at the same time achieving higher accuracy. In order to achieve this, we use a
new PEI representation which helps us to overcome segmentation problems that
arise when the RGB-D image is used directly. In order to improve computational
efficiency, we use multistage filtering detection which eliminates the doubtful
candidates at the earlier stages so that fewer candidates are passed on to the latter
stages where they can be further reduced in number to obtain significant accuracy.
Joint histogram of color and height is used for associating the detection results and
its combination with Kalman filter provides the 3D trajectory of the humans.

2.1 Point Ensemble Image Representation

Due to occlusion among humans, segmenting the original image domain often
causes under segmentation and oversegmentation. Plan view transformation from
original data helps in solving these to problems. In the camera coordinates, a 3D
cloud as in Fig. 2 can be created by back projecting the pixels with depth information into 3D space. We establish a new co-ordinate system with the XY plane as
the ground and the Z-axis as the upper vertical vector. The ground plan is dis-
credited with the help of a square grid. The point cloud that is mapped to the new
coordinate system is projected to the XY plane and the plan view on the floor is

Fig. 1 Block diagram of
tracking system using a single
RGB-D camera

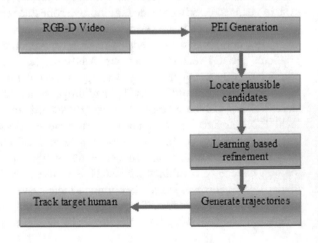

Fig. 2 3D point cloud of a
human body

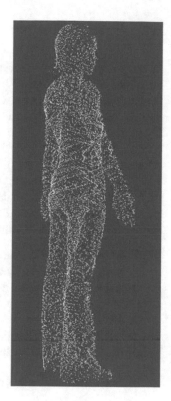

generated. This virtual view, segmentation can be simplified. The square grid for
discrediting the plan view is defined as

$$g^{i,j} : \begin{cases} (i-1)\varepsilon \leq x < i\varepsilon \\ (j-1)\varepsilon \leq y < j\varepsilon \end{cases} \tag{1}$$

where $g^{i,j}$ is the (i, j) cell of the grid, (x, y) is the point coordinates on the XY plane,
and ε is the resolution of the grid.

In the plan view representation, each cell holds the information of the highest
point in that cell as result lot of detail information is lost. Hence, we propose the
usage of PEI representation, to take advantage of plan view at the same time
preserving the information of all 3D points. In PEI (Point Ensemble Image) which
is denoted by E, each cell records a set of points projected into that cell and is
formulated as

$$E^{i,j} = \{p | p \in P, (x^p, y^p) \in g^{i,j}\} \tag{2}$$

where P is the new coordinate systems point cloud and $p = (x^p, y^p, z^p)$ is a 3D point
in P.

2.2   Human Detection

In most of the detection process, background subtraction is used, but in real-world situations, it does not prove to be efficient since background is dynamic in nature for most of the cases. The next option is to use depth cameras for human detection without background subtraction. The performance of depth cameras is limited since the shape is crucial in these methods prior to head and body and hence can be corrupted easily due to mutual occlusions and incomplete depth data. Thus, instead of imposing strong shape prior on target, we do it progressively. Our method includes a two-phase filtering method which is capable of extracting the advantage of both depth and RGB data. Most of the false positives that are not human physique wise probable are quickly rejected in the first phase. The responses are further refined by using a learning-based classifier in the second phase.

Phase 1: Detecting the locations that are human physique wise probable

In the initial phase, we find the 3D points which are higher than all the neighborhood points. For a point p in the point cloud, we draw a cylinder with radius $\omega/2$ where ω is the average width of the human torso. The height of the points within the cylinder is compared with the height value of p. The point ensemble image representation makes this computation quite efficient. The neighboring set of points of p is calculated on the PEI in the following manner:

$$n = \{x | x \in c', c' \in E, |c' - c| < \omega/2 \tag{3}$$

where c is the cell in E into which p is projected and c' is the set of neighboring cells of p including c. All the points with the maximum value of height in the neighborhood are considered to be possible crowns of human heads. The points having height value between minimum height threshold (h_{min}) and maximum height threshold (h_{max}) are considered in order to improve processing speed and also reliability. This helps in eliminating a large number of points that are closer to the floor and ceiling. This process ensures that only head crown locations are present in the resulting response even in complex and crowded situations. The space to be further searched has been reduced considerably in this phase, even though there are false positives.

Phase 2: Learning-based refinement

The second phase involves the process of determining whether the objects picked up in the previous phase are humans. We train a classifier using two types of features for encoding shapes and appearance information. Both the information's are obtained from the set of neighboring points in the PEI. The two features are based on the upper portion of the body only, since the lower part is prone to frequent occlusions and deformations. The features are described below.

2.3 HOHD—Histogram of Height Difference Measurement

As shown in the Fig. 2, it is seen that humans have their 3D shape of the upper body distinct from the remaining part of the body. We can see that there is a slight difference in height between the head crown and other points in the head and a large difference in height between the head crown and the other part of the body. Since the lower part of the body often suffers from occlusion, we only obtain points from the upper torso. Those points must satisfy the following two conditions.

(i) Among the points projected into the cell, the point that is to be selected should have the largest height value.

(ii) We set a threshold, γ which is the quarter of the average human height. The difference in height between the head crown and point must be less than the threshold γ. The normalized histogram is constructed using the data collection of height variations which is shown in the Fig. 3. The normalized histogram of 15 frames is been constructed. This feature alleviates noisy data and avoids hard assumptions as we construct this histogram.

2.4 JHCH—Joint Histogram of Color and Height

In human detection and tracking, the height and the color features are helpful and proved effective. And simply combining both the features are also proved to be efficient. To obtain a good output, we need both the features to be compared. Thus, we use joint histogram of color and head feature to describe the information about the human head. A human is detected in his front then the color information of face and hair are collected. But if the human is observed from the back, then only the information of the hair is collected. Those values are given the different height values. As the HSV (height, space, and value) color space model is less sensitive to illumination changes, we use this model.

We use separate frames to uniquely record the points of the black and white pixels. We build 2D JHCH with the help of the points that are smaller than that of

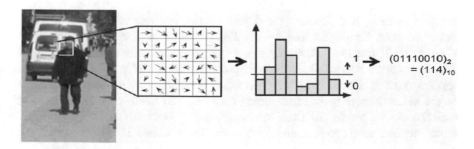

Fig. 3 Histogram of height difference

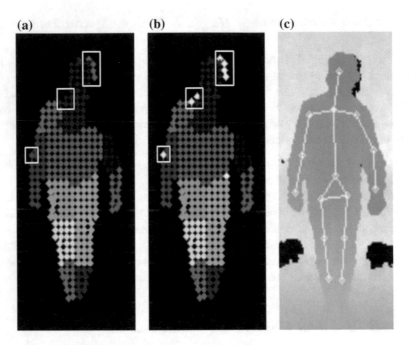

Fig. 4 Joint histogram of color and height representation

head crown as shown in the Fig. 4. JHCH shows eminent performance even in detecting people with different hairstyles and head postures.

2.5   3D Joint Histogram Color and Height Tracking Method

The points that satisfy the condition defined in the detection stage are then entered into the tracking stage. The Kalman filter model is used and we associate data in each frame. We use Kalman filter model because the results produced by this model is better than the measurement of the data as seen in Fig. 5. We assign a response to each track that is similar to the position predicted. In order to differentiate among different humans, it is necessary to define a similarity function. Again to differentiate between different humans, we use JHCH that we used at the detection stage. We use JHCH due to its efficiency more than the other color models. But in the detection stage, we used 2D JHCH descriptor. The 2D JHCH is fine to describe the appearance of the human but it is limited when we try to describe the characteristics of the entire human body. This limitation is due to the highly differing cloth varieties. So we go for 3D (hue, saturation, and height) JHCH. We divide hue, saturation, and height to 9, 5, and 5 intervals. To build this 3D JHCH, only those points which are in the cylindrical neighborhood of the head crown are helpful.

Fig. 5 Performance of Kalman filter

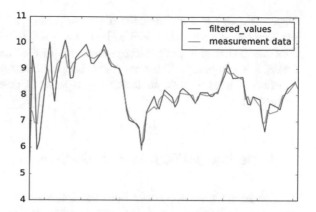

Let τ denote the number of frames in the 3D JHCH to be constructed. We go through these frames in the frame left to right, top to bottom, and front to back. Then, the histogram of the target Tg is given by $H_{Tg} = \left\{ H_{Tg}^i \right\}_{i=1,2,3...} \tau$. We need to assign different weights to the 3D points in the different height level. The weighting function can be given as follows.

$$wt(p_i) = \exp\{\alpha \cdot h(p_i)\} \tag{4}$$

where α is the weight, $h(p_i)$ is the height value of the acquired point. The JHCH of each frame can be calculated as follows.

$$H_{Tg}^i = B. \sum_{p_i \in n} \{wt(p_i) \cdot \delta[f(p_i) - i]\} \tag{5}$$

where δ is the Kronecker delta function, $f(p_i)$ is the frames of JHCH, I is the neighboring point set, $B = 1/\sum_{p_i \in n} wt(p_i)$ is the factor that ensures that sum of total histograms is unity.

Now, we define the similarity function between the targets A and B as follows:

$$\partial(A, B) = \gamma \cdot \rho(H_A, H_B) + (1 - \gamma)\varphi(S_a, S_b) \tag{6}$$

where S_a, S_b are the spatial locations of the target, $\rho(H_A, H_B)$ is the Bhattacharya similarity, $\varphi(S_a, S_b)$ is the spatial location similarity. $\rho(H_A, H_B)$, $\varphi(S_a, S_b)$ is given as below.

$$\rho(H_A, H_B) = \exp\left\{ -\mu \cdot \left(1 - \sum_{k=1}^{r} \sqrt{H_A^i H_B^i} \right)^2 \right\} \tag{7}$$

$$\varphi(S_a, S_b) = \exp\{-v \cdot D(S_A S_B)\} \tag{8}$$

where $D(S_A S_B)$ is the Euclidean distance, μ and ν are the weighing factors. The similarity function's values will be between 0 and 1. If the similarity function is 1, then the two targets are perfectly matched. In a frame detection, responses and tracks are compared. If the response has no matching track, a new object is detected. If a track with no matching response is observed, then the tracking is terminated.

3 Experimental Results and Outputs

The experiments are done on various datasets such as office dataset, mobile camera dataset, and clothing store datasets to analyze the proposed systems' performance and its accuracy. In an office dataset provided by Choi et al., people take different poses such as standing, moving, and sitting in a largely occluded environment. Choi et al., has also provided a mobile camera dataset where a robot is made to drive around a building. This dataset is subjected to different light conditions and discontinuous background. Along with these two dataset, clothing store dataset which has been captured in a clothing store is used. This dataset has a background more complicated than in the mobile store dataset and also this dataset includes walking in groups. Thus, this clothing store dataset helps us to analyze even more effectively.

The proposed system's performance is evaluated by comparing it with the previous works such as Edgelet-based part detector, Choi et al. detector and Zhang et al., detector. There is no source code available for Choi et al., system. So, we only evaluate the office and mobile camera dataset for Choi et al., detector. The office and mobile camera dataset uses four images per second and the clothing store dataset uses one image for 3 seconds. We provide hand-informative label along the bounding boxes in the upper bodies of people in assessment frame. The two assessing method used by Choi et al., is been used to find out the positive detection. The two methods are based on the degree of being common between the detected areas and the bounding boxes and on the 3D distance.

Our proposed system on experimentation shows that it excels the other three detectors mentioned before which are shown in the Fig. 6. In the Edgelet-based part detector, the performance is limited due to highly cluttered background and different poses of people. Whereas in the Zhang et al. detector, the recall rate is limited due to the partial occlusion of head that occurs more often. The proposed system unlike Choi et al., combines RGB and depth data and implements 3D shape and appearance details together. Thereby, the system works well even if the partial occlusion occurs. The results of the first detection phase of our detector are shown in Table 1 and the comparison chart is shown in the Fig. 6. From the result, it is observed that the detector at this phase yields a low miss rate. But this reduces the

Table 1 Detection error for our proposed detector against various other detectors

Detector	Miss rate	FPPI
Proposed	0.6	15
Choi et al.	0.8	17
Z. Zhang et al.	1.0	15
Edgelet-based part	1.2	16

Fig. 6 Comparison chart of detection error for the proposed detector against existing methods

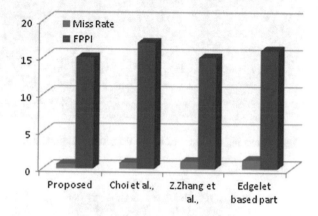

search space for the detector because only many false positives per image (FPPI) are contained in the detection responses. The miss rate in this phase is due to the data loss in depth.

In the three datasets, people are very far from the camera. The depth data of these far people are not available. Therefore, they get missed in the first detection phase. The quantitative performance of the two features is evaluated by turning off any one feature either JHCH or HOHD at a time. And, we compare the results on the datasets. Turning on the HOHD feature results in abrupt performance decrease. This is because JHCH is a combination of appearance and depth information and it yields higher discriminative details. But the HOHD has only depth data. The office environment's result and tracking of vehicles are shown in the Figs. 7 and 8, respectively.

It is seen that the tracking errors are significantly reduced. Particularly, the ID switch errors have been greatly reduced.

The tracking performance is analyzed by evaluating the tracking errors such as track lost errors and the ID switch error and the results are compared in the Fig. 9.

We implement RANSAC matching algorithm in this system as shown in Figs. 10 and 11. Using this algorithm, we can detect and track the human body in different poses as shown in Fig. 12.

Fig. 7 Tracking performance of our detector in different stages

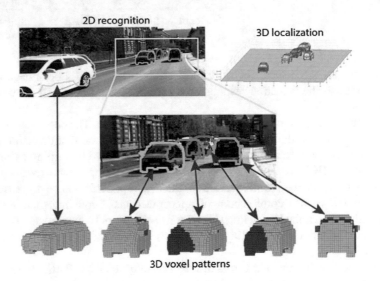

Fig. 8 Tracking performance on vehicles

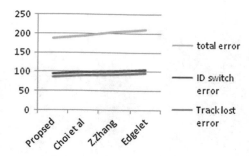

Fig. 9 Tracking performance of our proposed system against various other detectors

Fig. 10 RANSAC matching algorithm on street lights

Fig. 11 RANSAC matching algorithm

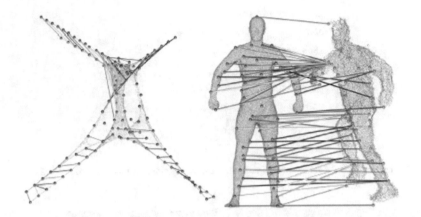

Fig. 12 Matching humans at different poses

4 Conclusion

We use a RGB-D camera to detect and track the target human using the enhanced PEI (Point ensemble image) representation. The proposed system can detect and track people in the various poses and in the dynamic background more effectively as shown in our results. We have combined the RGB and depth data to detect and track. We re-project the 3D points back to the original domain. We accomplish the tasks incrementally at separate stages such as locating, detecting and tracking. We have features such as HOHD and JHCH in the detecting stage and Kalman filter model in the tracking stage. The PEI representation makes the computation process easier in the detection and tracking stages.

References

1. Jun Liu, Ye Liu, Guyue Zhang, Peiru Zhu, Yan Qiu Chen, "Detection and tracking people in real time with RGB-D camera", Elsevier, 2014, pp. 16–23.
2. I. Haritaoglu, M. Flickner, D.Beymer, "Video-crm: understanding customers behaviors in stores, in: SPIE Proceeding", Video surveillance and transportation imaging application, 2013.
3. Zhenghao Xi, Dongmei Xu, Wanqing Song, Yang Zheng, "A* algorithm with dynamic weights for multiple objects tracking", Elsevier, 2015.
4. Qiang Zhu, Shai Avidian, Mei-Chen Yeh, and Kwang-Ting Cheng, "Fast human detection using a cascade of histogram of oriented gradients", IEEE Conf. on Computer vision and pattern recognition, 2006.
5. K.Kim, V.Lincent Lepetit, W.Woo, "Real time interactive modeling and scalable multiple object tracking for AR", Elsevier, 2012, pp. 945–954.
6. A.H.Maziran, "A powerful approach to real-time mobile objects tracking in crowded environments utilizing 3Dvideo analysis of fixed camera, particle filter and natural network", Elsevier, 2013, pp. 6357–6365.

7. Harish Bhaskar, Kartik Durivedi, Debi Prosad Drogra, Mohammed Al-Mualla, Lyudmila Mihaylova, "Autonomous detection and tracking under illumination changes occlusion and moving camera", Elsevier, 2015, pp. 343–354.

8. Z.He,Y.Cui, Hongpeng Wang, X.You, P.Chen, "One global optimization method in network flow model for multiple object tracking", Elsevier, 2015, pp. 21–32.

9. Bogdan Kwalek, Michal Kepski, "Human fall detection on embedded platform using depth maps and wireless accelerometer", Elsevier, 2014, pp. 489–501.

10. Olga Zoidi, Nikos Nikolaidis, Anastasio Tefas, Loannis Pitas, "Stereo object tracking with fusion of texture, color and disparity information", Elsevier, 2014, pp. 573–589.

11. A.H.Maziran, "A powerful approach to real-time mobile objects tracking in crowded environments utilizing 3Dvideo analysis of fixed camera, particle filter and natural network", Elsevier, 2013, pp. 6357–6365.

12. Houari Sabirin, Munchurl Kim, "Low complexity object detection and tracking with inter-layer graph mapping and intra-layer graph refinement in H.264/SVC bit streams", Elsevier, 2013, pp. 1531–1539.

13. W.Choi, C.Pantofaru, Silvio Savarese, "A general framework for tracking multiple people from a moving camera", IEEE Trans. Pattern Anal. Mach. Intell, 2013, pp. 1577–1591.

14. Kiyoung Kim, Vincent Lepetit, Woontack Woo,"Real time interactive modeling and scalable multiple object tracking for AR", Elsevier, 2012, pp. 945–954.

15. B.K. Dan, Y.S. Kim, Suryanto,J.Y. Jung, S.J. Ko, "Robust people counting system based on sensor fusion", IEEE Trans. Consum. Electron, 2012, pp. 1013–1021.

16. L.Xia, C.C.Chen, J.K Aggarwal, "Human detection using depth information by kinect", IEEE Conference on computer vision and pattern recognition workshop, 2011, pp. 15–22.

17. K.Levi, Y. Weiss, "Learning object detection from a small number of examples: the importance of good features", IEEE Conf. Computer vision and pattern recognition, 2013, pp. 53–60.

18. A.Ess,B.Leibe, K.Schindler, LV.Gool, "Robust multiperson tracking from a mobile platform", IEEE Trans. Pattern Anal. Mach. Intell, 2009, pp. 1831–1846.

19. M.Enzweiler, A.Eigenstetter, B.Scheile, D.M.Gavrila, "Multi cue pedestrian classification with partial occlusion handling", IEEE Conf. on computer vision and pattern recognition, 2010, pp. 990–997.

20. M.Luber, L.Spinello, K.O.Arras, "People tracking in RGB-D data with on-line boosted target models", IEEE Conf. Intelligent robots and system, 2011.

21. N.Dalal, B.Triggs, "Histograms of oriented gradients for human detection", IEEE Conf. on computer vision and patterns recognition, 2005.

22. F.Fleuret, J.Berclaz, R.Lengage, P.Fua, "Multicamera people tracking with a probabilistic occupancy map", IEEE Trans.Pattern Anal.Mach.Intell, 2008, pp. 267–282.

23. Q.Zhu, M.C.Yeh, K.T.Cheng, S.Avidan, "Fast human detection using a cascade of histograms of Oriented gradients", IEEE Conference on Computer Vision and Pattern Recognition, 2006, pp. 1491–1498.

24. P.Felzenszwalb, "Learning models for object recognition", IEEE Conf. on computer vision and pattern recognition, 2001, pp. 1056–1062.

Studies on Properties of Concrete Using Crumb Rubber as Fine Aggregate

P. Manoharan, P. T. Ravichandran, R. Annadurai and P. R. Kannan Rajkumar

Abstract The waste generated from the disposal of tyres, i.e., waste rubber, rubber products, and angle scrap is very high in volume raising serious environmental concerns. The disposal of this waste has become an environmental concern that needs to be attended to as quickly as possible. The effect of partial replacement of waste fine crumb rubber with sand are investigated on the cement concrete specimens. Due to the wide use of concrete in construction, the application of crumb rubber as a material of concrete would provide an efficient way of disposing the enormous amount of waste that is generated from scrap tyres. In this study, concrete with crumb rubber replacement 3, 6, 9, 12, and 15% is adopted. The objective of this paper is to establish the applicability of crumb rubber as a promising substitute to fine aggregate in conventional concrete. The potential advantage of using the crumb rubber in concrete is revealed through the tests such as compressive strength, split tensile strength and flexural strength. Even though the compressive strength of concrete is reduced to some extent in comparison to conventional concrete by replacing the crumb rubber by fine aggregate but the results showed a substantial increase compared to target compressive strength until 6% replacement.

Keywords Crumb rubber concrete · Impact · Toughness

1 Introduction

Concrete which is widely used construction material is increasing day by day and the use of main constituents such as Cement, Sand, and Coarse aggregate consumption is also increasing simultaneously. One of the global problems is the anthropogenic CO_2 emission is originated from cement production which poses 5% of environmental pollution [1]. On the other hand, health hazards and difficulty in

P. Manoharan · P. T. Ravichandran (✉) · R. Annadurai · P. R. Kannan Rajkumar
Department of Civil Engineering, Faculty of Engineering and Technology,
SRM Institute of Science and Technology, Kattankulathur 603203, Tamil Nadu, India
e-mail: ptrsrm@gmail.com

© Springer Nature Singapore Pte Ltd. 2019
M. A. Bhaskar et al. (eds.), *International Conference on Intelligent Computing and Applications*, Advances in Intelligent Systems and Computing 846,
https://doi.org/10.1007/978-981-13-2182-5_20

197

land filling also creates increased environmental effects due the large generation of waste rubber. The illegal dumping of waste rubber and the high cost of disposal pose a huge threat to the environment [2]. The flexibility and ductility of concrete is enhanced due to the addition of crumb rubber to the concrete as a replacement to fine aggregate which nullifies the above problem [3, 4]. More research work has been carried out by the researchers mainly to use recycled tyres by made into fine particles known as crumb rubber to mix into the cementitious based materials like concrete [5–7]. Previous experimental studies in crumb rubber aggregate concrete showed increase in its performance through ductility, impact, toughness, and reduced damping [8–10]. However, the results indicate that there is reduction in mechanical properties such as compression, tension, and modulus of elasticity [11, 12]. The interface adhesion between rubber and cement particles is very poor due to the smooth surface of the rubber particles is the main reason for the strength reduction [13, 14]. Studies by different researchers show that there is an enhancement of elastic behavior while using crumb rubber in concrete but on the other hand there is a reduction in compressive strength. The various research articles indicate that rubber mixed with concrete will enhance the deformability and durability [15]. The strain capacity of concrete improves the macro-crack localization when crumb rubber is used in the concrete [16].

The concrete with crumb rubber replacement (five replacement ratios are preferred 3, 6, 9, 12, and 15%) is proposed in this study. This research focusses on the mechanical properties and the density of concrete with crumb rubber. Wong and Ting [17] have studied the properties of high-strength concrete after replacing the amount of total aggregates by 25% of rubber aggregates and rubber chips. Normal and High Strength Concrete which had rubber aggregate in it has shown gradual ductile failure when subjected to high loads, whereas the conventional normal and high-strength concrete suffered a sudden brittle failure [18]. After partially replacing the fine aggregate, the concrete has been tested for Compressive Strength, Split Tensile Strength, and Flexural Strength. The durability properties of concrete were also studied. The various results obtained after performing the required tests were interpreted and analysed.

2 Experimental Methods

The target compressive strength of the control mix (M1) was 40 MPa with a constant water to binder (W/B) ratio 0.39. Silica Fume was used to replace 8% of the total amount of cement as binding material for both the mixes. The specific gravity of crumb rubber and coarse aggregate used was having a specific gravity of 1.20 and 2.65. A super plasticizer of Polycarboxylate ether was used to increase the workability of concrete. Concrete mix was obtained with substitution of sand volume with 0, 3, 6, 9, 12, and 15% of Crumb Rubber. Six mixtures with and without crumb rubber was tested for Compression, split tension and flexural strengths. All the ingredients such as cement, silica fume, sand, crumb rubber (with different percentage), coarse aggregate was dry mixed for 2 minutes and then water mixed with SP was added and mixed for another 3–4 min. Thereafter fresh concrete is cast into moulds and removed from the mould after 24 h and cured for 28 days. The various materials that were used in this investigation and their properties are explained below. The investigations were carried out in three series, each involving six different mixes of materials, thus having 18 different mixes.

3 Test Results and Discussions

The procedures that were followed in order to perform the various tests that give the characteristics of concrete upon the partial substitution of fine aggregate with crumb rubber are given below.

3.1 Compression Test

For this test, concrete cubes of dimensions 100 mm cubes were cast. The tests were performed in accordance to the standards provided in IS 516-1959 [19]. The compressive strength values for various replacements of crumb rubber were obtained by performing the test at 3, 14, 21, 28 days. The results obtained are mentioned in Table 1. After the performance of all the tests on the samples, values were obtained which confirm our speculations about the effect of crumb rubber on the performance properties of concrete. The results of compressive strength test show that the compressive strength of concrete increases with increase in crumb rubber up to 6% and thereafter further increase reduces the compressive strength reduces upon the addition of crumb rubber. The compressive strength values of concrete after 28 days for various crumb rubber percentages are shown in Fig. 1.

Table 1 Compressive strength of concrete

Mix ID	Compressive strength (N/mm^2)			
	3rd day	14th day	21st day	28th day
CR0	20.2	43.1	45.2	46
CR1	21	44.27	48	49.7
CR2	19.5	42.3	44.1	46.04
CR3	19	37.3	38.9	42.3
CR4	18.18	35.2	37.2	39.3
CR5	15.8	32.9	35.8	37.1

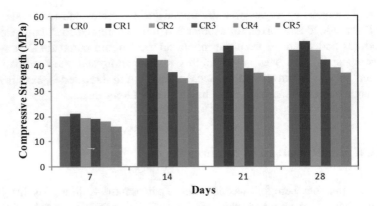

Fig. 1 Compressive strength of crumb rubber concrete

3.2 Split Tensile Strength of Concrete

To conduct the indirect tensile test, concrete cylinders of dimensions 100 mm
 200 mm were cast. The tests were performed in accordance to the standards
provided in IS 5816-1999 [20]. The results were obtained by performing the test at
7, 21, 28 days. The results thus obtained are mentioned in Table 2. The results were
compared with the codal provisions as per IS 456-2000 as shown in Eq. 1 [21].
Generally, engineers assume that the direct tensile strength of concrete is around
10% of its compressive strength; split tensile strength is around 1 percent more [22].
The value of split tensile strength increases with percentage increment of crumb
rubber in concrete till 6%, following which, any further increase in crumb rubber
goes on to reduce the value of tensile strength of concrete. The results obtained in
this test are also in similar pattern to those obtained by Ali I Tayeh et al. during their
research on a similar experimental investigation [23]. The split tensile strength of
crumb rubber with varying percentages in concrete at 28 days are shown in Fig. 2.

$$fcr = 0.7[fck]^{0.5} \tag{1}$$

Table 2 Split tensile strength of concrete and comparison with compressive strength of concrete

Mix ID	Split tensile strength (N/mm^2)			Compressive strength (N/mm^2) at 28 days	Split tensile strength (N/mm^2) at 28 days	Split tensile strength as per IS
	7th day	21st day	28th day			
CR0	2.87	3.57	4.2	46	4.2	4.74
CR1	3.1	4	4.7	49.7	4.7	4.93
CR2	3.15	3.86	4.55	46.04	4.55	4.74
CR3	2.07	3.61	4.3	42.3	4.3	4.55
CR4	1.98	2.63	3.2	39.3	3.2	4.38
CR5	1.9	2.28	3	37.1	3	4.26

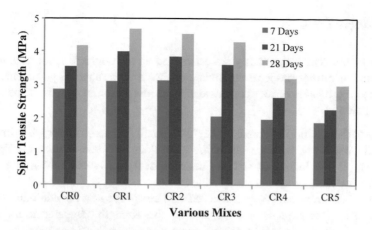

Fig. 2 Split tensile strength of crumb rubber concrete

3.3 Flexural Strength

The test for Flexural Strength was carried out by casting a beam of size 100 mm × 100 mm × 500 mm. The test was carried out in as per IS 516-1959 [19]. The results were obtained after performing the test after 28 days. The results thus obtained are mentioned in Table 3. It is well known that the concrete is a material with low energy absorption capacity due to the low resistance to tensile stresses which makes the concrete to be more brittle. The resistance to tensile loads using crumb rubber is in the range of two to three times higher than cement matrix resistance. The results show us that the value of flexural strength increases till addition of 9% crumb rubber after which, a decline in the strength has been noticed. The flexural strength under four-point loading was 8.30 MPa for specimens without crumb rubber. Whereas the specimens with 9% crumb rubber was 12.20 MPa with an increase of 46.90% compared with concrete without crumb rubber.

Table 3 Flexural strength of concrete

Mix ID	Ultimate flexural load (kN)	Flexural strength (N/mm²) at 28 days
CR0	20.75	8.3
CR1	23	9.2
CR2	28.55	11.42
CR3	30.5	12.2
CR4	23.1	9.24
CR5	22.9	9.16

4 Conclusions

All the required investigations were performed on the concrete specimens in order to obtain the various properties of concrete when crumb rubber is partially replaced with fine aggregate. A few conclusions were drawn based on the experimental investigation.

- Due to the addition of crumb rubber in concrete the flexural strength is increased until a certain extent after which, the value does not yield an increase. From the results, it concludes that with the addition of 9% of crumb rubber, the highest value for flexural strength is obtained.
- The results from the compressive strength test prove that the addition of crumb rubber in concrete increases the compressive strength compared to the normal conventional concrete till a certain percentage of crumb rubber, we can conclude that till the addition of 3% of crumb rubber, the compressive strength value is higher (8%) than conventional concrete; any further addition reduced the value.
- The value of split tensile strength of concrete also increases with increment in the quantity of crumb rubber, until a certain extent. The results show that the split tensile strength of crumb rubber concrete is higher by 11.2% than conventional concrete till 3% addition of crumb rubber; further addition of crumb rubber up to 9% increased the tensile strength by 2.3%; by further addition resulted in a decrement in split tensile strength.
- From the results obtained from all the tests performed, we can infer that the addition of crumb rubber in concrete mix increases the flexural strength, the compressive strength, the split tensile strength, of concrete until an addition of certain percentage, and it also increases the greenness.
- Even though the concrete with crumb rubber mixture reduces the strength properties, modification of concrete with crumb rubber is one of the best solutions for the waste tyre utilization. The concrete with crumb rubber produces a concrete with lesser density, good absorption of vibration, higher plasticity, etc.

References

1. K.L.Scrivener., R.J. Kirkpatrick.: Innovation in use and research on cementitious material. Cement and Concrete Research, vol. 38(2), pp. 128–136 (2008).
2. R. Siddique., T.R.Naik.: Properties of concrete containing scrap-tire rubber -An overview, Waste Management, vol. 24(6), pp. 563–569 (2004).
3. I.B. Topcu.: Assessment of the brittleness index of rubberized concretes. Cement and Concrete Research, vol. 27(2) pp. 177–183(1997).
4. M.M. Al-Tayeb., B.H. Abu Bakar, H.M.Akil., H.Ismail.: Effect of partial replacements of sand and cement by waste rubber on the fracture characteristics of concrete. Polymer Plastic Technology Engineering, vol. 51(6), pp. 583–589 (2012).

5. Garrick, G.M.: Analysis and testing of waste tire fiber modified concrete. Thesis (PhD), B.S Louisiana State University (2005).
6. Hernández-Olivares F., Barluenga G., Parga-Landa B., Bollati M., Witoszek B.: Fatigue behaviour of recycled tyre rubber-filled concrete and its implications in the design of rigid pavements. Construction and Building Materials, vol. 21, pp. 1918–1927. (2007).
7. Khatib Z.K., Bayomy F.M., 1999. Rubberized Portland cement concrete. Journal of Materials in Civil Engineering, Vol 11, pp. 206–213 (1999).
8. A.O. Atahan., A.Ö. Yücel.: Crumb rubber in concrete: static and dynamic evaluation. Construction and Building Materials, vol 36, pp. 617–622 (2012).
9. K.B. Najim., M.R. Hall.: Mechanical and dynamic properties of self-compacting crumb rubber modified concrete. Construction Building Materials, vol 27 (1), pp. 521–530 (2012).
10. M.M. Al-Tayeb., B.H. Abu Bakar., H. Ismail., H.M. Akil.: Effect of partial replacement of sand by recycled fine crumb rubber on the performance of hybrid rubberized-normal concrete under impact load: experiment and simulation. Journal of Cleaner Production, vol 59, pp. 284–289 (2013).
11. M. Bravo J., de Brito.: Concrete made with used tyre aggregate: durability related performance. Journal of Cleaner Production, vol. 25, pp. 42–50 (2012).
12. O. Youssf, M.A. ElGawady, J.E. Mills and X. Ma (2014), "An experimental investigation of crumb rubber concrete confined by fibre reinforced polymer tubes", Construction and Building Materials, vol. 53, pp. 522–532.
13. E. Ganjian, M. Khorami, A.A. Maghsoudi 2009, "Scrap-tyre-rubber replacement for aggregate and filler in concrete", Construction Building Materials, Vol. 23 (5) pp. 1828–1836.
14. Z.K. Khatib, F.M. Bayomy 1999, "Rubberized Portland cement concrete", Journal of Materials Civil Engineering, Vol. 11 (3) pp. 206–213.
15. Turatsinze, A., Bonnet, S., Granju, J.L., 2005. "Potential of rubber aggregates to modify properties of cement based-mortars: improvement in cracking shrinkage resistance." Construction and Building Materials, Vol. 21, pp. 176–181.
16. Ho, A.C., Turatsinze, A., Vu, D.C., 2008. In: Alexander, M.G., Beushausen, H.D, Dehn, F., Moyo, P. (Eds.), "On the Potential of Rubber Aggregates obtained by Grinding End-of-life Tyres to Improve the Strain Capacity of Concrete." Taylor & Francis Group, London, pp. 123–129.
17. Wong, S. F., Ting, S. K. Use of recycled rubber tires in normal- and high-strength concretes, ACI Materials Journal, Vol. 106(4):325–332 (2009)
18. Shubhada Ashok Gadkar. 2013. Freeze-Thaw Durability of Portland Cement Concrete Due To Addition of Crumb Rubber Aggregates, South Carolina, Clemson University, Doctoral Thesis.
19. IS: 516-1959.Method of Tests for Strength of Concrete, Bureau of Indian Standards.
20. IS: 5816-1999. Method of Test Splitting Tensile Strength of Concrete, Bureau of Indian Standards.
21. IS: 456-2000. Plain and Reinforced concrete—Code of Practice, Bureau of Indian Standards.
22. Lee, B. I., Burnett, L., Miller, T., Postage, B., Cuneo, J. 1993. "Tyre rubber cement matrix composites", Journal of Materials Science Vol 12 (13): pp. 967–968.
23. Ali I. Tayeh. 2013. Effect of Replacement of Sand by Waste Fine Crumb Rubber on Concrete Beam Subject to Impact Load: Experiment and Simulation, Civil and Environmental Research 3(13): 165–172.

Solar-Powered Multipurpose Backpack

S. Padmini and Md. Shafeulwara

Abstract Our main purpose of this project is to reduce the pain and suffering of travel. In everyday life, everybody is busy and time constrained. Sometimes it gets more difficult when there is not much power to use the electronic device. This project provides great comfort when travelling or on a hike. It has solar cells attached at the front of backpack with the output of 5 V which allows you to power your phones or other devices in daytime. A power bank stores the power received from solar plate which provides great flexibility to use that power even when you are not in daylight. The backpack has a Bluetooth speaker inbuilt with it to enjoy your moment. The speaker can be used by wire or by the Bluetooth which is compatible with any Bluetooth device. A magnet-based Compass which will give direction to your path when you need it. It has also a torch which can be powered by the same power bank or three AAA batteries. This torch is fixed at the shoulder hand of the bag. It has 14 led bulbs which is enough to guide you in the dark. One of the unique features of the backpack is that the backpack can be tracked down if it gets lost. It has an inbuilt GPS system which sends you the GPS coordinates whenever it gets the coded message from the registered phone number.

Keywords Gadget bag · Eco-G backpack · Solar bag · Smart bag
E-bag

S. Padmini (✉) · Md. Shafeulwara
Department of Electrical and Electronics Engineering, SRM Institute of Science
and Technology, Chennai, India
e-mail: padmini.s@ktr.srmuniv.ac.in

Md. Shafeulwara
e-mail: shafeulwara@gmail.com

© Springer Nature Singapore Pte Ltd. 2019
M. A. Bhaskar et al. (eds.), *International Conference on Intelligent Computing
and Applications*, Advances in Intelligent Systems and Computing 846,
https://doi.org/10.1007/978-981-13-2182-5_21

1   Introduction

Everyone needs a bag to carry their belongings. Since it is a smart world we live in, where everything we use in our regular day life is electronics, it is very helpful to have a smart bag which can collect solar energy and store it into a battery so that it can charge one's phones and tablets and power up your electronics devices. The backpack contains a 5 W foldable and removable solar panel which collects the energy of the sun and stores it into a power bank.

The power bank supplies 5 V to the phones and other electronic devices. It is of 20,000 mAh which means it has enough power for charging your devices multiple times [1]. In the backpack, it has some gadgets for which power is supplied by the same power bank. Like a Global Positioning System (GPS), an LED flashes light and a Bluetooth speaker [2, 3]. It happens sometimes that you forget your bag somewhere or it gets stolen, in that kind of situation you might find yourself into a trouble. But do not worry; the backpack contains a GPS system which will tell you the right location where it is. All one has to do is just send an SMS from your registered number. The GPS system has a GSM module with a GSM SIM card [4, 5]. Once the GSM module receives the massage, the microcontroller attached with it understands the message and send the GPS coordinates to the registered phone number then by searching it on Google map so that one can find the exact location.

Most of the people enjoy the music when they are travelling, hiking, or in a picnic. The backpack has inbuilt Bluetooth speaker which will allow any Bluetooth-enabled phone to connect and play the music. The speaker can be powered by 5 V supply which is provided by the power bank.

If you are hiking in the woods or any unknown place there is a chance someone get lost. To avoid this situation, the backpack has a compass which indicates the direction in which you are going and it will ensure that one never get lost. It also has an additional feature when someone does not have access to any light source. The backpack provides a solution in this situation. It has 14 (Light Emitting Diode) LEDs. An adjustable LED flashlight consumes very less power and energy but has the brightness enough to guide someone through the dark.

2   Solar Panel

Harnessing energy from sun is not a new thing. People were using sun to dry clothes and preserve their things. In the Vedic literatures of Indian flying machines are mentioned which were powered by the sun. Coming to twenty-first century, the solar technology has evolved by a great amount in advancing the solar cells which could be the main power source of the future. Solar panels consist of multiple solar cells connected together in parallel and series fashion [6, 7]. These cells are basically a p-n junction semiconductor devices with pure silicon wafer doped with 'n' type phosphorous layer on the top and 'p' type boron layer on the base.

If a Photo Voltaic cell is kept under the sunlight, photons hits at the p-n junction of cell and energise the electrons, force them to be move out from the valence band of the atoms. The electrons are repelled by negative charges of the p-type layer and attracted towards the positive charges of the n-type layer. Connection of wires across the junction contains a current in them. Solar cells have developed from efficiency of 6% to thin films and lightweight with the efficiency of almost 30%.[8–12].

Today, we have mono-crystalline, amorphous and polycrystalline thin film panels. Mono-crystalline has shown the highest efficiency so far, because of the presence of maximum silicon per unit area. Hence more current flows for the equal amount of photons received. They are made from a single silicon crystal as continuous lattice. But for polycrystalline panels, molten silicon poured into moulds and separate boundaries can be seen due to this. Lesser density of silicon per unit area means lesser efficiency of the cell.

The solar panel used in this project as shown in Fig. 1 is a 12 V with 5 W power output. It is a lightweight solar panel fixed at the front side of backpack as shown in the picture. The position of the panel is designed as such that it gets full exposure to the sun when someone wears it. The specifications of the solar plate are given in Table 1.

The limitation of solar panels is that it works only in broad daylight but we need energy when it is dark too. However, during the absence of sun light, it requires something which can store energy like batteries. The project requires to have a solar charger with a voltage regulator to charge a battery with the constant output voltage which is obtained through this IC LM2574.

Fig. 1 Solar panel used in this project

Table 1 Solar plate specification

Power (W)	5 W
Open circuit voltage (V)	21.6 Voc
Short circuit current (A)	0.34 Isc
Maximum power voltage (V)	17.0 Vmp
Maximum power current (A)	0.29 Imp
Cell type	Polycrystalline
Frame type	Silver
Junction box	Yes
Length	8.74″ (222) mm
Width	10.63″ (270) mm
Depth	0.67″ (170) mm
Weight	3.31 lb (1.50) Kg
Connector	J Box

3   Design of the Charging Circuit

Solar cells are inefficient of converting sunlight into electricity. This signifies that the solar panel should be efficient as much as possible. Unfortunately, the 7805 linear voltage regulator fails in this situation. Voltage regulator takes a variable voltage input (above 5 V) and outputs a relatively steady 5 volts. It simply burns the extra energy off as heat, which does not make much useful for any solar charger [13]; hence, a DC/DC converter, which is similar to a transformer. It receives a higher input voltage, and puts out power at a lower voltage, but carries more current. Ideally power would be conserved, but in reality, DC/DC converters are around 70–80% efficient.

A DC/DC step down converter with rating of 500 mA which is a USB limit and has 5 V output, the LM2574 chip as shown in Fig. 2. The datasheet for this chip is really helpful in designing a voltage regulation circuit. It just requires a couple of passive components (Fig. 3).

There is a problem with the solar cells that it can only work in daylight. If it is to use that solar panel at night, there is no way that it will work. There is a lot of power that goes waste unless you are very careful about when you charge. The best way to

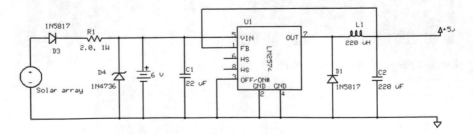

Fig. 2 Charging circuit

Fig. 3 5 V DC/DC stepdown converter to take the variable output from the solar panel and reduced to 5 V for charging USB devices

utilise the solar panel maximum is to use a chargeable battery to store the solar energy that you do not immediately use. In this project, a 20,000 mAh power bank is used to save the energy. The power bank consists of lithium ion battery which is very lightweight and portable and has high energy density. It has a protection circuit which protects it from over current, overvoltage and over charging. The power bank can also be charged manually through mains with any mobile charger with micro USB output. It has two USB outputs of 5 V. One with 1A and another with 2A. It can charge any device like mobile phones or tablet having 5 V input.

4 Global Positioning System (GPS)

It happens sometimes that your luggage bag gets lost or stolen. One of the common scenarios is when someone is travelling by plain, the incorporated GPS system fitted in the backpack will intimate the exact location of the bag (Fig. 4).

This GPS system is powered by the internal power source of the bag, and linked to a GSM phone. Whenever the location is required, an SMS needs to be sent to the GPS system to get the location. The GPS system has a GSM module which needs a SIM card to be functional. Whenever this module receives the message, it informs the sender with the geo location along with the time. This GPS system communicates over the GSM channel.

The GPS as shown in Fig. 5 is navigation system based on satellites consists of a network of 24 satellites located into orbit. The system is used to provide information to the military, civilians and the commercial users all around the world and

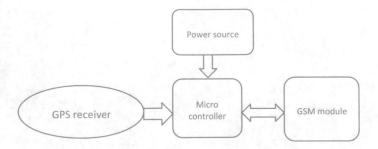

Fig. 4 Block diagram of GPS system

Fig. 5 GPS module and GSM based GPS tracker

it is accessible to anyone with a GPS receiver. GPS works all weather circumstances at everywhere in the world. Normally there are no subscription charges or system charges to use GPS.

A GPS receiver has to be locked on with the signals of minimum three satellites to estimate two-dimensional (2D) position (longitude, latitude) and to track the movements. To determine three-dimensional (3D) position (longitude, altitude and latitude) it needs connection with four satellites. Once the location is determined, the system can determine other information like, speed, distance and time. GPS receiver used in this project is to detect the backpack location and provide information to the owner through GSM technology.

5   GSM Module SIM900

The GSM modem as shown in Fig. 6 is a special kind of modem which has a SIM card and works with dedicated subscribers identification number over a GSM network, same as a mobile phones. It is a cell phone without display. SIM900 is a GSM/GPRS-compatible Quad-band cell phone and works on a certain frequency of 850/900/1800/1900 MHz. Further it may also be used to access Internet and as well

Fig. 6 SIM900 GSM module

as for voice communication and for SMS. Internally, the module is controlled by the AMR926EJ-S processor, which manages data communication, phone communication and the communication with the circuit interfaced with the cell phone itself.

6 Bluetooth Speaker

Most people like to listen to music, mostly youngsters. When a group of people gets together they like to have fun, they like to listen music especially when they are travelling or hiking they want to enjoy and listen to music loudly (Fig. 7).

In this project, a Bluetooth speaker is mounted on two sides of bag which is powered by inbuilt battery pack. The speaker operates at 5 V dc and can paired any

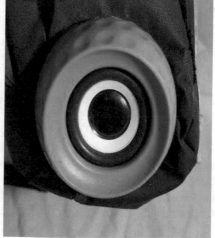

Fig. 7 Speaker mounted on the backpack

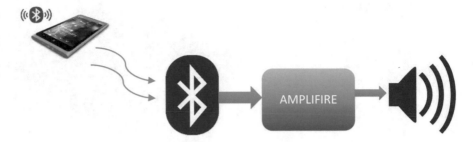

Fig. 8 Block diagram of bluetooth speaker system

Bluetooth-enabled phone. All user needs to do is just connect the phone and play the music. In future, a headphone splitter can be attached to output of Bluetooth receiver so that more than one person can get connected to same device to avoid the problems that may be caused to the people around.

In this project, speaker has been incorporated with built-in amplifier. The two speakers are fixed in both the bottom side of the bag. The specifications are given below (Fig. 8).

Specifications

- 2.0 USB-powered multimedia speaker system.
- DC 5 V Power input via a USB port, 3.5 mm audio jack output.
- 2 speakers with output of 4 W RMS each, driver of each unit is 50 mm in diameter and the Frequency Response is 60 Hz–20 kHz.
- Wired and wireless volume control.

7 Adjustable LED Flashlight

Everyone needs light in the dark. The bag has its own flashlight as shown in Fig. 9. A flashlight is an important piece of Every Day Carry (EDC) gear. It is very useful at night while camping and hiking. In this backpack a 14 LED flash light is mounted on the shoulder sleeve so that when someone hangs the bag he flashlight goes right to the shoulder.

The flashlight has 14 bright LED lights with high intensity (8000–10,000 MCD). It is also water-resistant and has adjustable strap, Adjustable inclination light in weight and it will feel comfortable and does not need to hold it to focus suitable for camping, hunting, fishing, caving, hiking, and climbing powered by three x AAA batteries. It has four modes of operation, four LED mode, seven LED mode, 14 LED mode, 14 LED Flashing (Fig. 10).

The bag has also a magnetic compass. It is very useful when you are travelling or hiking. It works according to the earth's magnetic field. Sometimes it happens that

Fig. 9 Adjustable LED flashlight

Fig. 10 Magnetic compass

someone gets lost and they cannot find the direction or cannot recognise the direction, the magnetic compass on the bag helps you in this situation.

8 Conclusion

Combining technologies together and put it into a travel bag to full fill the needs of travel was the main goal of this project. In today's scenario, everyone is very busy with their work. Most of the times they cannot find the time to charge the electronic devices. It can be very helpful for them that they have a backpack which can provide power for their devices without having electricity. It is eco-friendly, the solar cell converts solar power into electrical energy and store it into battery in the bag.

The backpack does not only hold one's necessities, laptop, tablet, bottle gadgets but also provide power for these devices without electric supply. It can power smart phones, tablets, external battery packs, mp3 players, portable speakers and many more that can be connected with a USB cable by harnessing energy straight from the scintillating sun.

The inbuilt GPS can be described as antitheft system in the backpack. It can provide the exact location when it gets lost. A Bluetooth speaker in the bag can be used for partying also. It can be paired up with any compatible phone. It can be a perfect companion for camping, hiking, biking trip and all kinds of travel as well as many other lifestyle activities.

References

1. System Design", Journal of Electronic Systems Volume 2 Number 2, (June 2012).
2. Arnquist Sarah, "In Rural Africa, a Fertile Market for Mobile Phones," *New York Times*, (2009).
3. Bluetooth Special Interest Group. Advanced Audio Distribution Profile Specification, URL https://www.bluetooth.org/docman/handlers/DownloadDoc.ashx?doc_id=260859. Jul 2012.
4. Asaad M. J. Al-Hindawi, Ibrahim Talib, "Experimentally Evaluation of GPS/GSM Based.
5. Chen, H., Chiang, Y. Chang, F., H. Wang, H. Toward Real-Time Precise PointPositioning: Differential GPS Based on IGS Ultra Rapid Product, SICE Annual Conference, Taipei, Taiwan. (2010).
6. Schuss Christian, Eichbergert Bernd, "Design specifications and guidelines for efficient solar chargers of mobile phones", 11th International Conference on Systems, Signals & Devices (SSD), Feb. 2014.
7. Schuss Christian, Rahkonen Timo, "Solar energy harvesting strategies for portable devices such as mobile phones", 14th Conference of Open Innovations Association (FRUCT), 11–15 Nov. 2013.
8. Bellis Mary, "Definition of a Solar Cell, History of Solar Cells," About.com Inventors, http://inventors.about.com/od/sstartinventions/a/solar_cell.htm, (1997).
9. Boreham Ray, "The Potential Importance of the Solar Energy for the Future," Ezine Articles, (2008).
10. Gal Jonathan."The Importance of Solar Energy in Our Everyday Lives," PRLog, http://www.prlog.org/10171090-the-importanceof-solar-energy-to-our-everyday-lives.html, (2009).
11. Kho Jennifer, "Are Thin-Film Solar Efficiency Standards Unfair?" (2008).
12. Digital Opportunity Channel, "Solar Mobiles Speeding Up Cell phone Revolution," http://www.digitalopportunity.org/comments/solar-mobiles-speeding-up-cellphone-revolution, (2009).
13. Chih-HaoHou Chun-Ti Yen Tsung- Hsi Wu Chin-Sien Moo, "A Battery Power Bank of Serial Battery Power Modules with Buck-Boost Converters", International Power Electronics Conference, 2014.

Optimal Load Frequency Control of an Unified Power System with SMES and TCPS

**Priyambada Satapathy, Manoj Kumar Debnath
and Pradeep Kumar Mohanty**

Abstract TCPS (Thyristor-controlled phase shifter) and SMES (Super Conducting Magnetic Storage) are implemented in this research work to regulate the frequency of an interlocked two area power system. Two distinct PD + PID double-loop controller is recommended in each area for frequency regulation in the scrutinized system. The optimum controller gains are obtained using firefly algorithm considering ISE (Integral Square Error) as objective function. The system responses of this suggested model is examined by introducing a rapid load variation of 0.1 pu in area 1. The supremacy of the recommended PD + PID controller is established over PID controller considering several time response indices like peak overshoots, settling time and minimum undershoots. The toughness of the proposed system is also verified by amplifying the loading of the system.

Keywords Firefly algorithm · PID controller · Load frequency control
FACTS devices · TCPS · SMES

1 Introduction

AGC plays so many different roles in an interconnected restructured power system network such as maintaining stability and to suppress the oscillations and fluctuations in system constraints during disturbances, etc. [1]. The mismatch between total generation and demand loads leads to fall in frequency that in turn increase the speed of governor. In this case AGC provides suitable and continuous regulation over load frequency and exchange of tie-line power to keep area control error (ACE) to zero [2].

Currently various classical and modern technologies have been executed to face all the challenges regarding AGC. In 2011 Sinha, Patel and Prasad applied FACTs devices in AGC using Fuzzy controller to reduce the oscillations occur in frequency

P. Satapathy · M. K. Debnath (✉) · P. K. Mohanty
Siksha 'O' Anusandhan University, Bhubaneswar, Odisha, India
e-mail: mkd.odisha@gmail.com

© Springer Nature Singapore Pte Ltd. 2019
M. A. Bhaskar et al. (eds.), *International Conference on Intelligent Computing
and Applications*, Advances in Intelligent Systems and Computing 846,
https://doi.org/10.1007/978-981-13-2182-5_22

and generated power [3]. Literature survey reveals that S. Sridhar and M. Govindara proposed a model based mitigation algorithm to get best optimized solution for AGC [4]. Dated back CR network has been suggested by on-off switch along with stopover times to analyze the automatic generation of smart grid [5]. Regulation of ancillary services is mandatory in any interconnected power system network. So device statement mechanism is applied in ref. [6] to keep the system frequency as per prescribed value. Gravitational search algorithm (GSA) optimized classical proportional-integral-derivative (PID) controller has been implemented to investigate a two area non-reheat thermal system [7]. Recently penetration of irregular renewable energy sources brings more complications in stability and reliability of grid power system. In [8] gravitational wolf optimization technique (GWO) optimized PID controller has been implemented to examine a two area thermal system incorporation with solar system. To achieve stability in transmission line during load disturbances FACTs devices like TCPS and TCSC are considered in [9]. Fractional order PID (FOPID) controller is used to optimize a two area interconnected power system with electrical vehicle (EV) loading [10]. The impact of TCPS and TCSC are illustrated on AGC of a two area thermal-thermal system under disturbances [9]. An attack impact model is derived which obtained a series of false data injections (FDI) to minimize the remaining time until the onset of disorderly remedial action [11]. An analysis has been done on the basis of efficiency and utilization of battery energy storage system (BESS) for regulation and grid integration [12]. The objective of model predictive model is to maintain steady value and avoid disturbances during tie –line interactions. This method is examined over a four area power system for satisfactory responses in AGC [13].

Literature survey shows that recently various advance FACTs devices like TCPS, SSSC, etc., and energy storage devices are proposed in many research area of AGC to keep the deviations of system constraints within the specified limit. Thus in this research paper a two area interconnected thermal system with TCPS and SMES are designed and examined with firefly algorithm tuned PD + PID double-loop controller.

2 System Investigated

A linearized model of two control areas connected through tie-line is shown in Fig. 1 used for Automatic Generation Control. Both the area comprises of reheat thermal power generation with SMES and TCPS is placed in series with tie-line. The torque of thermal turbine controls the frequency of the power system. For the better dynamic response of suggested system with AGC the PID/PD + PID controller gains has to be optimized. To obtain optimum integral gain setting integral square error is used which is defined by

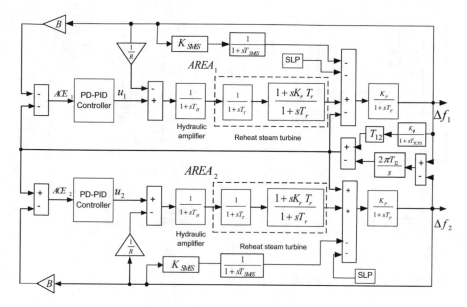

Fig. 1 Transfer function model of two area system with SMES and TCPS

$$J = \int \left(\Delta f_1^2 + \Delta f_2^2 + \Delta P_{\text{tie}12}^2 \right) \mathrm{d}t$$

Here the incremental change in frequency is Δf and ΔP_{tie} is the incremental change in tie-line power. The details about TCPS and SMES are described in upcoming section.

2.1 Thyristor-Controlled Phase Shifter (TCPS)

TCPS control technique is used to determine the phase shift angle as a nonlinear function of rotor angle and speed. However, it is not very meaningful to measure the rotor angle of a single generator with respect to system reference because in power station huge number of generators is used for generation purpose. In this paper for research purpose a two area reheat thermal interconnected power system has been proposed along with TCPS in series with tie-line. A schematic diagram of two area interlocked power system along with TCPS has been shown in Fig. 1. Basically the tie-line resistance does not have significant influence on the dynamic response of a system due to its high reactance to resistance ratio.

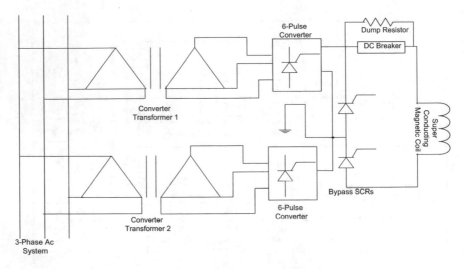

Fig. 2 Connection diagram of SMES

2.2 Super Conducting Magnetic Energy Storage (SMES)

The connection diagram of SMES unit is shown in the Fig. 2. In the scrutinized system SMES is placed in both the areas to improve the frequency stability of the system. Inverter or rectifier is used to link to the grid with a DC superconducting coil. The conducting coil is excited to a fixed value from grid and immersed it in a helium containing tank during stable operation. A line commutated converter helps to exchange the energy among the superconducting coil and suggested system. The SMES allowed flow within the converter and power system during rise or fall in load demand. After that the governor and other control mechanism starts to fix the new equilibrium condition of power system and the super conducting coil again starts to charge from its initial value.

3 Proposed Method

3.1 Controller Architecture

The PD + PID double loop controller is basically combination of two sequential inner and outer processes. Any disturbance occurs in the internal process is attenuated by the inner process where the quality of final output is controlled by outer process. The main motive of using this double loop controller is to reject the initial disturbances before it spread over to all over the plant. Figure 3 represents the modest PD + PID controller having inner and outer loops.

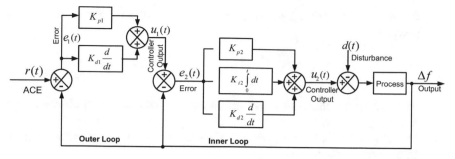

Fig. 3 Configuration of PD + PID controller

Outer loop:
This primary outer loop generally is known as master loop. The output from the inner process becomes the input to the outer process. The output of process is controlled to meet a set point signal. A load disturbance of is applied to entire system.

Inner loop:
The secondary inner loop also known as slave loop. An inner comparator is located inside the inner loop to control the inner process and the output of inner loop denotes as $u_2(t)$. Here $u_1(t)$ is the input of inner process. The disturbances generated in system are first corrected by inner loop before it harms the outer process. So to satisfy the above statement the inner loop should have much faster response as compare to outer loop. The inner loop plays a great role to minimize the effect of the performance of the controller due to gain variations which may change the operating set point.

3.2 Firefly Algorithm

Optimization technique is a very essential tool for decision making and to analyze physical conditions of a system. The firefly algorithm was developed by Xin-she Yang at Cambridge University in 2007. Fireflies are one of the families of insects who produce yellow, green and pale-red light in tropical environment. As fireflies are unisex in nature, they attracted to each other regardless of their sex. Attractiveness and brightness are proportional to each other and both decrease with the increase in distance between the fireflies. The objective function simply depends upon the brightness of a firefly and in the case of maximization problem it is proportional to the brightness. Following steps has to follow for a firefly algorithm.

i. Based upon inverse square law, the light intensity $I(r)$ varies and that is described by

$$I(r) = \frac{I_s}{r^2}$$

where I(r) is the intensity at the source and r is the distance of observer from source.

ii. The intensity of light I changes with the square of distance by considering the co-efficient of absorption

$$I = I_0 e^{-\gamma r^2}$$

iii. The population of fireflies can be initialized by following the below equation:

$$x_{i+1} = x_t + \beta e^{-\gamma r^2} + \alpha \varepsilon$$

Here the second and third term indicates the attraction and randomization respectively.

iv. The light intensities of each firefly can be determined from the brightness of every firefly.

$$I = I_0 e^{-\gamma r^2}$$

v. Evaluate the attractiveness of fireflies

$$\beta = \beta_0 e^{-\gamma r^2}$$

vi. The below equation is used to determine the Movement of firefly to another brighter firefly

$$x_i = x_i + \beta_0 e^{-\gamma r_{i,j}}(x_j - x_i) + \alpha \varepsilon$$

vii. We have to update the intensity of light and also rank them to find the present best among the fireflies.

For implication purpose most of the cases we take $\beta_0 = 1$, $\alpha \in [0, 1]$ and $\gamma = 1$. The parameter γ characterizes the variations of attractiveness and at the time iterations the value of γ can also be adjusted to get various optimal values.

4 Result and Analysis

To illustrate the effectiveness of this proposed controller the dynamic model was developed using MATLAB Simulink under different operating conditions. A disturbance of 0.1 pu is subjected on area 1. In this proposed model TCPS is connected with the tie-line and SMES is placed on both the control areas. The simulation results are performed under various circumstances. Initially the system is simulated with PD + PID controller with/without TCPS and SMES. The frequency oscillation in area 1 with and without TCPS and SMES is shown in Fig. 4. This figure clearly states that by using TCPS and SMES the frequency stability of the system improves significantly. In the later case it is proved that the implemented controller over conventional PID controller the system is simulated with PD + PID controller and PID controller. The optimum gains of different controllers obtained by firefly algorithm are listed in Table 1. The dynamic responses for Δf_1, Δf_2 and ΔP_{tie} are shown in Figs. 4, 5, 6 and 7 respectively with above-mentioned optimized gains. Various time response indices are calculated and tabulated in Table 2. The response indices like minimum undershoot, settling time, and peak overshoot bear the least values for PD + PID controller as compared to PID controller. From the Figs. 4, 5, 6 and 7 it is evident that the system dynamic performances have been improved with PD + PID as compared to PID controller. Finally we concluded that the above responses with TCPS-SMES combination can effectively improve system stability during sudden perturbation.

In the later study the system behavior is analyzed by increasing the loading of the system. The p.u. load of the area 1 is increased by 10% to verify the robustness of the system. The system frequency oscillations in area $1(\Delta f_1)$, area $2(\Delta f_2)$ and tie-line power fluctuations between area 1 and 2 (ΔP_{tie}) are shown in the Figs. 8, 9 and 10 respectively. These figures confirm that the suggested control approach preserves its supremacy even we have amplified the loading of the system.

Fig. 4 Frequency deviations in area 1 without and with TCPS and SMES

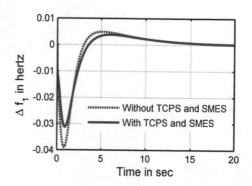

Table 1 Optimum values of parameters of PD + PID double controllers tuned by firefly algorithm

Controller	Area 1					Area 2				
PD + PID	K_{p1}	K_{d1}	K_{p2}	K_{i2}	K_{d2}	K_{p1}	K_{d1}	K_{p2}	K_{i2}	K_{d2}
	1.735	1.512	1.445	1.116	0.346	1.178	1.392	1.021	1.572	0.154
PID	K_p		K_i		K_d	K_p		K_i		K_d
	1.8186		1.6615		1.2861	1.7802		1.4436		1.098

Fig. 5 Frequency deviations in arca 1

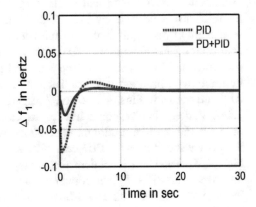

Fig. 6 Frequency deviations in area 2

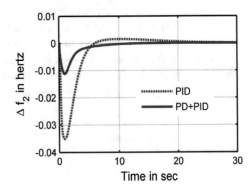

Fig. 7 Tie-line power
deviations

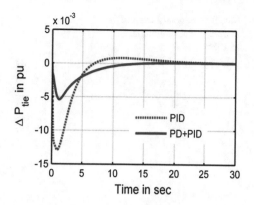

Table 2 Response specifications of different controllers

Deviations	Settling time (T_s), peak overshoot (O_{sh}) and undershoot (Ush)	PD + PID	PID
Δf_1	O_{sh} in Hz	0.0038	0.0121
	U_{sh} in Hz	−0.0312	−0.0804
	T_s in s	16.5500	17.3100
Δf_2	O_{sh} in Hz	0.0001	0.0016
	U_{sh} in Hz	−0.0112	−0.0351
	T_s in s	8.5100	21.3700
ΔP_{tie}	O_{sh} in PU	0.0001	0.0008
	U_{sh} in PU	−0.0054	−0.0129
	T_s in s	9.5500	16.9400

Fig. 8 Frequency deviations
in area 1 with augmented
loading

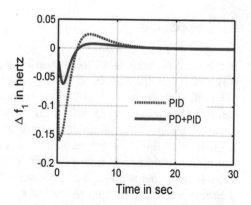

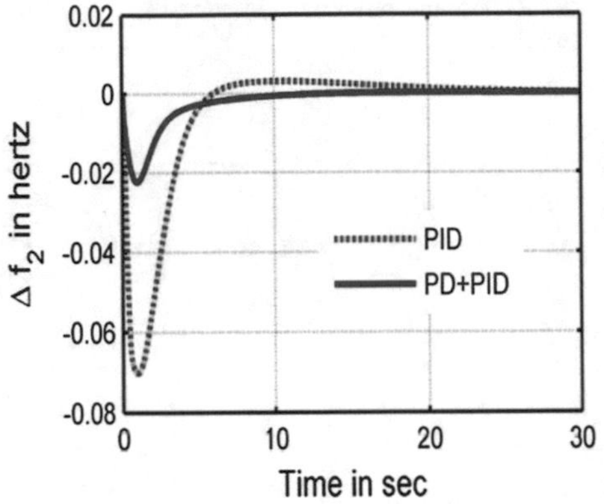

Fig. 9 Frequency deviations
in area 2 with augmented
loading

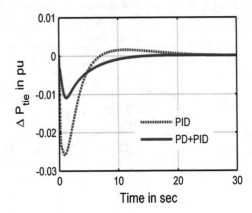

Fig. 10 Tie-line power
deviations with augmented
loading

5   Conclusion

In this research analysis a synchronized combination of SMES and TCPS has been
executed for an interconnected thermal power system for Load frequency Control.
PD + PID double loop controller has been successfully implemented as secondary
controller for frequency regulation. The finest values of the suggested controller are
obtained using firefly algorithm considering ISE as evaluative function. The
dominance of the PD + PID controller is established over PID controller in consideration with different time response indices like peak overshoot, settling time,
and minimum undershoot. The toughness of the suggested methodology is verified
with increase of the loading pattern of system. The dynamic responses reveal that
firefly algorithm optimized PD + PID double-loop controller along with the SMES
and TCPS can effectively control the frequency oscillations when there is a change
in system load levels.

Appendix

$$T_H = 0.8, T_r = 10, K_r T_r = 5, T_t = 0.3, T_P = 20, B = 0.425,$$
$$R = 0.425, K_P = 120, T_{12} = 0.01, T_{\text{TCPS}} = 0.01, K_\phi = 1.5,$$
$$\phi_{\max} = 13°, \phi_{\min} = -13°, K_{\text{SMES}} = 0.12, T_{\text{SMES}} = 0.035$$

References

1. Kundur, Prabha. Power system stability and control. Eds. Neal J. Balu, and Mark G. Lauby. Vol. 7. New York: McGraw-hill, 1994.
2. Padiyar, K. R. Power system dynamics. BS Publ., 2008.
3. Sinha, S. K., R. N. Patel, and R. Prasad. "Applications of FACTS devices with Fuzzy controller for oscillation damping in AGC." Recent Advancements in Electrical, Electronics and Control Engineering (ICONRAEeCE), 2011 International Conference on. IEEE, 2011.
4. Sridhar, Siddharth, and Manimaran Govindarasu. "Model-based attack detection and mitigation for automatic generation control." IEEE Transactions on Smart Grid 5.2 (2014): 580-591.
5. Liu, Shichao, Peter X. Liu, and Abdulmotaleb El Saddik. "Modeling and stability analysis of automatic generation control over cognitive radio networks in smart grids." IEEE Transactions on Systems, Man, and Cybernetics: Systems 45.2 (2015): 223–234.
6. Pai, Vilina, and Manju Gupta. "Deviation price based Automatic Generation Control for a two area power system." Electrical Power and Energy Systems (ICEPES), International Conference on. IEEE, 2016.
7. Mohapatra, Sangram Keshori, Bhabani Sankar Dash, and Pabitra Mohan Dash. "Application of GSA optimized PI controller parameters in automatic generation control for interconnected power system." Signal Processing, Communication, Power and Embedded System (SCOPES), 2016 International Conference on. IEEE, 2016.
8. Mate, Nidhi, and Sandeep Bhongade. "Automatic generation control of two-area ST-thermal power plant optimized with grey wolf optimization." Power India International Conference (PIICON), 2016 IEEE 7th. IEEE, 2016.
9. Pandey, Kamlesh, S. K. Sinha, and Ashish Shrivastava. "Impact of FACTS devices in automatic generation control of a deregulated power system." Power Electronics (IICPE), 2016 7th India International Conference on. IEEE, 2016.
10. Kachhwaha, Aditya, et al. "Interconnected multi unit two-area Automatic Generation Control using optimal tuning of fractional order PID controller along with Electrical Vehicle loading." Power Electronics, Intelligent Control and Energy Systems (ICPEICES), IEEE International Conference on. IEEE, 2016.
11. Tan, Rui, et al. "Modeling and Mitigating Impact of False Data Injection Attacks on Automatic Generation Control." IEEE Transactions on Information Forensics and Security 12.7 (2017): 1609–1624.
12. Chakraborty, Tapabrata, David Watson, and Marianne Rodgers. "Automatic Generation Control using Using an Energy Storage System in a Wind Park." IEEE Transactions on Power Systems(2017).
13. Monasterios, Pablo R. Baldivieso, and Paul Trodden. "Low-Complexity Distributed Predictive Automatic Generation Control with Guaranteed Properties." IEEE Transactions on Smart Grid(2017).

2DOF-PID Controller-Based Load Frequency Control of Linear/Nonlinear Unified Power System

Nimai Charan Patel, Manoj Kumar Debnath, Binod Kumar Sahu and Pranati Das

Abstract This paper describes about the Load Frequency Control (LFC) of an interconnected linear/nonlinear power system having two degree freedom of PID controller (2DOF-PID). The reheat thermal applied on a two-area system is considered for frequency control analysis by applying an abrupt load perturbation of 0.1 pu in area 1. Two systems are scrutinized for studying the load frequency control namely a linear system and a nonlinear system. The nonlinearities of the system are included by implementing governor with dead band (GDB) in each of the areas. The parameters of 2DOF-PID controllers are tuned on basis of Cuckoo Search Algorithm (CSA) considering time domain based fitness function, i.e., ITAE (Integral Time Absolute Error). For both types of system the dominance of 2DOF-PID controller is proved over conventional PID controller in terms of responses like minimum undershoot, settling time and peak overshoot. The supremacy of the recommended methodology is also established over some pre-published results.

Keywords Cuckoo search algorithm · 2DOF-PID controller · Load frequency control · Nonlinear system

1 Introduction

In a supply system to achieve acceptable range of power quality there should be a well balance between active power demand and output power generation [1]. Load frequency control has a serious responsibility to keep system frequency in their

N. C. Patel
Government College of Engineering, Keonjhar, Odisha, India

M. K. Debnath (✉) · B. K. Sahu
Siksha 'O' Anusandhan University, Bhubaneswar, Odisha, India
e-mail: mkd.odisha@gmail.com

P. Das
Indira Gandhi Institute of Technology, Sarang, Odisha, India

© Springer Nature Singapore Pte Ltd. 2019
M. A. Bhaskar et al. (eds.), *International Conference on Intelligent Computing and Applications*, Advances in Intelligent Systems and Computing 846,
https://doi.org/10.1007/978-981-13-2182-5_23

prescribed value under small disturbances. The area control error is formed by the suitable summation of changes occurred in frequency and tie line power during perturbation. For continuous stable operation we have to bring back the ACE to zero [2].

To improve performances in automatic generation control many control strategies are implemented in an interconnected power system in the last few decades. In the paper [3] two load frequency controllers have been proposed to control the speed by fuel rack position control of the generator that in turn maintain the nominal value of frequency. A new dynamic model for the load frequency control of two areas is proposed in article [4] without including the integral controller and error while controlling the area, i.e., the area control error to quench the fluctuations in system constraints. PSO optimized proportional integral derivative controller being proposed for AGC & AVR to control the real power & reactive power respectively [5]. The paper [6] proposed FO controller in LFC and AGC to improve system stability which reduces the steady state error. To reduce the LFC problem in hybrid power system a decentralized control scheme is implemented to improve dynamic performances in a closed-loop system [7]. The craziness based PSO is used in article [8] for AGC of non-reheat thermal on a two area system. Article [9] presented a GA optimized integral controller that helps the system to function in stable situation under various disturbances with inclusion of time delay, GRC along with speed governor with dead band constraints. Gains are optimized for PI and PID controllers in paper [10] by the population-based algorithm called as Artificial Bee Colony (ABC) optimization technique for automatic generation control. Yang and Deb (2009) introduced a nature-inspired optimization algorithm which is called as cuckoo search algorithm basing upon the reproduction system of cuckoos [11]. An approach of Teaching-Learning Based Optimization (TLBO) algorithm is made to achieve global solutions for nonlinear optimization problems [12]. To search the optimal gains of proportional controller differential algorithm is implemented for AGC in an interconnected system [13]. A 2DOF-FO-PID controller is being employed for AGC with accurate GRCs on a three unequal thermal systems in paper [14]. In article [15] PID controller along a derivative filter optimized by JAYA algorithm was used for AGC to reduce the unwanted harmonics in the input signals.

From the research background it is seen that maximum research work focused on linear system for load frequency control. But in our present analysis we have considered both linear and nonlinear system for load frequency control with the help of CSA tuned 2DOF-PID controller.

2   System Investigated

For load frequency control analysis here two types of system are considered. Initially a linear system is considered with reheat steam turbine [10]. The parameters of this linear system are considered from article [10]. In the second case the system nonlinearity is implanted in terms of Governor Dead Band. The model of

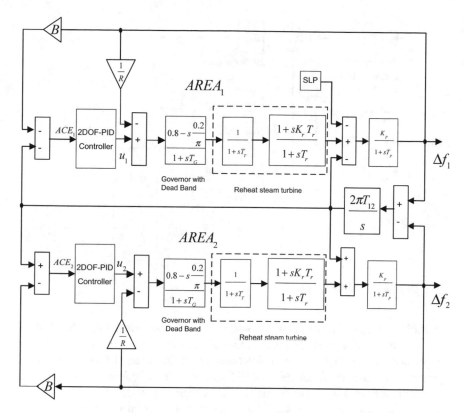

Fig. 1 Model of the two-area nonlinear interconnected power system

the two-area nonlinear system with GDB is displayed in Fig. 1. Appendix contains the parameters of the system. In both types of system in each area thermal generating unit with reheat steam turbine is implemented. Both PID and 2DOF-PID controllers are placed in each system for observing the frequency oscillation in each area. The parameters of these controllers are tuned by CSA methods. For examining the frequency control, a SLP (Step-Load Perturbation) of 0.1 pu is applied in area 1. T_G, T_T, T_r and T_P represent the time constant of governor, turbine, reheat and power system respectively. Δf_1 and Δf_2 denote frequency oscillations in each of the areas.

3 Proposed Method

3.1 Controller Architecture

The total number of closed-loop transfer functions in the control system which can be attuned separately is known as degree-of-freedom of the control system.

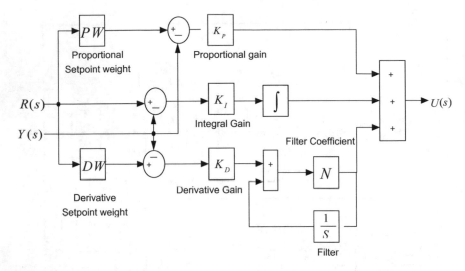

Fig. 2 Outline of 2DOF-PID controller

A 2DOF system generally has more benefits than the traditional 1-degree or SDOF system as there are different parameters that are needed to be controlled for meeting the performance criteria or the desired output of the control system. Figure. 2 represents the diagram of parallel 2DOF-PID controller where the reference signal is denoted as R(s), Y(s) is the feedback signal from the whole system output, the term U(s) denotes the output signal, the proportional set point weight is denoted by PW, the derivative set point weight is denoted by DW, the filter coefficient of derivative mode is denoted by N and the proportional, integral as well as derivative gains are represented by K_P, K_I and K_D respectively.

3.2 Optimization Method (Cuckoo Search Algorithm)

Yang and Deb (2009) introduced a nature-motivated optimization technique which is called cuckoo search algorithm basing on the reproduction system of cuckoos [11]. Basically it is inspired by obligate brood parasitism of some cuckoo species that they produce their eggs in the nests of the host. Each of the cuckoos lay one egg at a single time and randomly chooses a nest where the host bird has produced its own egg. High quality of eggs will be considered as the best nest and it will be carried on to the upcoming generation. There are fixed amount of host nests available and the probability of discovering of an alien eggs by a host bird is [0,1]. Sometimes when the host bird detects that the eggs are not belonging to them, then they will either destroy it or simply build a new one. Female parasitic cuckoo are very specialized in imitating the colors as well as patterns on the eggs of the few selected species. From the research Yang and Deb concluded that Levy flights performed better than simple random walk in case of discovering the search space. Naturally while searching foods, animals are offering an arbitrary March for the subsequent move which depends upon the current position and transition due to the probability of the subsequent position. As the step-length of the Levy flight is much longer in the long run therefore it is more efficient than others in exploring the search space.

The following steps describe the basic behavior of Cuckoo search algorithm:

Step-1: Initialization of population having number of host nest as 'n' and 'm' as number of cuckoos to search the nest in a defined area.

Step-2: Randomly search 'k' nest among 'n' nest by Levy flight method where the female cuckoo laid its egg. Suppose cuckoo 'j' lays its egg at 't' position along 'u' which is considered as the dimension of search space. Then,

$$x_j^t = \left(x_{j1}^t, x_{j2}^t, \dots, x_{ju}^t\right) \tag{1}$$

Suppose egg of mother cuckoo 'j' grows into adult, then it will seek the new nest for new generation by following below expression:

$$x_{jk}^t = x_{jk}^t + 0.01.\alpha.L.R.\left(x_{jk}^t - p_{gk}^t\right), \tag{2}$$

where α = parameter of controlling the moving step $(0 < \alpha > 1)$
L = Levy distribution
R = random number sampling with standard Gaussian distribution

$p_{gk}^t = \left(p_{g1}^t, p_{g2}^t, \dots, p_{gu}^t\right)$ best nest found by the swarm of cuckoo

Step-3: The fitness of cuckoo egg is compared by the fitness of host egg.

Step-4: Replace the cuckoo egg by host egg in 'k' nest, if the fitness of cuckoo egg is found better than that of host egg.

Step-5: If the host bird detects an alien egg, subsequently the host bird will build a new one to avoid local optimization $(p < 0.25)$. If it happened, new egg should be laid by cuckoo in another nest by following phenomena:

$$x_{jk}^{t+1} = x_{jk}^t + R \cdot \left(x_{sk}^t - x_{uk}^t\right) \tag{3}$$

Step-6: If the termination criterion is not satisfied then repeat the process from step 2 to step 5.

4 Result and Analysis

To examine both the frequency as well as tie line power control, PID and 2DOF-PID controllers are implemented separately in each of the areas of the scrutinized system. Both types of system are examined separately. The model of the two-area system is examined with the help of MATLAB/SIMULINK environment. To optimize the gains of the both types of controllers CSA method is used with ITAE as fitness function (Eq. 4). 10% of an abrupt load disturbance is applied on area 1 to investigate the frequency along with the inter-line power oscillations. Initially the linear system [10] is considered. The optimum results which are obtained by the implementation of the CSA method to tune the gains of the proposed 2DOF-PID controller are listed in Table 1. The frequency and interline power oscillations, i.e., Δf_1, Δf_2 and ΔP_{tie} are shown in the 3, 4 and 5 respectively. The response indices like minimum undershoot, settling time and peak overshoot are tabulated in Table 2. Figures 3, 4 and 5 along with Table 2 clearly portrays that the suggested 2DOF-PID controller exhibits better performance as compared to pre-published result of ABC optimized PID controller [10].

$$J = \text{ITAE} = \int\limits_0^t (|\Delta f_1| + |\Delta f_2| + |\Delta P_{\text{tie}}|)t \cdot \mathrm{d}t \qquad (4)$$

In the further analysis the system nonlinearity is considered in terms of governor with dead band as shown in the Fig. 1. Here also the frequency variations are examined with PID and 2DOF-PID controllers separately. The finest controller gains obtained by CSA method is tabulated in Table 1. The frequency and interline power oscillations (Δf_1, Δf_2 and ΔP_{tie}) of this nonlinear system are shown in

Table 1 Optimum values of different controller gains tuned by cuckoo search algorithm

2-DOF PID Controller					
	Area 1			Area 2	
Linear	PW = 1.871, DW = 0.1, K_P = 4, K_I = 4, K_D = 3.6534, N = 109.99			PW = 1.1984, DW = 1.5, K_P = 2.6, K_I = 0.01, K_D = 2.345, N = 200.52	
Nonlinear	PW = 0.6229, DW = 2, K_P = 1.356, K_I = 1.8656, K_D = 0.4436, N = 194.66			PW = 0.54, DW = 0.1, K_P = 1.1644, K_I = 0.01, K_D = 0.4741, N = 135.99	
PID controller					
	K_P	K_I	K_D	K_P	K_I	K_D
Linear [10]	1.966	9.5902	3.9320	0.710	0.6827	0.7419
Nonlinear	2.0000	2.0000	1.2787	0.8028	1.6092	0.4443

Fig. 3 Frequency
oscillations in area 1 (linear)

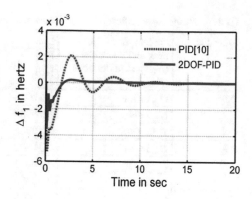

Fig. 4 Frequency
oscillations in area 2 (linear)

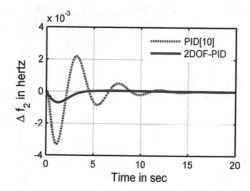

Fig. 5 Tie-line power
oscillations (linear)

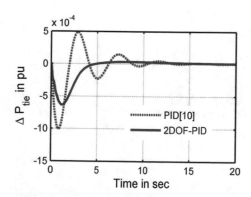

Figs. 6, 7 and 8 respectively. The response indices like minimum undershoot settling time and peak overshoot are tabulated in Table 2. Figures 6, 7 and 8 along
with Table 2 again proves the dominance of suggested 2DOF-PID controller over
conventional PID controller.

Table 2 Response specifications of different controllers

Deviations	Settling time (T_s), peak and overshoot (O_{sh}) and undershoot (U_{sh})	Linear		Nonlinear	
		PID [10]	2DOF-PID	PID	2DOF-PID
Δf_1	O_{sh} in Hz	0.0021	0.0002118	0.0014	0.0009061
	U_{sh} in Hz	−0.0052	−0.0035	−0.0099	−0.0127
	T_s in s	12.3700	8.9100	16.070	14.2000
Δf_2	O_{sh} in Hz	0.0022	0.0000588	0.0008	0.0003841
	U_{sh} in Hz	−0.0033	−0.0007	−0.0075	−0.0039
	T_s in s	12.9300	9.4400	17.460	14.7400
ΔP_{tie}	O_{sh} in PU	0.0005	0.0000325	0.0002	0.0000699
	U_{sh} in PU	−0.0010	−0.0006	−0.0022	−0.0022
	T_s in s	8.3400	3.8700	14.690	8.2000

Fig. 6 Frequency oscillations in area 1

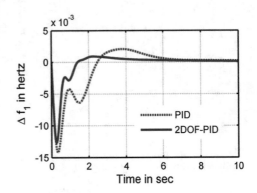

Fig. 7 Frequency oscillations in area 2

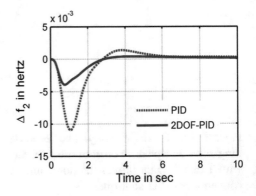

Fig. 8 Tie-line power oscillations

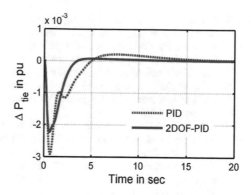

5 Conclusion

Here the CSA was successfully applied to calibrate the gains of the proposed controller, i.e., 2DOF-PID and the conventional PID controller. With these optimum gains the system was found to be performing well in terms of regulating system frequency and interline power variations in linear/nonlinear power system. The system nonlinearity is considered by incorporating governor with dead band in each area of the reheat type of thermal power system having a two-area system. The 2DOF-PID controller proved to be performing well as compared to PID controller in consideration with minimum undershoots, settling time, and peak overshoot. The results obtained in linear system were also related with the previously established result like ABC tuned PID controller to prove the dominance of suggested 2DOF-PID controller. Here we have limited our analysis in a two-area system but the research analysis can furthermore be extended to a multi-source system containing numerous areas.

Appendix

$$T_G = 0.8, T_r = 10, K_r = 0.5, T_r = 10, T_t = 0.3, T_P = 20,$$
$$K_P = 120, B = 0.425, R = 0.425, T_{12} = 0.086$$

References

1. Fosha, Charles E., and Olle I. Elgerd. "The megawatt-frequency control problem: A new approach via optimal control theory." IEEE Transactions on Power Apparatus and Systems 4 (1970): 563–577.
2. Kundur, Prabha. Power system stability and control. Eds. Neal J. Balu, and Mark G. Lauby. Vol. 7. New York: McGraw-hill, 1994.

3. Karnavas, Y. L., and D. P. Papadopoulos. "AGC for autonomous power system using combined intelligent techniques." Electric power systems research 62.3 (2002): 225–239.
4. Prasanth, B. Venkata, and SV Jayaram Kumar. "Load frequency control for a two area interconnected power system using robust genetic algorithm controller." Journal of Theoretical and Applied Information Technology 4.12 (2008): 1204–1212.
5. Soundarrajan, A., S. Sumathi, and C. Sundar. "Particle swarm optimization based LFC and AVR of autonomous power generating system." IAENG International Journal of Computer Science 37.1 (2010): 37–1.
6. Alomoush, Muwaffaq Irsheid. "Load frequency control and automatic generation control using fractional-order controllers." Electrical Engineering (Archiv fur Elektrotechnik) 91.7 (2010): 357–368.
7. Alrifai, Muthana T., Mohamed F. Hassan, and Mohamed Zribi. "Decentralized load frequency controller for a multi-area interconnected power system." International Journal of Electrical Power & Energy Systems 33.2 (2011): 198–209.
8. Gozde, Haluk, and M. Cengiz Taplamacioglu. "Automatic generation control application with craziness based particle swarm optimization in a thermal power system." International Journal of Electrical Power & Energy Systems 33.1 (2011): 8–16.
9. Golpira, H., and H. Bevrani. "Application of GA optimization for automatic generation control design in an interconnected power system." Energy Conversion and Management 52.5 (2011): 2247–2255.
10. Gozde, Haluk, M. Cengiz Taplamacioglu, and Ilhan Kocaarslan. "Comparative performance analysis of Artificial Bee Colony algorithm in automatic generation control for interconnected reheat thermal power system." International Journal of Electrical Power & Energy Systems 42.1 (2012): 167–178.
11. Yang, Xin-She, and Suash Deb. "Cuckoo search via Lévy flights." Nature & Biologically Inspired Computing, 2009. NaBIC 2009. World Congress on. IEEE, 2009.
12. Rao, R. Venkata, Vimal J. Savsani, and D. P. Vakharia. "Teaching–learning-based optimization: an optimization method for continuous non-linear large scale problems." Information sciences 183.1 (2012): 1–15.
13. Rout, Umesh Kumar, Rabindra Kumar Sahu, and Sidhartha Panda. "Design and analysis of differential evolution algorithm based automatic generation control for interconnected power system." Ain Shams Engineering Journal 4.3 (2013): 409–421.
14. Debbarma, Sanjoy, Lalit Chandra Saikia, and Nidul Sinha. "Automatic generation control using two degree of freedom fractional order PID controller." International Journal of Electrical Power & Energy Systems 58 (2014): 120–129.
15. Singh, Sugandh P., et al. "Analytic hierarchy process based automatic generation control of multi-area interconnected power system using Jaya algorithm." Engineering Applications of Artificial Intelligence 60 (2017): 35–44.

Modified Shunt Active Line Conditioner Using Enhanced Self-restoring Technique with Step Size Error Elimination Algorithm

Gopalakrishnan Muralikrishnan and Nalin Kant Mohanty

Abstract This paper reveals an enhanced control algorithm by employing a novel idea known as step size error elimination in order to achieve improved control over DC-coupled capacitor voltage in modified Shunt active line conditioner (MSALC). Earlier works on self-restoring algorithms were reported solely under operation for steady-state condition. But self-restoring algorithms were employed by any one of the controllers such as conventional proportional—integral (PI) or adaptive fuzzy logic control (FLC). But, power system will be subjected to dynamic operation also. Therefore, by proposing step size error elimination (SSEE) as a merit feature to the self-restoring algorithm, all the conditions such as steady-state and dynamic operations in the power system can be enhanced. For analysis and evaluation, self-restoring with SSEE algorithm was developed and MATLAB–Simulink environment tool was employed for stimulation along with MSALC. From the result outcomes, it was proved that the proposed self-restoring with SSEE stands to be remarkable with high accuracy, high frequency response, and minimum overshoot and undershoot feature. It responds excellently under conditions such as steady-state and dynamic operations.

Keywords ADALINE control approach · DC electric potential regulation Shunt active line conditioner · Step size error elimination · Total harmonic distortion

G. Muralikrishnan (✉)
Department of Electrical and Electronics Engineering, Panimalar Engineering College, Chennai, India
e-mail: muralitkg@yahoo.com

N. K. Mohanty
Department of Electrical and Electronics Engineering, SVCE, Kanchipuram, India

© Springer Nature Singapore Pte Ltd. 2019
M. A. Bhaskar et al. (eds.), *International Conference on Intelligent Computing and Applications*, Advances in Intelligent Systems and Computing 846,
https://doi.org/10.1007/978-981-13-2182-5_24

1   Introduction

Power system quality is a serious concern of sensitive load that considers the voltage and current at a particular instant and in the specified zone of power distribution system [8]. The Power system parameter variations manifested in terms of voltage, current, or frequency deviations which are likely to prevail, in turn leads to failure or mal-operation of end-user loads that are identified as power quality issues [11]. Integer multiples of fundamental frequency, fractional multiples of fundamental frequency, unbalance of electric potential, changes in the electric potential amplitude and deviation in fundamental frequency tend to be the predominant Power quality issues [8, 11].Profoundly the harmonic current intensity on the end-user is detailed in various articles [5, 10, 19]. As an enhancement to passive LC filter, active power line conditioner is employed as an advanced solution to eliminate the harmonics [9, 19]. Harmonics in power distribution system can be reduced by the use of Active Line Conditioner (ALC). The goal is to set the upstream current of power distribution system to be sinusoidal under the circumstances such as downstream current is unbalanced and/or distorted. Of all the methods, the instantaneous power theory by Akagi stands as one of the outstanding feature by its control and design [4]. But, due to conditions such as distortion and/or unbalance in voltage profile, the instantaneous PQ theory is not supposed to perform well.

Harmonics which is an integral multiple of fundamental frequency can be further classified as electric potential and current harmonics or high current intensity harmonics. Nonlinear load operations contribute to current harmonics exploited due to power electronic switches and appliances that influences the upstream topology coupled at point of common coupling (PCC) [24]. The almost outcomes of high current intensity can be condenser failure, devices and machines excess heating, mechanical distortion of the rotating parts and excessive neutral currents [7]. A number of line conditioners and filters have been designed and employed to enhance the power quality at the distribution side [4, 5]. But shunt active line conditioners too preferred as the best one for various industrial processes [16]. To reduce the current harmonics, shunt active line conditioner (SALC) is employed superior to series active line conditioner (SeALC) which reduces electric potential harmonics. These conditioners tend to provide the merits of wattless power compensation, current harmonics elimination, and equalization of downstream currents. In turn provide an idea for us to ensure an economic and an alternate path of dominance strategy with much compatibility to provide solution for exceed limit of PQ issues.

Dual functions that need to be carried by a SALC are the generation of exact current instructions and its propensity to rapidly follow the instruction current signals. For generating instruction current for compensation, harmonic constituents of downstream currents need to be identified which is a vital feature needed in the control structure of a SALC. This will be spotted by correctly notifying the harmonics constituents, or alternately by subtracting the fundamental element of the downstream current. In case of obtaining, the efficient conditioner function, alternate method stands to be simple and have low execution time [21]. In SALC, harmonic current is injected in direction which is just opposition in order to guarantee sinusoidal nature in the upstream of the power distribution system. Traditional design procedure of single-phase SALC employs full-bridge inverter wherever it contains four thyristor switches with a condenser named as electric potential condenser for DC-coupled source.

Primary role for DC-coupled potential condenser will be saving the DC potential to be constant in order to provide compensation or mitigation current when compared with the upstream current. Traditional design procedure for commanding the condenser potential of DC-coupler is based on the operating values including the instant value of electric potential and condenser potential of specified DC-coupler. But, due to deployment of the design procedure, the condenser potential of DC-coupler is a distorted voltage due to no restriction and results to unregulated value [8, 15]. Capacitor blow out and high frequency distortions are some listed drawbacks as a result of unbalanced compensation current [8, 24].

In the modern era, self-restoring algorithm has gathered peculiar focus from the research analysts towards its merits in comparative to established design feature associated for DC-Coupled condenser electric potential control [11]. Out of its merits, crest perfect DC potential and unpolluted restricted electric potential with exactly no power frequency variations are significant [25]. The self-restoring algorithm employs electric conversion principle for regulating restoring and charge dispense of the DC-Coupled condenser potential in comparison to the traditional design procedure for solely handling the change in values of preferred electric potential and subsequent restoring DC-Coupler capacitor potential to be the foremost variable in regulating the condenser potential. Traditional PI-based supervise [3, 12, 13] and fuzzy rule based supervise (FLC) [12] can be considered as best known present technical supervising methodology for regulating the change in potential procreated by the self-restoring concept.

Self-restoring employing PI design procedure will be the most familiar and understand able too. But, little demerits of PI such as voltage amplitude variations, DC-Coupler potential imbalance, and maximum overshoot, low frequency response, presence of unwanted DC constituents, other frequency variation constituents in the supervised DC-Coupler potential. Unreliable outcomes due to specification changes, non-determinism and end-user distortions as PI perform solely for steady-state condition [2, 3].

Conventional PI controller is too stiff for adjusting the frequency and intended principally while concerning SALC, as PI controller desires providing specific numerical specifications for obtaining the variables of proportional k_p and integral k_i constants. With advancement in soft computing techniques, FLC to be the major technical control method provides greater performance, in terms of stability, preciseness; speedy response and outcomes are good while employed for complicated system [3]. FLC control method performance is independent of desired, accurate numerical specifications for developing the model and adjusting the frequency variations. FLC performance is good for vague inputs, but precisely tackles non-deterministic system. FLC is too reliable and simple when compared with conventional PI controller [3, 17]. FLC has self-regulating control technique too and performs based on the group of understandable and specific intelligible linguistic fuzzy commands [6].

Since FLC control method has superior performance compared to PI control method, they too even share some important demerits such that both working procedures consider no parameter variables, nonlinearity, and load distortions. Earlier research articles reported solely examines for steady-state operation and research on dynamic operation was not carried out yet [3, 6]. But, dynamic operating condition tend to prevail and is peculiar in the case of DC-Coupled capacitor, which in turn lead to blow out under the conditions of over voltage and possibly mal-operation of compensation current prevails under the condition of under voltage [20]. If there is any deviation on load side, then comparable deviation also occurs in form of electric potential of DC-Coupled capacitor [3]. At present PI or FLC are the technical procedures employed to self-restoring algorithm, in order to respond voltage deviation precisely, which results incredible disquiet for obtaining the design procedure learning rate. Regulation against over voltage deviation, tend to be a straight-forward regulation procedure (major instruction pulse), as they will not be too flexible when either the voltage deviation parameter deviates or will be stiff, when further proceeded to restrict the deviation.

Hence, this paper confers the research task that is capable of enhancing the regulation of DC-Coupled capacitor under both operating condition. By applying step size error elimination as supplementary characteristic for self-restoring algorithm, SALC need to tackle both under the condition such as steady-state and dynamic state. The advanced self-restoring algorithm will be justified using Step size error elimination control method. To discuss in detail about the research, Sect. 2 of the paper includes advancement made in single-phase SALC such as harmonic extraction method, DC-Coupled capacitor electric potential regulation. This is preceded using merits of self-restoring algorithm employing PI, FLC and more enhancements performed by step size error elimination. Simulation carried out and its outcomes are explained in Sect. 3. Ultimately, Sect. 4 derives the inference from the outcomes of the proposed research.

2   Single-Phase MSALC

Figure 1 depicts a complete MSALC that consists of a full-bridge inverter with four thyristor switches, a DC-Coupled capacitor and controller. Harmonic source considered is a rectifier-deployed circuit that introduces dominant THD level in power distribution system [23]. The system is associated by non-deterministic current drawn appliances such as uncontrolled diode bridge rectifier along with parallel RLC load [1].

Figure 2 details about the MSALC regulation scheme which includes harmonic extraction model, DC-Coupled capacitor (condenser) electric potential supervise model, synchronizer model, current supervising model and switching state model.

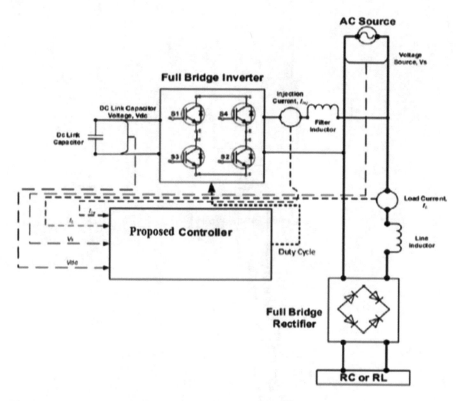

Fig. 1 System model of single-phase MSALC with typical distribution system

For the proposal, a prominent control procedure is the DC-Coupled condenser electric potential supervise model. For harmonic extraction, the supervising scheme employed is improved Windrow–Hoff (W–H) Adaline algorithm [2, 6, 22]. But, regarding current regulation, PI regulation model is employed [6, 22]. A synchronizer is implemented for generation of instruction pulse. For generation of switching sequence, pulse-width modulation is employed.

From Sect. 1 it is hinted clearly that, DC-Coupled condenser electric potential supervise model is a significant regulation scheme for MSALC. For best performance, DC-Coupled condenser electric potential needs to be greater than two-third of the upstream electric potential so that properly ensuring the desired compensation current can be produced. Least capacitance to be designed for the condenser can be examined as below [15]

$$C \geq \frac{\max \left| \int_0^t I_{inj}(t) \right|}{\Delta V_{C\max}}, \tag{1}$$

where I_{inj} indicates compensation current and $\Delta V_{C\max}$ indicates the crest unwanted electric potential of DC-Coupled condenser.

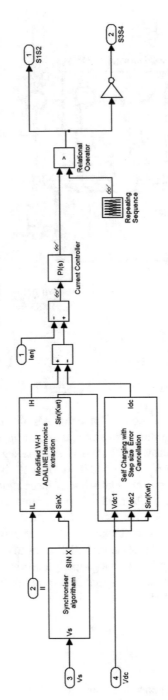

Fig. 2 Block diagram of regulation scheme for MSALC

2.1 Harmonic Extraction

ADALINE control approach is executed for harmonic extraction, separating sine, and cosine constituents which results in evaluating harmonic constituents persisting in power distribution network. First-order constituent and integral multiple constituents are interrelated with the non-sinusoidal downstream current I_L by sampling every k and for sampling time t_s employing discrete operation under nominal fundamental frequency w [14]. Unwanted downstream current which will be expressed as

$$I_L(k) = \sum_{n}^{N} = 1, 2, \ldots [W_{an} \sin(nkwt_s) - W_{bn} \cos(nkwt_s)], \qquad (2)$$

where W_{an} and W_{bn} are sine and cosine constituents magnitude of downstream current I_L. n indicates integral multiple constituent index up to N crest index.

ADALINE control approach is applied for deriving equal significance of $I_L(k)$. Most important characteristic deployed by the extraction control scheme is that weight upgrading control model employing W–H control procedure [6, 22]. Compensation current I_{inj} that need to be applied ensuring the cancelation of harmonic distortion, that will be in opposition compared to integral multiple constituent current I_H, which is depicted in Fig. 3. An advancement presented for improved W–H ADALINE is the application of only the foremost index in the integral multiple constituent compared to n numeral in the integral multiple constituent that exists in the case of W–H ADALINE control approach. Due to the application of the control methodology, this had underneath the issues that exist in W–H ADALINE control approach. Since numerical weight n can are revised that involves too long time to respond [2]. For simplifying the weight to be constant with respect to the numerals of integral multiple constituent indices, solitary revise

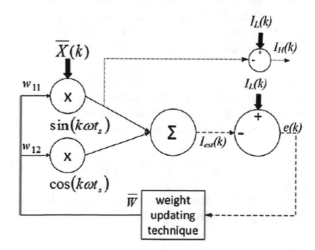

Fig. 3 Block diagram of improved W–H ADALINE control approach

all weight of first order constituent. The development will be obtained from numerical representation as the constituents are of perpendicular to each other. With this alteration, the computation time is very much reduced, leading to high speed evaluation.

But, upgrading solely both the weights leads to high value of mean square error *e*. So learning factor α need to be included in (3) [13]. For optimal generation of specified first-order constituent from the integral multiple constituent current, appropriate learning factor is vital needed. Integral multiple constituent current I_H (*k*) is generated based on end-user current identification from (4) [18].

$$\overline{W}(k+1) = \overline{W}(k) + \frac{\alpha e(k)Y(k)}{\overline{Y}^T(k)\overline{Y}(k)} \tag{3}$$

$$I_H(k) = I_L(k) - W \sin(kwts) \tag{4}$$

2.2 DC-Coupled Condenser Electric Potential Control

For the generation of compensation current, the DC–AC power converter draws power from the DC-Coupled condenser which realizes to be a stable DC utility. Traditional control approach employed for regulating voltage of DC-Coupled condenser through forward mode variations comparing instant value of electric potential and preferred DC-Coupled electric potential [15]. Dynamic operation always tends to happen in distribution network and in particular for DC-Coupled condenser, that might lead to blow out condition if excess electric potential appear. If less electric potential below the standard occurs, that can results for compensation current to mal-operate [20]. Enhanced self-restoring control approach must enable to improve the performance level by having control over the DC-Coupler capacitor under both operating conditions. Hence, at steady-state and dynamic conditions that are to be tackled, merit aspect termed as step size error elimination need to be employed for self-restoring control approach.

Surplus active power is needed for the control of DC-Coupled condenser electric potential drawn from utility side in order to restore condenser. In the due of restoring period, electric potential of the DC-Coupled condenser in most cases deviates from the specified voltage which leads to deviation of DC-Coupled condenser energy storage alone. Therefore, deviation of energy storage ΔE in DC-Coupled condenser can be expressed as (5).

$$\Delta E = \frac{c}{2}\left[(V_{dc2})^2 - (V_{dc1})^2\right] \tag{5}$$

c represents condenser estimation for DC-Coupled condenser. V_{dc1} represents DC-Coupled condenser specified electric potential and V_{dc2} represents DC-Coupled condenser instant value of electric potential. In an alternative concern, condenser is restored from 1 Φ distribution network E_{ac} which can be expressed as

$$E_{ac} = Pt_c = V_{rms} I_{dc,rms} \cos \theta t_c, \qquad (6)$$

where P represents surplus active power essential for charging. t_c represents condenser restoring period. V_{rms} represents root mean square of upstream electric. $I_{dc,\ rms}$ represents root mean square of restoring condenser current $I_{dc.}$ θ represents angle deviation comparing upstream electric potential and restoring condenser current. But, t_c is expressed equivalent to $T/2$ as restoring event of the condenser solely requires half cycle. T represents fundamental constituent time duration that need to be nominal 50 Hz.

Electric Potential deviation e as given by Eq. (7) in the self-restoring control approach delivers largest performance for the evaluation of DC-Coupled condenser restoring current I_{dc}. DC-Coupled condenser electric potential regulation is preferred for regulating restore level in terms of controlling I_{dc}.

$$e = \left[(V_{dc2})^2 - (V_{dc1})^2 \right] \qquad (7)$$

2.2.1 Self-restoring Scheme Using PI

PI was dominantly employed for the voltage control of DC-Coupled condenser. Figure 4 depicts functional set of self-restoring scheme using design procedure of PI. Lower limit variables of PI design procedure will be evaluated from [3].

$$k_p \geq 2C \xi \omega \qquad (8)$$

$$k_i \geq Cw, \qquad (9)$$

where k_i indicates integral gain. k_p indicates proportional gain. C denotes condenser estimation for DC-Coupled condenser. ξ denotes under damped ratio that need to be typically 0.707 ω indicates angular frequency.

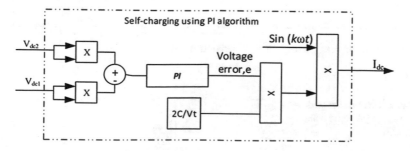

Fig. 4 Functional set of self-restoring scheme using PI

2.2.2 Self-restoring Scheme Using FLC

Self-restoring scheme employing FLC design procedure is depicted in Fig. 5. Fuzzification, defuzzification, fuzzy rule-base, and fuzzy inference engine are the sub-processes involved in FLC design procedure. During fuzzification, mathematical entry elements being transformed as fuzzy entry elements with respect to membership functions during fuzzification. Different fuzzy values are employed regarding entry and response elements [12, 17]. Delusion E and change of delusion CE are applied to be entries having sampling time k for self-restoring scheme employing FLC design procedure from (10) and (11). But, response for FLC design procedure will be fuzzified electric potential error e_f.

Behind the computation of E and CE, those entries will be transformed as fuzzy entry elements. Fuzzy response will be determined using look-up table having a set of fuzzy laws. FLC design gives the results considering fuzzy prime law 'If X and Y, Then Z' (Table 1).

$$E(k) = \left[(V_{dc2})^2 - (V_{dc1})^2 \right] \tag{10}$$

$$CE(k) = e_f(k) \tag{11}$$

For examining FLC response, fuzzy inference engine can be employed. Extrapolation methodology will be Mamdani [8]. Normally, weight need to be summed with the rules for enhancing in order to ensure precise reasoning and acceptable logical effects. During defuzzification, fuzzy responses are retrieved again to crisp elements using fuzzy entry elements. Major standard methodology

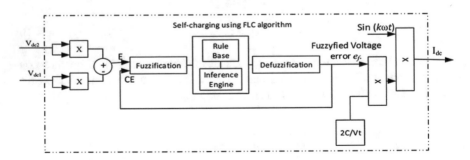

Fig. 5 Functional set of self-restoring scheme using FLC

	CE/E	N_{Min}	ZE	P_{Min}	PC	P_{Max}
Table 1 Base law regarding self-restoring scheme using FLC [25]	N_{Min}	N_{Min}	N_{Min}	ZE	P_{Min}	PC
	ZE	N_{Min}	ZE	P_{Min}	PC	P_{Max}
	P_{Min}	ZE	P_{Min}	PC	P_{Max}	P_{Max}

applied in the case of defuzzification need to be centroid of area due to the desired mean value effects, produces high precise results [13].

2.2.3 Self-restoring Scheme Using Step Size Error Elimination

Since, it is necessary to focus on dynamic operating condition; enhancement for self-restoring scheme needs to be performed in terms of applying step size error elimination from (12).

$$e_{\text{new}} = e + \Delta e, \tag{12}$$

where e_{new} represents updated electric potential deviation. e represents electric potential deviation. Δe represents proposed step size variation. Hence, updated restoring current I_{dc} from (13).

$$I_{dc} = \frac{2C\left[\left[(V_{dc2})^2 - (V_{dc1})^2\right] + \Delta e\right]}{VT} \tag{13}$$

If condenser electric potential need to be regulated within a specified level, so that restoring DC-Coupled condenser current I_{dc} need to be or nearly equivalent to zero. Instead of forwardly regulating electric potential deviation, self-restoring control scheme with an annexed feature step size error elimination design procedure that need to be included is depicted in Fig. 6. Its significant objective is to provide flexibility for the control approach with regard to overshoot and undershoot in order to eliminate any deviation of electric potential variation. Since, the possibility for disquiet to occur instantly will be less. Step size error elimination control method presents another way for commanding electric potential deviation for self-restoring scheme; instead of self-restoring using PI and FLC design procedure that instantly commands electric potential deviation. The optional feature results in quick responses and enhanced functionality. When there exist any variation (overshoot) above or below (undershoot) the standard of deviation in electric potential,

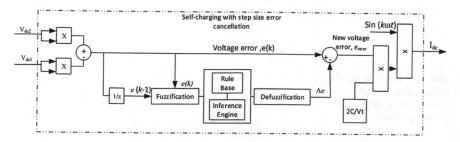

Fig. 6 Block diagram of self-restoring technique with enhanced step size error elimination algorithm

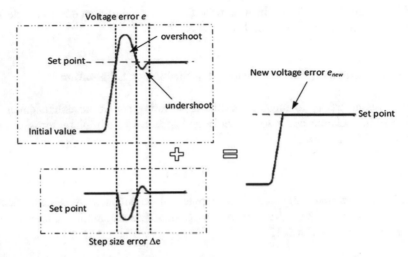

Fig. 7 Enhanced step size error elimination scheme based waveform depiction [25]

enhanced self-restoring effect can eliminate all effects with respect to variation of electric potential, resulting it equivalent to zero. In case of undershoot or overshoot, Δe results in numerically of positive or negative value respectively; as they need balancing overshoot and undershoot of electric potential deviation.

The step size error can solitary provide prominent effect to have solution when there is overshoot or undershoot likely to be present in electric potential deviation particularly as condenser electric potential beyond reaching the specified voltage. To control Δe, FLC control method is preferred as it had more merits over PI technique as it was obviously stated previous. FLC can be employed for regulating step size as elimination needs to be correct and desired one in order to ensure much preferred outcomes. Number of alternative procedures that improves FLC technique includes fuzzification entries and standard fuzzy laws [25] (Fig. 7; Table 2).

Present control scheme of self-restoring using FLC control approach applies deviation E and variants of deviation CE to be key entries regarding fuzzification. But for, Fuzzification entries can be optioned as electric potential deviation $e(k)$ and prior electric potential deviation $e(k-1)$ in the case of self-restoring scheme using step size error elimination design procedure, so as to derive the step size error (SSE).

Table 2 Base law regarding self-restoring scheme using step size error elimination design procedure [25]

$e(k)/e(k-1)$	N_{Min}	ZE	P_{Min}	PC	P_{Max}
N_{Max}	ZE	N_{Min}	N_{Max}	N_{Max}	N_{Max}
N_{Min}	P_{Min}	ZE	N_{Min}	N_{Max}	N_{Max}
ZE	P_{Max}	P_{Min}	ZE	N_{Min}	N_{Max}
P_{Min}	P_{Max}	P_{Max}	P_{Min}	ZE	N_{Min}
P_{Max}	P_{Max}	P_{Max}	P_{Max}	P_{Min}	ZE

3 Simulation Design and Results

Simulation model is developed using MATLAB software in Simulink environment and waveforms are analyzed using scope and harmonic spectrum is analyzed using FFT window [7]. Step Size Error Cancelation (SSEC) Controller based single-phase SALC which is modified is termed as modified Shunt Active Line Conditioner (MSALC).

It is connected to a test model system which consists of supply of 230 V, 50 Hz connected to nonlinear loads. The nonlinear load includes an H-bridge uncontrolled diode rectifier, Capacitor load and RL load all together as depicted in Fig. 8. The load parameters are C = 10 μF, RL = (23 Ω, 1 mH).

Simulation is carried out using SSEC controller. The vital needed performance factors considered to analyze the performance of DC-Coupled condenser electric potential regulation procedure are preciseness, period of responses, THD of supply current across different loads.

Figures 9, 10 and 11 represents the simulation block of control strategies for MSALC, simulation block of improved W-H ADALINE design procedure and simulation block regarding self-restoring scheme using enhanced step size error elimination design procedure respectively designed using MATLAB Simulink environment.

Figures 12, 13 depicts membership functions with respect to entry and response elements. Entry variables, (e) error voltage and e(k − 1) previous error voltage are applied as input parameter for the evaluating the output (de) step size error which is the deviation in error values which is employed to command the corresponding dc voltage.

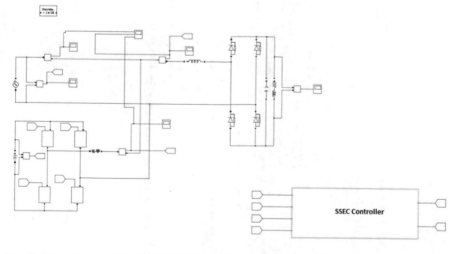

Fig. 8 Simulation model of 1Φ MSALC using SSEC controller with test distribution system

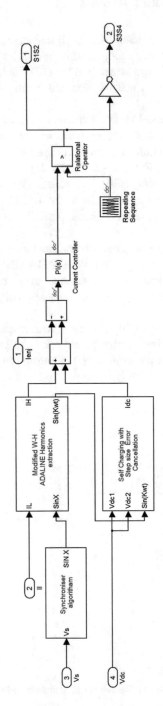

Fig. 9 Simulation block of control strategies for MSALC

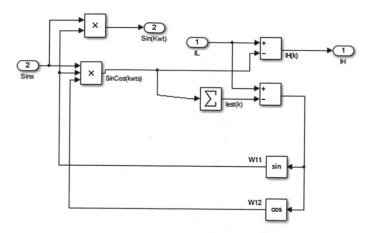

Fig. 10 Simulation block of improved W–H ADALINE design procedure

From Fig. 14, it is clear that the source voltage is maintained stiff at 230 V in spite of nonlinear load with the nominal frequency of 50 Hz from the source side.

By comparing the source current of 20 A with load current of 10 A, it is understandable that load current is supposed to be dropped to 10 A due to non-linear load. So, we require injecting of 10 A current as the compensation current through controller so as to compensate or cancel the effect of nonlinear load in the load current to have the precise source current value of 20 A. So, we are injecting compensation current of 10 A which is 180° phase opposite to that of load currents to ensure minimization of the harmonic indices within the prescribed IEEE standard regarding harmonic regulation in power distribution network depicted from Fig. 15. THD concerning supply current and load current without (before compensation) MSALC in the power distribution system are 14.67 and 29.17% respectively. But, THD concerning supply current and load current with (after compensation) MSALC in the power distribution system are 1.65 and 3.28% respectively which is under the control level of less than 5% as prescribed by IEEE standard regarding harmonic regulation in power distribution network [11].

From Fig. 16, obviously the thyristor switches S1 and S2 are ON during the positive half cycle. While the thyristor switches S3 and S4 are ON during the negative half cycle. So each thyristor pairs (S1, S2) and (S3, S4) will operate for 180°.

FFT analysis is carried out for overall non- linear load including the rectifier, RL load and capacitor. By analyzing Fig. 17, it can be explained that THD of supply current and load current are 1.65 and 3.28% respectively which is less than 5% as prescribed by IEEE standard regarding harmonic regulation in power distribution network [11].

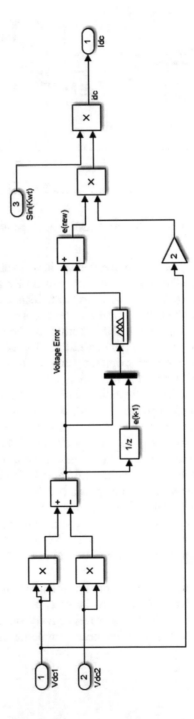

Fig. 11 Simulation block of self-restoring scheme using enhanced step size error elimination

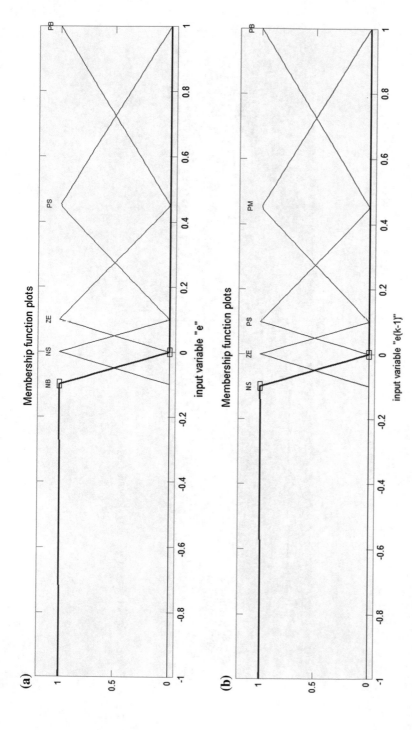

Fig. 12 **a, b** Membership functions of entry elements ($e(k)$ & $e(k-1)$))

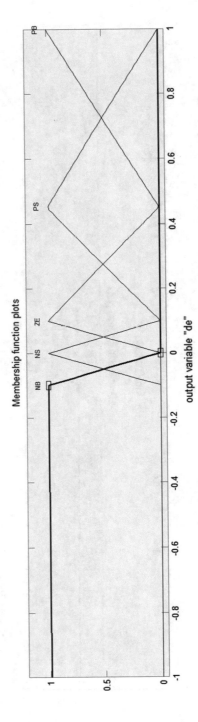

Fig. 13 Membership functions of response elements (de)

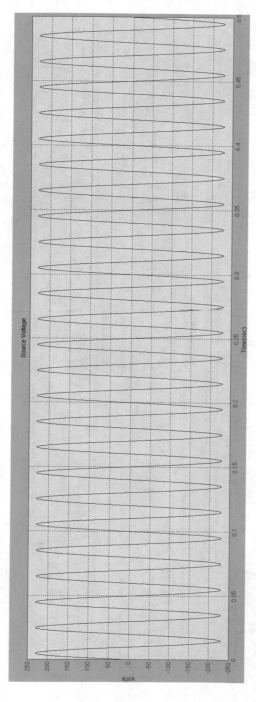

Fig. 14 Waveforms of source voltage

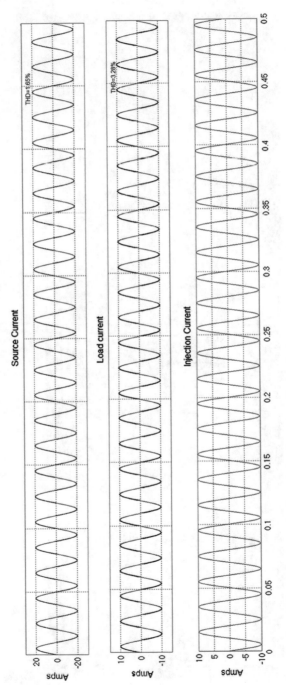

Fig. 15 Waveforms of source current, load current and compensation current

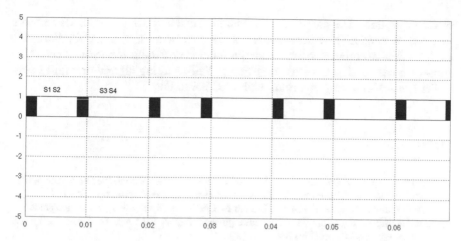

Fig. 16 Switching sequence of thyristor switches

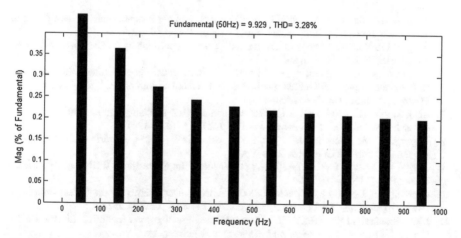

Fig. 17 FFT analysis of supply current

4 Conclusion

Both self-restoring scheme using PI and self-restoring scheme using FLC are designed to control the voltage deviation directly under steady-state resulting to slow response time. Due to slow tracking process, there is possibility of high overshoot and undershoot with PI and FLC controllers under dynamic operation. But enhanced self-restoring algorithm with step size error elimination controller is to be designed to command the voltage error in alternate path (indirectly) resulting to high preciseness, less (overshoot, undershoot) THD value and quick responses under dynamic condition. Analysis is carried out based on the performance parameter such as accuracy, (overshoot, undershoot) THD value and response time of the controllers. The proposed algorithm is stimulated using MATLAB software to check the validity of scheme results together with 1 Φ MSALC. From the analysis, Self-restoring algorithm with Step size error elimination (SSEE) controller shows the best performance as compared to Self-restoring algorithm with PI controller and self-restoring algorithm with FLC controller.

References

1. Abdel Aziz MM, Zobaa AF, Hosni AA (2006), "Neural network controlled shunt active filter for nonlinear loads," *Eleventh Int. Middle East Power Systems Conf.*, pp. 180–188.
2. Abdul Rahman NFA, Radzi MAM, Mariun N, CheSoh A, Rahim NA (2013),"Integration of dual intelligent algorithms in shunt active power filter," *IEEE Conf. on Clean Energy and Technology (CEAT),* pp. 259–264.
3. Afghoul H, Krim F (2012), "Comparison between PI and fuzzy DPC control of a shunt active power filter," *Second IEEE ENERGY CON Con. &Exhibition Advances in Energy Conversion Symp.*, pp. 146–151.
4. Akagi ANH, Kanagawa Y (1983) Generalized theory of the instantaneous reactive power in three-phase circuits. In: International conference on Power Electron.
5. Akagi H (1996), "New trends in active filter for power conditioning," *IEEE Trans Industrial Application,* Vol.32, No.6, pp. 423–435.
6. Bhattacharya A, Chakraborty C (2011), "A shunt active power filter with enhanced performance using ANN-based predictive and adaptive controllers," *IEEE Trans. Ind. Electron.*, Vol. 58, No. 2, pp. 421–428.
7. Bhim Singh AC, Al-Haddad K (1999), "Computer-aided modeling and simulation of active power filters," *Electr. Mach. Power Syst.*, Vol. 27, pp. 1227–1241.
8. Dugan RC, McGranaghan MF, Santoso S, Beaty HW (2003), *Electrical power systems quality*, 2nd edn, McGraw-Hill, New York.
9. Gyugyi L, Strycula E (1976) Active ac power filters. In: Proceedings IEEE Ind. Appl. Ann. Meeting 19-C (NA), pp 529.
10. Habrouk ME, Darwish Mk, Mehta P(2000), "Active power filters a review," *IEEE Proc Elect Power Application,* Vol.147, No.5, 403. doi:10. 1049/ip-epa: 20000522.
11. IEEE Standard 1159: (2014)"IEEE Recommended Practice for Monitoring Electric Power Quality," *Institute of Electrical and Electronics Engineers,* New York.
12. Jain SK, Agrawal P, Gupta HO (2002), "Fuzzy logic controlled shunt active power filter for power quality improvement," *IEEE Proc. Electrical Power Applications*, No.149, pp. 317–328.
13. Mikkili S, Panda AK(2013), "Types-1 and -2 fuzzy logic controllers-based shunt active filter I_d–I_q control strategy with different fuzzy membership functions for power quality improvement using RTDS hardware," *IET Power Electron.*, Vol. 6, No. 4, pp. 818–833.
14. MorenoVM, Lopez AP, Garcías RID (2004), "Reference current estimation under distorted line voltage for control of shunt active power filters," *IEEE Trans. Power Electron.*, Vol. 19, No. 4, pp. 988–984.
15. Orfanoudakis GI, Yuratich MA, Sharkh SM (2013), "Analysis of DC-link capacitor current in three-level neutral point clamped and cascaded H-bridge inverters," *IET Power Electron.*, Vol. 6, No. 7, pp. 1376–1389.
16. Petit JF, Robles G, and Amaris H (2007), "Current reference control for shunt active power filters under non sinusoidal voltage conditions," *IEEE Transaction on Power Delivery,* vol. 22, no. 4, pp. 2254 2261.

17. Ponpandi R, Durairaj D (2011), "A novel fuzzy-adaptive hysteresis controller based three phase four wire-four leg shunt active filter for harmonic and reactive power compensation," *Energy Power Eng.* Vol. 3, pp. 422–435.
18. Singh B, Verma V, Solanki J (2007),"Neural network-based selective compensation of current quality problems in distribution system," *IEEE Trans. Ind. Electron,* Vol. 54, No. 1, pp. 53–60.
19. Singh ACB, Al-Haddad K (2014), "A review of active filters for power quality improvement," *IEEE Trans Industrial Electron,* Vol. 46, No. 5, pp. 960–972.
20. Suresh Y, Panda AK, Suresh M (2012),"Real-time implementation of adaptive fuzzy hysteresis-band current control technique for shunt active power filter," *IET Power Electron.,* No. 5, pp. 1188–1195.
21. Trinh QN and. Lee HH (2013), "An advanced current control strategy for three phase shunt active power filters," *IEEE Trans. Ind. Electron.,* vol. 60, No.12, pp. 5400 –5410.
22. Vasumathi B, Moorthi S (2011), "Harmonic estimation using modified ADALINE algorithm with time-variant Windrow–Hoff (TVWH) learning rule," *IEEE Symp. on Computers & Informatics,* pp. 113–118.
23. Wenjin D., Taiyang H (2007), "Design of single-phase shunt active power filter based on ANN," *IEEE Int. Symp. on Industrial Electronics,* pp. 770–774.
24. YongtaoD, Wenjin D (2008), "Harmonic and reactive power compensation with artificial neural network technology," *Proc. 7th World Congress on Intelligent Control and Automation,* pp. 9001–9006.
25. Zainuri MAAM, Radzi MAM, Che Soh A, Mariun N, Rahim NA (2016), " DC-coupled capacitor voltage control for single-phase shunt active power filter with step size error cancellation in self-charging algorithm," *IET Power Electron.,* Vol. 9, No. 2, pp. 323–335.

A Simplified Pulse Generation Control Algorithm Based upon the Concept of Synchronverter

K. Subha Sharmini

Abstract This paper discusses development of a suitable m—file code for generating required pulse signals for a power electronic interface (PEI)—a single-phase voltage source inverter. The code is composed based upon the conceptual idea of synchronverter. Synchronverter is one of the voltage controlled strategy used for controlling voltage and power, when these PEI circuits are used as the coupling circuit between micro-source and the grid. The data required for writing the code were taken from the results obtained from individual simulations of the power circuit and electronic circuit of synchronverter. The drafted code is then embedded onto a PIC micro-controller and is used to control the switching pattern of the single-phase inverter thereby validating the proposal.

Keywords Coding · Control · PWM · Power electronic interface
SIMULINK · Synchronverter

1 Introduction

Energy and its sustainability are the two major issues in the present scenario. Using locally available renewable energy resources to feed medium-/low-voltage power to the local consumers in small scale seems to be the plausible answer. Despite being beneficial in terms of overall performance improvement, such micro-resources cannot be directly put in connection with grid. The reason is the dubious nature of the yield outcome from such renewable energy resources (PV, wind, etc.,). Thus, it becomes indispensible to identify a proper interfacing technology that functions in an analogous fashion to a synchronous generator. Inverters in conjunction with a suitable control thus gains its place as a fitting substitute.

The control strategies are classified into voltage control and current control. As part of this article, a VCS (Voltage Control Strategy) technique, i.e., synchronverter

K. S. Sharmini (✉)
Department of EEE, SRM IST, Kattankulathur, Chennai, India
e-mail: subha_kannan@srmuniv.edu.in

© Springer Nature Singapore Pte Ltd. 2019
M. A. Bhaskar et al. (eds.), *International Conference on Intelligent Computing and Applications*, Advances in Intelligent Systems and Computing 846,
https://doi.org/10.1007/978-981-13-2182-5_25

is used. It overcomes several drawbacks of CCS (Current Control Strategy). P, PI, PID, H infinity repetitive and deadbeat predictive control strategies are all categorized under CCS. But they necessarily require one or the other transformation techniques and voltage, frequency cannot be controlled directly. VCS techniques especially synchronverter are best suited for stand-alone applications as no change over of control technique is required to be done between grid connected and grid forming modes of operation.

2 Synchronverter—Literature Review

As stated in the above section, synchronverters in a LV micro-grid are an analogous equivalent to synchronous generator in conventional HV power system. Synchronverters is the working combo of an inverter (power part) and controller (electronic part). The droop control behavior, i.e., electro-magnetic characteristics, rotor inertia, and voltage and frequency regulation of the synchronous generator is inducted into the control part/electronic part of the synchronverter [1–10].

Articles [9–19] substantiate that synchronverter model and control mechanism could be used to synthesize various FACTS controllers (STATCOM, SVC), HVDC converters, hybrid MTDC systems. It is also proven that a dedicated PLL unit is not mandatory [15].

3 Modeling of Synchronverter

Design Equations

Given below are the equations that are introduced into the heart of the electronic part [1]:

$$T_e = M_f i_f \langle i, \sin \theta \rangle \tag{1}$$

$$e = \dot{\theta} M_f i_f \sin \theta \tag{2}$$

$$Q = \dot{\theta} M_f i_f \sin(\theta - \varphi) \tag{3}$$

$$\ddot{\theta} = (1/J)(T_m - T_e - D_p \dot{\theta}) \tag{4}$$

$$\text{where,} \ i = i_0 \sin \varphi \tag{5}$$

$$J = D_p \tau_f \tag{6}$$

$$K = \dot{\theta}_n D_q \tau_v \tag{7}$$

T_e Electro-magnetic torque,
T_m Mechanical torque,
J Moment of inertia,
K Dual of inertia/Gain,
M_f Mutual inductance of field coil,
i_f Field current,
e Back EMF,
θ Virtual angle,
$\dot{\theta}$ Virtual angular speed,
$\dot{\theta}_r$ Angular frequency reference,
$\dot{\theta}_n$ Nominal angular speed,
D_p Damping factor,
D_q Voltage droop co efficient,
τ_f Time constant—Speed/Frequency loop,
τ_v Time constant—Voltage loop.

Synchronverter—Control Part

With the aid of the above equations the controller part is developed as shown in Fig. 1. The partial schematic that is found within the dashed lines is run in SIMULINK and virtual angular speed value in radians is noted (Fig. 2). The other design parameters are as shown in Table 1.

Synchronverter—Power Part

A two-legged four switch common voltage source inverter is used as the power part. The circuit (Fig. 2) is also simulated in SIMULINK and the statistics of the load current waveform is observed. As it follows the sinusoidal rule it can be mathematically expressed as given below (Fig. 4):

$$i_o = I_m \sin \omega t \tag{8}$$

$$\theta = \omega t \tag{9}$$

$\dot{\theta}$ $= 417$ rad (from simulation, Fig. 2)
I_m $= 0.3$ A (from simulation, Fig. 4)

The inverter/load voltage is also observed (Fig. 4). The same is also validated as part of hardware results. A suitable m—code is written using the statistical/numerical details obtained with simulations of power and control circuits.

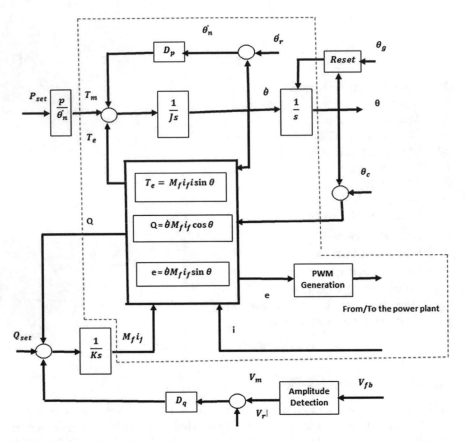

Fig. 1 Synchronverter—electronic part

4 Hardware Implementation

Figure 3 depicts the general block schematic of the hardware along with the practical hardware model implementation. A 230:15 V step-down transformer is used to give input supply to the power circuit. A full-wave diode bridge rectifier— rectifies the 15 V AC output of transformer and a capacitor filter eliminates the presence of ripples. A pair of analog IC three terminal fixed voltage regulators (7812 and 7805) is used to provide constant DC power to pulse amplifier (12 V) and PIC micro-controller (05 V).

The control program for PWM generation is embedded onto the PIC micro-controller—"PIC 16F877A". PIC microcontroller is a RISC processor with reduced instruction set. The PIC16F877A has a host of features intended to maximize system reliability, minimize cost through elimination of external components, provide power saving operating modes and offer code protection. TLP250 plus IC

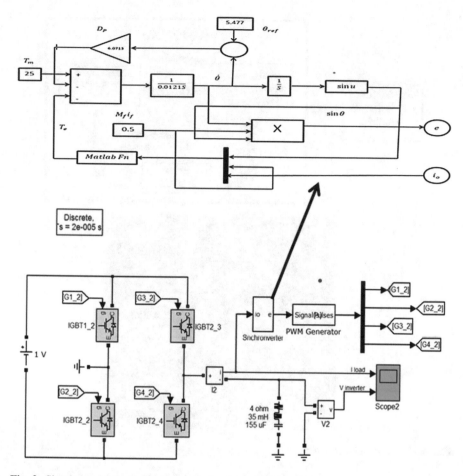

Fig. 2 Simulation implementation—control and power part

<table>
<tr><td>Table 1 Design specifications</td><td>J</td><td>0.0121</td></tr>
</table>

J	0.0121
K	2430.1
D_p	6.0713
D_q	386.7615
$M_f i_f$	0.5
T_m	25

IR2110 driver circuit is included as part of the hardware to gate and amplify the pulses generated by the PIC micro-controller.

The IR2110 is a high voltage, high speed power MOSFET driver with independent high and low side referenced output channels. It is fully operational to +500 V or +600 V and tolerant to negative transient voltage. The output drivers

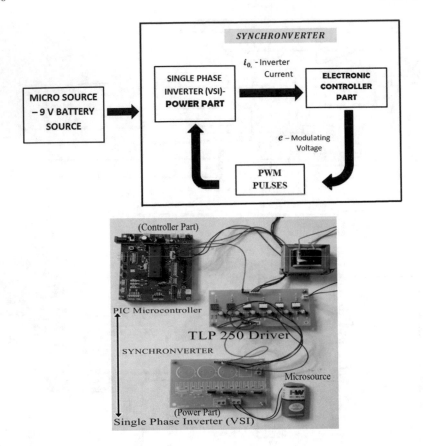

Fig. 3 Hardware implementation

feature a high pulse current buffer stage designed for minimum driver cross-conduction.

Propagation delays are matched to simplify use in high frequency applications. The floating channel can be used to drive an N-channel power MOSFET or IGBT in the high side configuration which operates up to 500 or 600 V. The four switches of the single phase voltage source inverter are being catered with the pulse signal output from this driver circuit. IRF840 MOSFET switches are chosen for their ruggedness, switching speed, low on state resistance and expense.

5 Conclusion

This article proposes synchronverter voltage control strategy in the form of a MATLAB code, to generate the necessary trigger pulse for the switches belonging to a single phase voltage source inverter. The embedded code is then dumped onto a

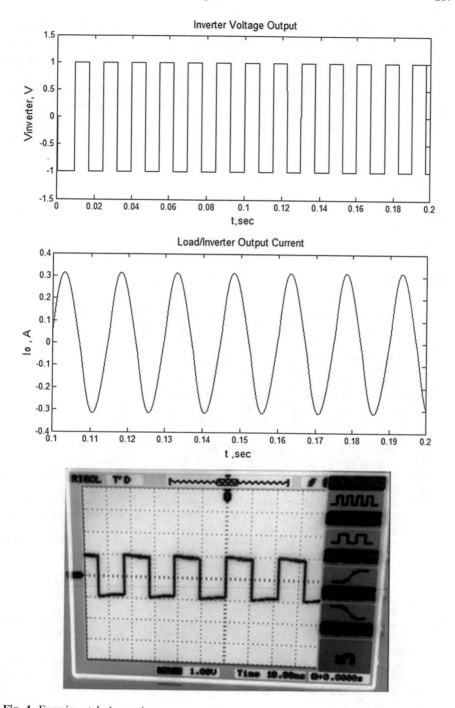

Fig. 4 Experimental observations

PIC microcontroller and the micro-chip is used to control the switching pattern of the VSI. The inverter output voltage matches that of the simulation results thus corroborating the proposal. It is also observed that the m—code is well suited for carrier with fundamental as well as higher switching frequency.

Future scope-The PEI is to be connected to a grid in which case fullest implementation of the entire control structure is to be done. This will result in power control as well.

Appendix

See Table 1.

References

1. Qing – Chang Zhong and Tomas Hornik, "Control of power inverters in renewable energy and smart grid integration,"Wiley –IEEE Press, 1st Edition, 2013.
2. Qing Chang Zhong, Zhenyu Ma Wen Long Ming, George C Konstantopoulos, "Grid friendly wind power systems based on the synchronverter technology," Elsevier-Energy Conservation and Management, Vol. 89, No. 4, pp. 719–726, Mar. 2015.
3. Q C Zhong and G Weiss, "Synchronverters: Inverters that mimic synchronous generators" IEEE Transactions on Industrial Electronics, Vol. 58, No. 4, pp. 1259–1269, Apr. 2011.
4. L Zhang, L Harnefors and H P Nee, "Power synchronization control of grid connected voltage source converters," IEEE Transactions on Power Systems., Vol. 25, No. 2, pp. 809–820, May. 2010.
5. H P Beck and R Hesse, "Virtual synchronous machine," in Proceedings of 9th Intl. Conference on Electrical Power Quality and Utilisation,, pp. 1–6, 2008.
6. J Driesen and K Visscher, "Virtual synchronous generators," in Proceedings of IEEE Power and Energy Society General Meeting, pp. 1–3, 2007.
7. Y Chen, R Hesse, D Turschner and H P Beck, "Improving the grid power quality using virtual synchronous machines," in Proceedings of Intl. Conference of Power Engineering, Energy and Electrical Drives, pp. 1–6,. 2011.
8. M Torres and L A C Lopes, "Frequency control improvement in an autonomous power system: An application of virtual synchronous machines," in Proceedings of 8th Intl. Conference of Power Electronics (ICPE & ECCE), pp. 2188–2195, 2011.
9. Raouia Aouini, Bogdan Marinescu and Mohamed Elleuch, "Synchronverter based emulation and control of HVDC transmission," IEEE Transactions on Power Systems, Vol. 31, No. 1, pp. 278–286, Jan. 2016.
10. Lin Yu Lu and Chia Chi Chu, "Consensus based distributed droop control of synchronverters for isolated micro grids," in Proceedings of the IEEE Intl. Symposium on Circuits and Systems, pp. 915–916, 2015.
11. P L Nquyen, Q.C Zhong, F Blaabjerg and J M Guerrero, "Synchronverter based operation of STATCOM to mimic synGrid friendly wind power systems based on the synchronverter technology," Elsevier-Energy Conservation and Management, Vol. 89, No. 4, pp. 719–726, Mar. 2015.

12. Raouia Aouini, Bogdan Marinescu, Khadija, Ben Kilani and Mohamed Elleuch, "Improvement of transient stability in an AC/DC system with synchronverter based HVDC,"in Proceedings of the 12th Int. Multi Conference on Systems, Signals and Devices, pp. 1–6, 2015.
13. Shanglin Mo, Bin Peng, Zhikang Shuai, Jun Wang, Chunming Tu, Z John Shen, Wen Huang, "A new self- synchronization control strategy for grid interface inverters with local loads," IEEE Transactions on Power Systems, Vol. 31, No. 1, pp. 278–286, Jan 2016.
14. Q C Zhong, Phi – Long Nguyen, Zhenyu Ma, Wanxing Sheng, "Self – synchronized synchronverters; Inverters without a dedicated synchronisation unit," IEEE Transactions on Power Electronics", Vol. 29, No. 2, pp. 617–630, Feb. 2014.
15. Chang- Hua Zhang, Qung- Chang Zhong, Jin – Song Meng, Xin Chen, Qi Hunag, Shu- heng Chen, "An improved synchronverter model and its dynamic behavior comparison with synchronous generator," IET. 2013.
16. Shuan Dong, Yong Ning Chi and Yan Li, "Active voltage feedback control for hybrid multi terminal HVDC system adopting improved synchronverters," IEEE Transactions on Power Delivery, Vol. xx, No. x, Jan. 20xx.
17. George C Konstantopoulos, Qing Chang Zhong, Beibei Ren and Miroslav Krstic, "Boundedness of synchronverters,"in Proceedings of the ECC, pp. 1050–1055, 2015.
18. Zhou Wei, Chen Jie and Gong Chunying, "Small signal modeling and analysis of synchronverters," IEEE. 2015.
19. P Kundur, "Power system stability and control," McGraw Hill-New York, 1994.

A Self-Sustained Solar Power for Energy-Efficient- and Power-Quality Improvement in Grid Connected System

S. Poongothai, S. Srinath and S. Joyal Isac

Abstract Renewable Energy based electricity is a requisite to convene power scarcity. Solar PV can support the countries to get unpolluted, safe, trustworthy and inexpensive energy. In the modern hi-tech world have directed to custom power electronics-based switching devices for all residential loads like LED lamps, BLDC fans and charging devices comes into the existence and hence power quality problems like harmonics become visible. The elimination of the above stated problem can be made incredible to an extend by using a reliable and sustainable off grid solar dc system for dc loads. The cost of inverter and custom power devices used can be obviated, leading to a good economy. This paper attempts to illustrate the benefit of a standalone solar dc power for a multi-stored buildings, low power residential loads and educational institute in order to economize cost and improve the quality of power and energy saving. A new transformerless buck boost converter with simple structure is proposed in this paper.

Keywords Distributed generation · Power quality · Total harmonic distortion
Light emitting diode · Compact fluorescent lamp · Balance of system

1 Introduction

Energy technology is the backbone of modern civilization and national economy. The speedy growth of renewable energy technologies can help countries to avoid the severe electricity shortage in near future. For customer benefit of back up generation and improving safe and reliable power supply Distributed Generation PV is used possibly at lower cost. Installation cost is substantial and running cost is negligible due to the availability of abundance sunlight free of cost [1].

S. Poongothai (✉) · S. Srinath
Velammal Engineering College, Chennai, India
e-mail: kothaisudhan@gmail.com

S. Joyal Isac
Saveetha Engineering College, Chennai, India

© Springer Nature Singapore Pte Ltd. 2019
M. A. Bhaskar et al. (eds.), *International Conference on Intelligent Computing
and Applications*, Advances in Intelligent Systems and Computing 846,
https://doi.org/10.1007/978-981-13-2182-5_26

Now a days there are various types nonlinear loads work on low voltage dc supply like personal computer, laptop charger, LED lamps, BLDC fans etc. introduces harmonics into the system. These new loads become advanced and increasing in number in the past few years. However momentous number of such loads will have a harmful influences yields harmonic in electrical grid [2, 3].

AC of course won out over DC as the power distribution of choice, mainly because of the ability to have large generators in a central location and then transmit the power efficiently over power lines to homes and businesses. DC would have required local generators on every street or even every home, which is possible now with the help of off grid solar PV.

With the company of inverter, the solar PV is integrated with grid. This switching device directs harmonics and voltage flicker in distribution system which in turn grid. Extensive performance of grid integrated PV systems shows a astonishing deprivation of efficiency due to the variant of source and performance of inverter [4]. This intermittent solar energy with power electronic converters and nonlinear loads, make the PV system is unsteady is in terms of grid integration.

2 Standalone Solar System

Drastic price reduction of solar PV system is optimal for many remote and rural areas are far from the power grid. Many photo voltaic systems operate in standalone mode and it is suitable for low power DC applications. Apart from PV panels and loads, the Balance Of System (BOS) enables the PV generated electricity to be properly applied to the load. It adjusts and converts the electricity to the proper magnitude suitable for loads. The BOS can also include storage devices, such as batteries as well as charge controller to controls the charges in battery. So, PV-generated electricity can be used during cloudy days or at night.

A MPPT is used for extracting the maximum power from the solar PV module and transferring that power to the load. A dc/dc converter (step up/step down) serves the purpose of transferring maximum power from the solar PV module to the load. A dc/dc converter acts as an interface between the load and the module. To get a squared voltage gain and positive output a transformerless buck boost converter is used. Apart from normal buck boost converter, an extra inductor capacitor diode and switch is used to get a wider range of positive output. Compared to other buck boost converter the proposed transformerless buck boost converter has the best performance [5]. The proposed Buck-Boost converters are used in situations where array voltages and battery voltages are nearly matched. This converter provides to have either higher or lower output voltages compared to input voltages. The basic operating principles and analysis has been done. By changing the duty cycle the load impedance as seen by the source is varied and matched at the point of the peak power with the source so as to transfer the maximum power [6].

3 Low Power Residential Loads

When low power residential loads are operated with AC supply, harmonic distortion occurs on the line which in turn to other loads and grid. With a view to decrease this effect, the use of standalone DG to supply low power residential dc loads is being researched.

Simulation is carried out using MATLAB/Simulink with few residential low power DC appliances with solar photovoltaic as input. The simulation model shown in Fig. 1 includes low power devices used in the present scenario to provide significant amount of energy savings in comparison with the traditional solution with a cost compromise in the view of power quality. Two loads via a LED lamp and a brushless DC motor are used with the simulation results.

The simulation results for the proposed standalone DC system have been presented in Figs. 2 and 3. It can be seen that the system works better than with AC mainly from the aspect of injecting nonlinear behavior in voltage and current into the system.

4 Exterior Lightning—High-Rise Buildings

An electrical engineer progress an attractive and energy-efficient approach to lighting design so as to economize the cost. In order to verify the effectiveness, energy, and cost savings of LED an analysis has been done with 50 lamps for 12 h over 25 years duration supporting common area lighting in multi-leveled apartment (Table 1).

The analysis shows the significant effect of energy and cost savings in exterior electrical lighting system with LED supported by grid connection. Approximately the energy and cost can be saved is 32,850 kWh and Rs. 2,30,000 with LED lightings.

In the LED, current is sensitive to the input voltage and luminance is proportional to the current. Hence current should not vary too much with the input voltage, which is influenced by the ambient temperature. Therefore, constant current is the preferred choice to drive the LED. Most of the energy-efficient lighting devices are based on power electronics. LED is a low-power electronic load which pollutes harmonics in the system. When large amounts such appliances operate in power distribution systems, the collective effect on the feeder power quality has become a large concern to utilities. The presence of harmonics in electrical systems means that current and voltage are distorted and deviate from sinusoidal waveforms [4, 5] which will introduce harmonics to other loads in turn to Grid.

Customer can use the power at affordable rate by the applicable of illumination control of LED. To avoid the electricity wastage the intensity control of LED is relevant for different purpose. Power quality analyzer has been used for

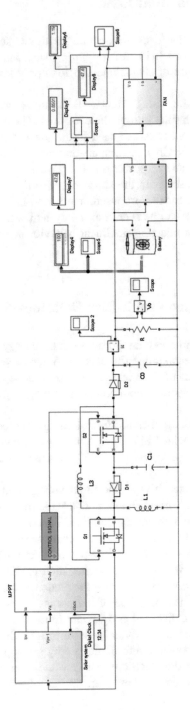

Fig. 1 Simulation model of the proposed system

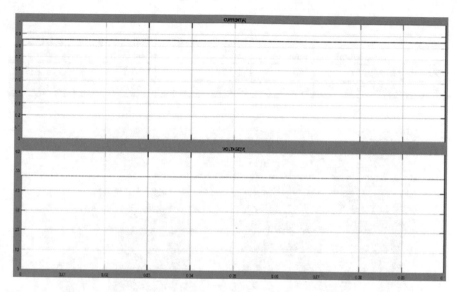

Fig. 2 Output waveform of LED

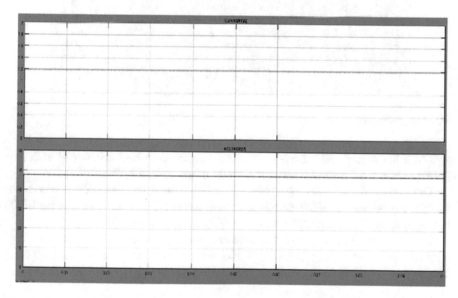

Fig. 3 Output waveform of BLDC motor

measurement of THD and distorted current of LED with different level of illumination shown in Fig. 4.

Application of illumination control in LED is to avoid power wastage. However, the following results show that the power quality issues (increase in distortion) in

Table 1 Cost analysis

Lamps	Lumen/Wattage/lamp	Total Wattage	Units Consumed (kWh)	Total Lamp cost (Rs)	Total Cost (Rs)
CFL	825/15 W	750 W	82,125	1,10,500	5,21,125
LED	825/9 W	450 W	49,275	45,000	2,91,375

Fig. 4 Experimental setup for LED with illumination control

aspect of harmonics and the power conversion losses can be incorporated with the presence of transformer and rectifier circuits (Figs. 5, 6, 7,8, 9, and 10).

The self-sustained solar dc system is suitable to drive the LED because its functions independent of the electric grid so as to provide constant voltage for LED lightings as well as there will not be a quality issues (Table 2).

The PV installation cost for standalone solar powered LED common area lighting for an apartment is Rs. 2,27,500 with government incentive of 30% it become Rs 1,72,000 for 25 years approximately (Fig. 11).

5 Educational Institute

The quality of power plays a vital role with the increase in number of equipment that is sensitive to distortions or dips in supply voltages. LED is a low-power electronic dc load which pollutes harmonics in the system. The biggest concern is

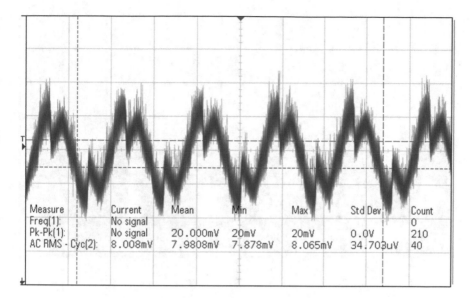

Fig. 5 Distortion in current with maximum intensity

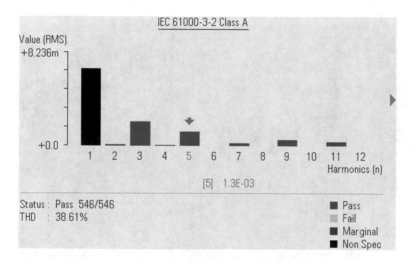

Fig. 6 Current harmonics with maximum intensity

the effect on the feeder power quality when more number of appliances are operated in the same network. Each LED reduces 230 V ac into 12 V dc. With this conversion and reduction, energy loss, and heat loss take place as well as life of lamp get reduced. Two times of energy conversion take place in solar inverter system

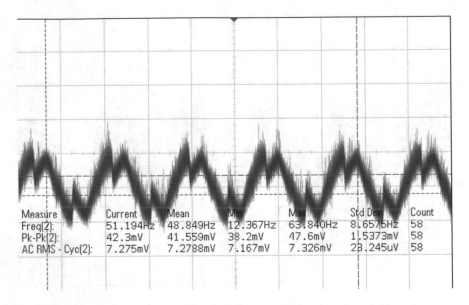

Measure	Current	Mean	Min	Max	Std Dev	Count
Freq(2):	51.194Hz	48.849Hz	12.367Hz	63.840Hz	8.6575Hz	58
Pk-Pk(2):	42.3mV	41.559mV	38.2mV	47.6mV	1.5373mV	58
AC RMS - Cyc(2):	7.275mV	7.2788mV	7.167mV	7.326mV	23.245uV	58

Fig. 7 Distortion in current with medium intensity

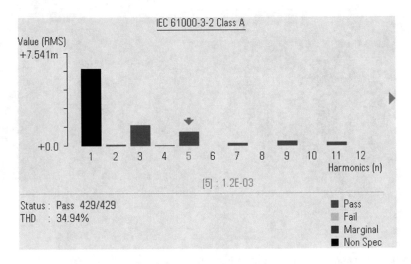

Fig. 8 Current harmonics with medium intensity

with AC loads. Due to this around 40–45% energy is getting wasted. Noiseless BLDC fans also injects harmonic into other loads which in turn to grid. To overcome the quality issues with LED lighting and BLDC fan as well as saving electricity and cost, a standalone inverter-less solar dc power suggested for fan and light loads in electrical laboratory.

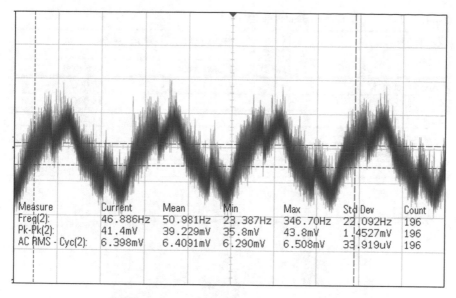

Fig. 9 Distortion in current with low intensity

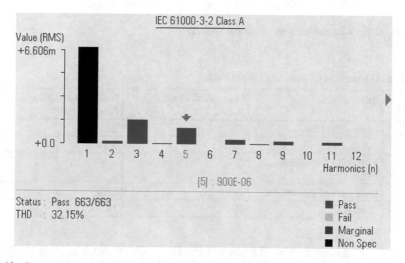

Fig. 10 Current harmonics with minimum intensity

In electrical laboratory 60 W fan and 36 W fluorescent lamp can be replaced with 32 W BLDC fan and 18 W LED lamp for solar dc system. The following calculation has been done for 7 h usage over 7 months with full load and for 5 months with 25% of load (Table 3).

Table 2 Total cost analysis—standalone PV system

Components	Common area lighting—50 lamps, 12 h, 25 years	
	No of time replacement	Price (Rs)
Solar panel (1 kW)	–	60,000
Converter (850 VA)	2	15,000
Battery (562.5 Ah)	2	97,500
Miscellaneous	–	10,000
Total lamp cost	5	45,000
Total cost	–	227,500

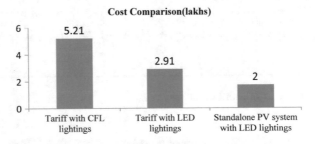

Fig. 11 Current harmonics with minimum intensity

Table 3 Comparison of costs of different loads

S. No	Requirements	Cost in AC system (Lakhs)	Cost in solar DC system (Lakhs)
1	Lamp (230 Nos)	5.5	2.3
2	Fan (85 Nos)	1.53	3.4
4	Electricity Bill with Rs 5/ unit	28	–
5	Solar system accessories	–	12.3
6	Total cost	35	18

Energy consumption over 25 years is 560,000 units whereas in solar dc system 280,000 units. As well as total cost of solar dc system is half of AC system supported by grid approximately.

6 Conclusion

An empirical cost analysis and power quality issues has been done with grid power supply and Standalone PV system. In the absence of transformer and rectifier circuits the power conversion losses can be greatly reduced. This has been inferred from the result. When compared with the traditional source, the cost of production of electricity by the off grid solar dc system is measurable and accountable. This system is easy to supply continuous and alternate emergency power if the grid power is knocked off or other disaster.

References

1. M. Ganga Prasanna, S. Mohammed Sameer, and G. Hemavathi, "Financial Analysis of solar Photovoltaic Power Plant in India." IOSR Journal and Finance., AP, India, pp. 9–15.
2. Sergey Yanchenko and Jan Meyer, "Harmonic Emission of household devices in presence of typical voltage distortion", pp. 1–6, IEEE, 29 June to 1 July 2015.
3. Lauri Kutt, Eero Saarijiarvi, Matti Lehtonen, Heigo Molder and Toomas Vinnal, "Harmonic load of residential distribution network – case study monitoring result", IEEE, pp. 93–98, 2014.
4. S.K. Khadem, M. Basu, M.F. Conlon, "Power Quality in Grid connected Renewable Energy Systems: Role of Custom Power Devices," International Conference on Renewable Energies and Power Quality, Granada (Spain), March 2010.
5. Shan Miao, Faqiang Wang and Xikui Ma, "A new transformerless Buck-Boost converter with positive output voltage," pp. 2965–2975, IEEE Transactions on Industrial Electronics, Volume: 63, Issue: 5, May 2016.
6. M.H. Taghvace, M.H.A. Radzi, S.M. Moosavaca, Hashim Hizam and M. Hamiruce Marhaban, "A Current and Future Study on non-isolated DC-DC converters for Photovoltaic applications." pp. 216–227, Renewable and Sustainable Energy Reviews, Volume 17, January 2013.

Use of Textile Effluent Treatment Plant Sludge as Sustainable Material in Brick Manufacturing

V. T. Ashok Kumar, P. T. Ravichandran and P. R. Kannan Rajkumar

Abstract The objective of this study is to use the unused of textile Effluent Treatment Plant (ETP) sludge in bricks. The physical and chemical and other engineering properties of the ETP sludge from Tirupur are studied. The tests were conducted as per Bureau of Indian Standards (BIS) specification to study the suitability of the ETP sludge for the use of bricks as a structural and non-structural application. The ETP sludge mixed with varying percentages of Flyash failed to meet the compressive strength requirements. The sludge mixed without varying the flyash requirement has met the requirement for the use of bricks. The strength and other durability properties of the bricks met the standards as prescribed by the Indian Standards. It is concluded that the use ETP sludge with quarry dust, flyash, lime and gypsum is possible to manufacture the bricks used for the non-structural applications. In particular, the 10% sludge mixed specimens with quarry dust 30%, flyash 40% after 28 days curing were found to have higher strength of 8.5% when compared with the companion specimens. Through the results obtained in this study, it has been found out that the mixing of ETP Sludge in bricks is beneficial and that it is one of the safest and efficient ways of disposing of the sludge

Keywords ETP sludge · Flyash · Quarry dust

1 Introduction

In the present scenario the entire globe has felt the increased awareness of the damage caused to the environment by the use of chemicals, some are carcinogenic and highly poisonous. The chemical waste produced by the textile industry pollutes the landfills and rivers since it demands a huge amount of water and chemicals [1]. The Tirupur city also named as textile city is polluted by water, land and air due to

V. T. Ashok Kumar · P. T. Ravichandran (✉) · P. R. Kannan Rajkumar
Department of Civil Engineering, Faculty of Engineering and Technology,
SRM Institute of Science and Technology, Kattankulathur 603203, Tamil Nadu, India
e-mail: ptrsrm@gmail.com

© Springer Nature Singapore Pte Ltd. 2019
M. A. Bhaskar et al. (eds.), *International Conference on Intelligent Computing and Applications*, Advances in Intelligent Systems and Computing 846,
https://doi.org/10.1007/978-981-13-2182-5_27

the industrial waste generated by the textile industries. The records with Tamilnadu state pollution control board (TNPCB) states that there are 830 textile units are engaged in Tirupur city. Sludge is generated through eight common effluent treatment plants (CETPs) established by these industries. The disposal of sludge in the non engineered landfills affects the ground water as well as the soil was found to be chemically polluted. Hence the disposal of sludge is a major problem in Tirupur city. The river Noyyal emerging from Velliangiri hills flowing through the Tirupur city is completely polluted affecting the irrigation around 16,000 acres of land [2]. The drinking water for the city is obtained from the river Cauvery from Erode district which is flowing 100 km away from the Tirupur City. Hence dumping of sludge in landfills creates an environmental problem to the city. Hence, sludge disposal is a never-ending process for any city. The disposal of sludge into the seawater has been banned by many countries. Filling the barren land with sludge waste is one of the common method of disposal of sludge. Incinerating the sludge waste is another solution followed by dewatering process is carried out by some countries. The land disposal of sludge is the cheapest way of disposing if it enables, then crops have to be grown on poor land for agricultural purposes. Some of these methods pose a different threat to environment. The sludge disposal problems will intensify when the sludge production increases. Landfilling these sludges will be a major problem. Hence, there is an urgent need to find the alternative way of disposing the sludge. The presence of high calcium and magnesium in ETP sludge indicates the potential use of this material in construction industry. The exact composition of these materials varies based on the industry or the placed from which waste water is originating and also based on the type of additives that are mixed with the sludge [3]. The depleting natural resources such as mud and clay in the lakes used for the preparation of bricks are demanding for the new building material. The sustainability of this product is difficult due to the possibility of leaching harmful constituents for prolonged period of usage, aesthetics of the building material, energy consumption during production, etc., [4]. This paper primarily focusses the issues of the Tirupur city attempting to use the ETP sludge in the manufacturing of bricks which is a dominant construction material in the country.

2 Materials and Methods

The various materials that were used in this investigation and their properties are explained below. The investigations were carried out in three series, each involving six different mixes of materials, thus having 18 different mixes.

2.1 ETP Sludge

The Effluent treatment plant (ETP) sludge collected from the sludge drying beds and landfilling areas from Tirupur town is used in this study. The moisture content in the sludge was 20–30% and was dried in a hot air oven for 24 h at 100°C. The sludge retained on a 90 μm sieve and passing through a 2.36 mm sieve was used in this study. The sludge passing through the 90 μm sieve was not used due to the handling problems of smaller size and lower density. The sludge has been employed as a part of the materials in different percentages according to the series of mixes being investigated. The sludge was mixed in three series. The sludge is varied from 0 to 30% at an increment of 5% per mix and the fly ash reduces in the same proportion in the first series. The second series shows the increment of ETP Sludge at the same rate but the decrement of material occurs in Quarry Dust. The third series has an increment rate of 10% per mix for ETP Sludge while the decrement per mix is shared equally by fly ash and quarry dust.

2.2 Quarry Dust

Quarry dust is a byproduct of the crushing process of the quarry stones which in turn used as aggregates for concreting works. During the process of quarrying the bouldered rock is crushed into various sizes and during this process dust is generated and this dust is generally termed as quarry dust. This dust material results in air pollution and it is a useless material. Hence, this material can be used in concrete which is also economical specially when it is used as a fine aggregate hence natural resources will be saved. Hence quarry dust could be used in concreting which will reduce the construction cost thereby depletion of natural resources could be reduced. Many developing countries use quarry dust to replace fine aggregate in concrete to some extent as an alternative material. The quarry dust of size 0–4.75 mm with a specific gravity of 2.63 is used in this study.

2.3 Flyash

Fly ash is a finely divided fuel dust that is usually obtained from the pulverization of coal. During combustion of coal, the mineral impurities of coal are carried away by the exhaust gases and once these impurities cool down, they solidify and form a material called Fly Ash. This material has often proven to be a proper substitute for soil in bricks and for fine aggregate in case of concrete. The material used has a specific gravity of 2.26. The size of a typical fly ash particle is usually less than 20 μm, although the sizes vary from less than 1 μm to more than 100 μm [5]. ASTM TYPE C Flyash was used for the mix purposes. [6]

Flyash is generated from electric power generating plants. Fly ash is a byproduct from burning pulverized coal in electric power generating plants. This product is collected from the bag filters of the power generating plants. This product resembles to a Portland cement but its chemical properties are entirely different. Flyash reacts with cement and water to provide additional cementitious products that improve many desirable properties in concrete. Various types (Class) of flyash are produced by the different power generating plants depending upon the physical and chemical properties of flyash. ASTM TYPE C Flyash was used for the mix purposes and the size is varying from 1 μm to more than 100 μm [5, 6].

2.4 Lime

Lime is one other material which has been used for the investigations. Lime is a material that has been employed in construction since more than 6000 years. Ancient Egyptians were among the first to use Lime as a binding material in the Giza Pyramid. The durability and binding properties of this material are the reason for its extensive use as a building material. The percentage of lime that has been used in all the mixes for the investigations is 18% [7–9].

2.5 Gypsum

Gypsum is a sedimentary mineral. It is found in layers that were formed under salt water millions of years ago. The water evaporated and left the mineral. Gypsum is composed of calcium sulphate ($CaSO_4$) and water (H_2O), i.e. Calcium Sulphate Dihydrate ($CaSO_4.2H_2O$). The specific gravity of Gypsum is found to be 2.3. The material is widely used in the manufacture of cement. However, the low hardness of the material limits the durability. Gypsum has been mixed with the other materials at a rate of 2% in all the mixes. Chemical composition of the materials used in the preparation of brick specimens given in Table 1.

Table 1 Chemical composition of materials	Description	Values (%)			
		Quarry dust	Fly ash	ETP	Gypsum
	SiO_2	62.48	65.29	17.47	3.8
	Al_2O_3	18.72	23.34	2.33	NIL
	Fe_2O_3	6.54	5.93	10.81	NIL
	SO_3	–	0.6	0.19	44.56
	MgO	2.56	1.05	4.56	NIL
	CaO	4.83	1.06	16.46	32.27
	LOI	0.48	1.22	46.08	16.5

3   Methodology

The various bricks required for the investigation were cast using rectangular moulds. The various mixes of materials were prepared according to the series to which they belong. The tests were carried on all these prepared specimens and the obtained results were tabulated. The method of manufacture of these bricks is explained below. Initially, a rectangular mould of size 230 × 110 × 80 mm was prepared. Then, the various materials, i.e. Fly Ash, Lime, Gypsum, ETP Sludge and Quarry Dust to be used for preparing the required mixes were prepared according to the proportions in which they are required for mixes. These materials are mixed properly and compacted in the mould in three layers. Each layer is sufficiently compacted by tamping bars. Finally, a trowel was used in order to provide a smooth finish to the moulds. The specimen was demoulded after 24 h and kept for curing. The dimensions of the bricks thus obtained are the same as the mould size, i.e. 230 × 110 × 80 mm. After proper curing of the bricks, various tests are performed to obtain the properties of the bricks; the effectiveness of these materials in bricks is thus concluded. The various tests performed on the bricks are compressive strength test and water absorption test. The various mixes prepared for the investigations are explained in the Tables 2, 3, and 4. Three series of mixes are prepared, each containing six different mixes, with a total of 16 different mix proportions of materials for the investigations.

Table 2 Details of variation of ETP sludge and fly ash in brick manufacturing-(S1)

Mix designation	ETP sludge (%)	Quarry Dust (%)	Fly ash (%)	Lime (%)	Gypsum (%)
M0	–	40	40	18	2
M1	10	40	30	18	2
M2	15	40	25	18	2
M3	20	40	20	18	2
M4	25	40	15	18	2
M5	30	40	10	18	2

Table 3 Details of variation of ETP sludge and quarry dust in brick manufacturing-(S2)

Mix designation	ETP sludge (%)	Quarry dust (%)	Fly ash (%)	Lime (%)	Gypsum (%)
M0	–	40	40	18	2
M6	5	35	40	18	2
M7	10	30	40	18	2
M8	15	25	40	18	2
M9	20	20	40	18	2
M10	25	15	40	18	2

Table 4 Details of variation of ETP sludge, fly ash and quarry dust in brick manufacturing-(S3)

Mix designation	ETP sludge (%)	Quarry dust (%)	Fly ash (%)	Lime (%)	Gypsum (%)
M0	–	40	40	18	2
M11	10	35	35	18	2
M12	20	30	30	18	2
M13	30	25	25	18	2
M14	40	20	20	18	2
M15	50	15	15	18	2

4 Results and Discussions

Compressive Strength and Water Absorption tests were performed on the sludge prepared bricks and the obtained results were used in order to draw conclusions. The procedures and the results are discussed.

4.1 Compressive Strength Test

The bricks of dimensions 230 × 110 × 80 mm were prepared for this test, and the bricks were subject to gradual compressive loads until the specimen fails. The test is conducted 28 days after the moulding of the brick specimen. The bricks of various mix proportions were tested and their final ultimate load values were tabulated as per IS 3495 – Part – I. [10]. From these tabulations, and using the dimensions of the bricks, the compressive strength values for various proportion bricks were obtained. The obtained values for this test are given in Table 5. According to the experimental results, the brick moulded with mix M6 gives the maximum compressive strength among all the mixes with ETP Sludge added, the proportion of materials in M7 is 5% ETP Sludge, 35% Quarry Dust, 40% Fly Ash, 18% Lime, 2% Gypsum. The results of the compressive test are also divided according to the series of the mix. Similar results were obtained by Raghunathan [11] and Balasubramanian [12] when the sludge was used in concrete.

4.2 Water Absorption Test

The water absorption test is performed in order to find out the difference of absorption between conventional bricks and the specially prepared sludge bricks. In this test, the brick specimen is dried at 115–120 °C until a constant weight occurs, and this weight is noted. After a constant weight is achieved, we then immerse the brick in water and let it soak till 24 h. After the stipulated time, the brick is removed

Table 5 Values of compressive strength and water absorption in brick

Mix ID	Density (kg/m³)	Compressive strength (N/mm²)	Water absorption (%)	Comp strength after sulphate intrusion (N/mm²)
M0	1626	5.1	13.1	4.8
M1	1648.2	3.9	12.3	3.6
M2	1655.13	3.39	12.9	3.2
M3	1631.66	2.44	14.2	2.2
M4	1639.57	3.155	14.2	2.9
M5	1630.18	2.675	16.3	2.3
M6	1511.36	4.24	12.7	4.1
M7	1618.82	4.065	12.3	3.8
M8	1559.04	3.85	12.2	3.7
M9	1568.42	3.01	17.7	2.85
M10	1376.73	2.265	17.9	2.0
M11	1368.57	3.5	12.6	3.2
M12	1472.82	2.95	16.2	2.65
M13	1516.55	2.7	17.2	2.6
M14	1496.04	2.55	17.3	2.4
M15	1666.75	2.0	17.8	1.82

from the water and weighed. Now, the difference in weight before immersion and after immersion gives us the absorption rate of that brick. The procedure is performed as given in IS 3495 – Part – II [13]. The absorption content is checked for safety, if the percentage is within limit, then the brick is safe for use. If the water absorption is less than 15%, the quality of the brick is high class, if the absorption is less than 20%, the brick is of medium quality. The results for the water absorption test are tabulated in Table 4. From the obtained results, it can be inferred that the brick mix ratio M8 provides the least water absorbent brick with an absorption value of 12.2%. The mix for this M8 brick contains 15% ETP Sludge, 25% Quarry Dust, 40% Fly Ash, 18% Lime and 2% Gypsum. The brick has the least water absorption value among all the mixes. The results of the water absorption test are also divided according to the series of the mix. The results obtained in this test are in line with those obtained by Baskar [14] where the bricks were subjected to elevated temperatures.

4.3 Sulphate Intrusion

Curing of the ETP Sludge bricks was done in a harsh sulphate environment in order to find out the durability of these bricks. After the curing is done, compressive test was performed on these bricks and the results are given in Table 4. For a time period of one week, the bricks were covered in wet gunny bags after being removed from their moulds. After this week, the bricks have achieved a strength that is sufficient for handling, and then, these bricks, instead of being transferred to water, they were transferred to curing tanks filled with sulphate solution. The sulphate solution was prepared by mixing 14.79 g of Na_2SO_4 in 1 L of water in the laboratory and this solution has a sulphate concentration equal to 10,000 PPM [15].

5   Conclusion

The results of the investigations thus performed have been studied and the characteristic properties of these sludge bricks have been obtained. These conclusions have been mentioned below:

1. The investigation for the compressive strength of the bricks resulted in a mix that produced a compressive strength of more than 4 N/mm^2. The M6 mix designated brick tested at 4.24 N/mm^2, which is considerably larger than the average 3.5 N/mm^2 that a conventional brick has.
2. The water absorption test upon the specimens shows us that the absorption value is less than 20% for all the specimens except three. The results thus obtained are promising, because a brick with absorption of less than 20% is deemed as suitable for construction.
3. The bricks with the mix with ETP Sludge 15%, quarry dust 25%, flyash 40%, lime 18% and Gypsum 2% has the least water absorption with 12.2% graded in the category of high class bricks.
4. From the results, it can be concluded that there are six mixes with a compressive strength that is greater than or equal to 3.5 N/mm^2. All these bricks hence have a compressive strength that is greater than the compressive strength of a conventional brick, and can, hence, be used for construction purposes.
5. It can be inferred from the obtained test results that Sulphate intrusion into brick mixes has reduced the compressive strength of that brick mix slightly.
6. The conclusions hence show that ETP Sludge and Fly Ash are applicable in brick compositions in specific mixes as they improved the compressive strength and reduced the absorption of water. Application of these in bricks helps reduce the amount of construction material used and also helps in reducing the amount of waste material that needs to be disposed of. The leachability and economic studies are to be studied for the next step of research.

References

1. S.S Muthu (eds.).: Environmental Implications of Recycling and Recycled Products, Environmental Footprints and Eco-design of Products and Processes. https://doi.org/10.1007/978-981-287-643-0_9 (2015).
2. Thomson Jacob. C., Azariah., J. and Viji A.G.R.: Impact of textile industries on river Noyyal and riverine groundwater quality in Tirupur, India. Environment Monitoring Assess, Vol. 18, pp. 359–368 (1999).
3. Chen Maozhe., Blanc Denise., Gautier Mathieu., Mehu Jacques., Gordon Remy.: Environmental and Technical Assessments of the Potential Utilization of Sewage Sludge Ashes (SSAs) as Secondary Raw Materials in Construction. Waste Management, vol. 33, pp. 1268–1275 (2013).
4. Ariful Islam Juel., Al Mizan an Tanvir Ahmed.: Sustainable use of tannery sludge in brick manufacturing in Bangladesh. vol. 60, pp. 259–269. http://dx.doi.org/10.1016/j.wasman.2016/12/041.
5. Vijaya Kishore K.: Utilisation of Sludge Concrete in Paver Blocks. International Journal of Emerging Trends in Engineering and Development, vol. 4, pp. 509–516 (2012).
6. ASTM C 469 – 10, "Standard Specification for the Coal Fly Ash and Raw or Calcined Natural Pozzolan for Use in Concrete" American Society for Testing and Materials, pp. 1–4.
7. IS: 712. Specification for building limes. New Delhi: Bureau of Indian Standards; 1973.
8. IS: 1514. Methods of sampling and test for quick lime and hydrated lime. New Delhi: Bureau of Indian Standards; 1959.
9. IS: 6932. Methods of Test for building limes. New Delhi: Bureau of Indian Standards; 1973.
10. IS: 3495-(Part 1). Methods of tests of burnt clay building bricks- Determination of compression strength. New Delhi: Bureau of Indian Standards; 1992.
11. Raghunathan T., Gopalasamy. P., Elangovan. R.: Study of strength of Concrete with ETP Sludge from Dyeing Industry. International Journal of Civil and Structural Engineering, vol. 1, pp. 379–389 (2010).
12. Balasubramanian J., Sabumon P. C., John U. Lazar., Ilangovan. R.: Reuse of Textile Effluent Treatment Plant Sludge in Building Material. Waste Management, Vol. 26, pp. 22–28 (2006).
13. IS: 3495-(Part 2). Methods of tests of burnt clay building bricks- Determination of water absorption. New Delhi, Bureau of Indian Standards; 1992.
14. Baskar R, Meera Sherrifa Begum. K.M and Sundaram S "Characterization and Reuse of Textile Effluent Treatment Plant Waste Sludge in Clay Bricks", Journal of the University of Chemical Technology and Metallurgy, Vol. 4, 2006, pp. 473–478.
15. Kannan Rajkumar P.R, Divya Krishnan K, Sudha C, Ravichandran P.T and Vigneshwaran T.D, "Study on Use of Industrial Waste in Preparation of Green Bricks", Indian Journal of Science and Technology, Vol. 9, February 2016, pp. 1–6.

Brain and Pancreatic Tumor Classification Based on GLCM—k-NN Approaches

D. Jithendra Reddy, T. Arun Prasath, M. Pallikonda Rajasekaran and G. Vishnuvarthanan

Abstract Diagnosis diseases at an untimely phase are a challenging task due to the lack of unfitted segmentation process. This paper focuses on developing an automatic recognition of the brain tumor and pancreatic cancer with precised segmentation and classification process. The proposed k-NN classifier composes of the three stages, namely, (a) Median filtering model for image preprocessing (b) Fuzzy C-segmentation model for accurate segmented image and (c) Gray Level Co-occurrence Matrix (GLCM) for selecting relevant features. The refined features are then given as input to k-NN classifier. The determination of k value clearly emancipates the classes. The proposed classifier tests on the images from the Harvard Medical School database and the Cancer Imaging Archive repositories. Experimental computation is done using metrics like precision, accuracy, specificity, and recall. The results have proved that our proposed classifier outperforms better than prior classifiers SVM, Naïve Bayes, and Probability Neural Network.

Keywords MFM · GLCM · MRI images · CT images · Improved K-NN classifier Fuzzy C means

D. J. Reddy (✉) · T. Arun Prasath · M. Pallikonda Rajasekaran · G. Vishnuvarthanan
Department of Instrumentation and Control Engineering, Kalasalingam Academy of Research and Education, Virudhunagar 626126, Tamil Nadu, India
e-mail: jithendrareddy.d@gmail.com

T. Arun Prasath
e-mail: arun.aklu@gmail.com

M. Pallikonda Rajasekaran
e-mail: m.p.raja@klu.ac.in

G. Vishnuvarthanan
e-mail: gvvarthanan@gmail.com

© Springer Nature Singapore Pte Ltd. 2019
M. A. Bhaskar et al. (eds.), *International Conference on Intelligent Computing and Applications*, Advances in Intelligent Systems and Computing 846,
https://doi.org/10.1007/978-981-13-2182-5_28

1   Introduction

Intelligent classification is a vital task in the medical images that imposes different modalities in anatomical structures. There has been a great opportunities on the medical imaging field by the advancements in computer technology which enhances the task of data acquisition, analysis, dispensation, and revelation. The necessity of the digital image processing on medical images is of two reasons, one is to enhance the pictorial information for data interpretation and the second is to process the relevant segment for an autonomous machine perception. Research onto the medical imaging application and interpretation preserves the radiology and medical condition leads to the investigation area. Most of the introduced techniques are widely adopted for scientific and industrial applications [1]. A few cells develop in an unusual way which all things considered frames a mass of tissue known as 'tumor or cancer'. As of late, a large portion of the general population experiences the ill- effects of the brain and pancreatic cancer. Brain Tumor identification [2] is hard in starting stage since it can't locate the precise estimation of tumor. In any case, once it gets recognized the brain tumor it provides for begin the best possible treatment and it might be repairable. Pancreatic cancer is known to be the most normal reason for cancer demise because of the trouble in diagnosing pancreatic cancer at an early state and halfway because of the detachment of the pancreas and its encompassing organs. Henceforth, accomplishing a great segmentation and classification result is a fundamental objective because of impending preprocessing techniques. Manual segmentation [3] and programmed segmentation procedures are utilized for discovering tumor in medical imaging. In the recent days, automatic segmentation strategies are investigated for accomplishing exact location of tumor in medicinal pictures. At times, some of computerized segmentation techniques do not give great location result [4]. Feature extraction [5] is useful in recognizing tumor where is precisely found and aides in foreseeing the upcoming stages. This proposal addresses the problems in classifying the tumor affected region using segmentation and classification model. Though the medical image analysis has assisted the technicians to rapidly make the decision over the diagnosis system, still it lacks in.

2   Related Work

This section depicts the prior works processed in the analysis of brain and pancreatic images. In [6], Brain tumor classification adaptive segmentation technique and classification method in which self-Organizing Map is one of the most popular neural network model. This automated analysis process fails to adopt the machine learning algorithms. The author in [7] suggested improved CEEMDAN and the Hilbert transformation technique. They converted the difference signal into the variant intrinsic mode functions that captures the textural details of the brain images. The author in [8] discussed about the recognition of metastasis and micro-calcifications using segmenting the brain images. Then, Convolutional neural networks for grouping the abnormal tissues of brain images were studied by [9] Due to the accumulation of the gliomas, the brain cells are affected and thus causes brain tumor. The author in [10] studied about the glioma labeling using patch based models. Multi-atlas model was used achieving appropriate segmented region. The learning stage of the test image throws risk in deriving objectives. The author in [11] discussed about Lateral Ventricular (LaV) deformation feature extraction component for brain tumor segmentation which detects the textual features. The author in [12] studied about the uncovering of brain metastases from MRI images. To enhance the speed of the detection process, model-based methods were defined by its high degree of similarity. The author in [13] discussed about the SVM model for feature extraction process. Their method fails to predict the similar accuracy when the training set is modified. The authors in [14, 15] presented pancreatic tumor detection in CT images. They suggested image classifier models to detect the affected regions for the pancreatic images. Distance-based classifier is used for classifying the affected and non-affected regions of the pancreatic images. The author in [16] studied about ANN model that investigates the pancreatic cancer. Finally, the classification is done by the neural networks. This system aided the technicians to make better decision process.

3 Proposed Methodology

This section describes the intelligent segmentation and classification model for tumor analysis in brain and pancreatic images. The prime aim of the proposed methodology is to effectively segment and classify the tumor cells of brain and pancreatic medical images. The proposed model is processed as follows:

Step 1: The Fig. 1 shows medical images are acquired from the Harvard Medical School database and the Cancer Imaging Archive repositories. Step 2: The acquired images are preprocessed using Median Filtering Model (MFM) that preserves the

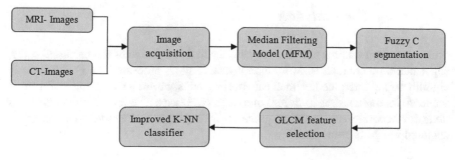

Fig. 1 Proposed method

edge information of the image. Step 3: Feature selection is a significant step of the proposed study. Gray Level Co-Occurrence Matrix (GLCM) adopts to accurately select the features for better classification accuracy. Step 4: With the outputs obtained from the GLCM approach, fuzzy C-segmentation model adopts for segmenting the appropriate region. It has been to facilitate the better computational time. Step 5: The segmented region is further given to improved k-NN classifier and studied the metrics like precision, recall, accuracy, and specificity. Step 6; Finally, it is compared with existing classifiers, SVM, Naïve Bayes, and PNN to prove the better performance of the proposed classifier. Image Acquisition is the first step of the proposed model. The brain MRI images are collected from the Harvard Medical School (HMS) database and the pancreatic CT images from the Cancer Imaging Archive repositories. These two repositories contain variants of Brain MRI and Pancreatic CT images that enhance effectiveness of the research study.

3.1 Image Preprocessing

It is the second step which eliminates the irrelevant details like noise, uncertainties of the collected images. The reason behind the delayed decision in diagnosis system is due to improper preprocessing techniques. Most of the medical images are prone to salt and pepper noise, i.e., possesses dark pixels in bright regions and bright pixels in dark region. This noise greatly affects the visualization of the image. Thus, median filtering model is used for preserving the sharpness of the image. This model computes the average of the pixel based on its current neighboring pixels. By doing so, the outliers of an image are removed without compromising the acuteness of an image. It is given as

$$y(a, b) = \text{median}\{x[i,j], (i,j) \in w \tag{1}$$

where, w represents the neighborhood defined by the user which is centered on the coordinates a and b in an image.

3.2 Image Segmentation

Since classification time and accuracy are the main objective of the proposed study, segmentation technique is an important task. This is processed by Fuzzy C-mean algorithm. To interpret the medical images, an appropriate segmented image is required. Its main task is to fit into one or more classes. The following are the steps done for accurate segmentation process. To begin, the objective function of an acquired image is computed (2)

$$J_m = \sum_{i=1}^{N}\sum_{j=1}^{C} u_{ij}^m ||x_i - c_j||^2 \quad 1 \le m < \infty \tag{2}$$

where, m is any real exceeds the value 1, u_{ij} is the member scale of x_i in the cluster j, x_i is the ith of d-dimensional measured data, c_j is the d-dimension center of the cluster; and $||*||$ is any norm expressing the similarity between any measured data and the center. Fuzzy partitioning is carried out as follows, Begin the variable $U = [u_{ij}]$ as matrix, $U^{(0)}$, For each vector $U^{(p)}$ in matrix, estimate mid-value $m^{(p)} = [c_j]$. It is given in Eq. (3), Revise $U^{(p)}$, $U^{(p+1)}$, If $||U^{(p+1)} - U^{(p)}|| < s$, then Stop the process else proceed the step again.

$$c_j = \frac{\sum_{i=1}^{N} u_{ij}^m \cdot x_i}{\sum_{i=1}^{N} u_{ij}^m} \tag{3}$$

$$\mu_{ij} = \frac{1}{\sum_{p=1}^{C}\left(\frac{||x_i - c_i||}{||x_i - c_p||}\right)^{\frac{2}{m-1}}} \tag{4}$$

3.3 Feature Extraction

It is the fourth step of our proposed model which assists us to extract the relevant features required for easy classification of the affected regions. It is done by Gray Level Co-occurrence matrix (GLCM) which extracts the texture features from the images. Since the extracted features are of second order, an abstract detail is obtained. The gray level information is portrayed in the rows and columns format for the specified image. The matrix element includes $P(i, j|\Delta x, \Delta y)$ which act as virtual frequency with two pixels and separated by a pixel distance $(\Delta x, \Delta y)$, occur within a given neighborhood, one with intensity 'i' and the other with intensity 'j'. The image correlation is given in (5), Angular second moment helps to preserve the homogeneity of an image, Where i and j are the spatial coordinates of the image with gray tone N_g, which is given in (6), Inverse Difference Moment measures the local homogeneity of an image which is given in (7), Entropy depicts the required information which is given in (8)

$$\text{Correlation} = \frac{\sum_{i=0}^{Ng-1}\sum_{j=0}^{Ng-1}(i,j)p(i,j) - \mu x \mu y}{\sigma x \sigma y} \tag{5}$$

$$\text{ASM} = \sum_{i=0}^{N_g-1}\sum_{j=0}^{N_g-1} P_{ij}^2 \tag{6}$$

$$\text{IDM} = \frac{\sum_{i=0}^{N_g-1} \sum_{j=0}^{N_g-1} P_{ij}^2}{1 + (i-j)^2} \tag{7}$$

$$\text{Entropy} = \sum_{i=0}^{N_g-1} \sum_{j=0}^{N_g-1} -P_{ij} * \log P_{ij} \tag{8}$$

3.4 Image Classification

It is the final step which classifies the tumor regions for the two inputs, Brain MRI images and Pancreatic CT images using proposed k-NN classifier. The procedure are described as calculate the Euclidean distance for the test images. It is given as

$$\text{Distance}_{(i,j)} = \sqrt{\sum_{i=1}^{k} (x_i - y_i)^2} \tag{9}$$

Reorder the trained images by its distance value. Then, find the no. of pixels in row and column sectors. Create the number of clusters for the training images. Then, optimal k–center value is computed. Again, compute the distance from the current pixel to the cluster center. Relied upon the cross validation, the optimal k value is found. Estimate the converse weighted average distance for the k-nearest values. Then, the process is continued for all pixels in an image.

4 Experimental Evaluation

This section reveals the experimental analysis processed in MATLAB program-ming language. Initially, the input images, Brain MRI images and Pancreatic CT images from the public repository, Harvard Medical School [17] that consists of T2-weighted MR brain images in axial plane and 256 * 256 in-plane resolutions and Cancer Imaging Archive [18] that composes of pancreatic CT images 512 * 512, shown in Fig. 2.

(a)   (b)   (c)   (d)   (e)   (f)

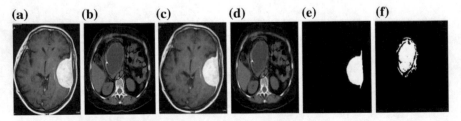

Fig. 2 a Test image of brain MRI **b** test image of pancreatic CT **c** preprocessed image of brain **d** preprocessed image of pancreatic **e** brain segmented image **f** pancreatic segmented image

Figure 2 explains proposed model (a) depicts the brain MRI images and Fig. 2b depicts the pancreatic CT images. In Fig. 2 portrays the preprocessed images using median filtering model. Thus, we have eliminated the different modalities in the images and preserved the sharpness of the image. From the Fig. 2e, f, it is inferred that the appropriate affected region of the image is segmented to achieve better classification process. The textural features are retrieved from the segmented region of brain and pancreatic images. Using GLCM approach, extract the features like, (a) Angular second moment describes the homogeneity factor of the segmented image. (b) Inverse difference moment describes the local homogeneity of the gray level images. (c) Entropy describes the amount of information required for the trained samples. (d) Correlation estimates the gray levels dependency of the adjacent pixels. (e) Contrast: It measures the pixel strength of the images (Table 1).

Using the above extracted features, the k-NN classification is preceded. In this case, we have considered 100 images which split into 60 training and 40 testing images. The nomination of k values decides the classification. The $k = 2$ which denotes that each image depicts two class labels, i.e., benign and malignant. We have analyzed all the metrics in classification systems. (a) Accuracy measures the analysis of true positive (TP) and true negative (TN) to the total no. of test images. (b) Precision estimates the analysis of true positive to the aggregate value of true positive and false positive rate. It is given in Eq. (11). (c) Recall estimates the analysis of true positive rate to the aggregate value of the true positive and false negative rate. It is given in Eq. (7). (d) Specificity measures the analysis of true negative to the aggregate value of false positive and true negative. It is given in Eq. (8)

$$Accuracy = \frac{TP + TN}{TP + TN + FP + FN} \tag{10}$$

$$Precision = \frac{(TP)}{(TP + FP)} \tag{11}$$

Table 1 GLCM features extraction values [13]

Feature extraction	Brain benign	Brain malignant	Pancreatic benign	Pancreatic malignant
Angular second moment	0.8393	0.8772	0.5587	0.8568
Inverse difference moment	0.9995	0.9993	0.9979	0.9978
Entropy	0.3104	0.2574	0.6804	0.3095
Correlation	0.9845	0.9721	0.9752	0.9187
Contrast	0.0025	0.0033	0.0107	0.0108

$$\text{Recall} = \frac{(TP)}{(TP + FN)} \tag{12}$$

$$\text{Specificity} = \frac{(TN)}{(FP + TN)} \tag{13}$$

Table 2 shows the comparative analysis of the proposed k-NN classifier with existing SVM, Naïve Bayes and PNN (Fig. 3 and 4).

Table 2 k-NN classifier with the existing classifier—comparison [1]

Classifiers	Images	Accuracy (%)	Precision (%)	Recall (%)	Specificity (%)
Proposed k-NN classifier	Brain MRI image	97.3	94.87	92.5	95
	Pancreatic CT image	94.6	92.1	87.5	90
Support vector machines (SVM)	Brain MRI image	90.3	95.1	91.6	92
	Pancreatic CT image	84.5	91.5	90	89.87
Naïve Bayes	Brain MRI image	90	94.3	90.5	91.6
	Pancreatic CT image	84	90.2	89.5	88.5
Probability neural networks (PNN)	Brain MRI image	88.87	92	89.1	90
	Pancreatic CT image	83.3	89.87	88.3	87.5

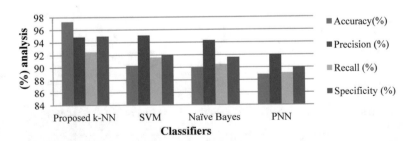

Fig. 3 Comparison analysis for the brain MRI images

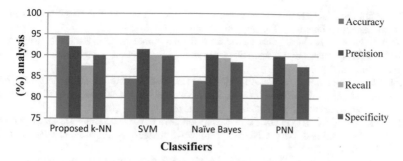

Fig. 4 Comparison analysis for the pancreatic CT images

5 Conclusion

This paper develops an intelligent classification model for an effective decision making process in medical diagnosis system. The proposed k-NN classifier accurately classifies the relevant classes, i.e., benign or malignant via better preprocessing and segmentation techniques. It has been tested on the Harvard Medical School and the Cancer Imaging Archive databases and evaluated on the Brain MRI images and Pancreatic CT images for prediction on early diagnosis. Table 3 depicts the efficiency of our proposed classifier than the prior classifiers like SVM, Naïve Bayes, and PNN. In future work, the similar classification model will be evaluated for other sorts of high-level medical images by reducing the image modalities.

Acknowledgements The author would like to thank the Kalasalingam University for providing financial help under the University Research Fellowship. We also would like to thank the Department of Electronics and Communication Engineering of Kalasalingam University, Tamil Nadu, India for permitting to use the computational facilities available in the Centre for Research in Signal Processing and VLSI Design which was setup with the support of the DST, New Delhi under FIST Program in 2013 (Reference No: SR/FST/ETI-336/2013 dated November 2013).

References

1. Sahu, Sanjib Kumar, Pankaj Kumar, and Amit Prakash Singh. "Modified K-NN algorithm for classification problems with improved accuracy." International Journal of Information Technology, pp 1–6, 2017.
2. Anila S, Sivaraju SS, Devarajan N. A new contourlet based multiresolution approximation for MRI image noise removal. National Academy Science Letters. 40(1):39–41, Feb 2017.
3. Zhang, Y., Ye, S., & Ding, W. Based on rough set and fuzzy clustering of MRI brain segmentation. International Journal of Biomathematics, 10(02), 1750026, 2017.
4. El Abbadi, Nidahl K., and Neamah E. Kadhim. "Brain Cancer classification Based on Features and Artificial Neural Network." Brain 6.1, 2017.
5. Usman, Khalid, and Kashif Rajpoot. "Brain tumor classification from multi-modality MRI using wavelets and machine learning." Pattern Analysis and Applications, pp 1–11, 2017.

6. V. Anitha, S. Murugavalli.: Brain tumor classification using two-tier classifier with adaptive segmentation technique. IET Computer Vision. 10 (1), 2016.
7. Taranjit kaur et al.: Quantitative metric for MR brain tumor grade classification using sample space density measure of analytic intrinsic mode function representation. IET image processing. 11(8), 2017.
8. Al-Rifaie, Mohammad Majid, Ahmed Aber, and Duraiswamy Jude Hemanth.: Deploying swarm intelligence in medical imaging identifying metastasis, micro-calcifications and brain image segmentation.IET systems biology. 9(6), pp 234–244, 2015.
9. Pereira, Sérgio, Adriano Pinto, Victor Alves, and Carlos A. Silva.: Brain Tumor Segmentation Using Convolutional Neural Networks in MRI Images. IEEE transactions on medical imaging. 35 (5), 2016.
10. Cordier, Nicolas, Hervé Delingette, and Nicholas Ayache: A patch-based approach for the segmentation of pathologies: Application to glioma labeling. IEEE transactions on medical imaging, 2015.
11. Jui, Shang-Ling, Shichen Zhang, Weilun Xiong, Fangxiaoqi Yu, Mingjian Fu, Dongmei Wang, Aboul Ella Hassanien, and Kai Xiao. : Brain MRI Tumor Segmentation with 3D Intracranial Structure Deformation Features. IEEE intelligent systems. 31(2). 2016.
12. Perez, Ursula, Estanislao Arana, and David Moratal.: Brain Metastases Detection Algorithms in Magnetic Resonance Imagin. IEEE Latin America Transactions, 14 (3). 2016.
13. Nanthagopal, A. Padma, and R. Sukanesh: Wavelet statistical texture features-based segmentation and classification of brain computed tomography images. IET image processing. 7(1), pp 25–32, 2013.
14. Shah, Jeenal, Sunil Surve, and Varsha Turkar.: Pancreatic Tumor Detection Using Image Processing. Procedia Computer Science (Elsevier), 2015.
15. Farag A, Lu L, Roth HR, Liu J, Turkbey E, Summers RM.: Automatic Pancreas Segmentation Using Coarse-to-Fine Super-pixel Labeling. Deep Learning and Convolutional Neural Networks for Medical Image Computing, pp 279–302, 2017.
16. Sanoob MU, Madhu A, Ajesh KR, Varghese SM: Artificial neural network for diagnosis of pancreatic Cancer: International Journal on Cybernetics & Informatics (IJCI). 5(2). 2016.
17. http://www.med.harvard.edu/aanlib/.
18. https://wiki.cancerimagingarchive.net/display/Public/Pancreas-CT.

Design of Flexible Pavement With Geosynthetic Reinforcement

Namrata Singh Solanki, Deepak Pabbisetty, Rupesh Majety,
P. T. Ravichandran and G. Priyanga

Abstract Problematic soils are of many types in which expansive clay covers vast area of our country. These types of soils are having the characteristics of low CBR value and low shear strength which leads to many problems both in super structures and sub structures. In this paper, we have mainly concentrated on pavement construction on such type of poor soils. Two types of Geosynthetic reinforcement namely, non-woven Geotextile and Geogrid has been adopted to improve the soil CBR value. The reinforcements are placed at single and multiple (2 and 3) layers from the top surface of the sample where the load is applied. This helps in determining the optimum layer combination that gives the most effective result. From the results, the rate of increase the CBR value for double layer combination is more than the triple layer and also the layers which are close to the top surface takes more load than the layers placed at deeper depth. The CBR value of soil with Geogrid is more compared to the geotextile. For the best combination adopted pavement design, quantity estimation and cost analysis has been determined for both reinforced and unreinforced soil. Using the Geosynthetic reinforcement in soil, increases the CBR value which in turn reduces the pavement thickness leading to cost benefits in the Design of Pavements.

Keywords Geogrid · Geotextile · CBR · Pavement design

1 Introduction

Pavement design is the procedure of combining the most economical and reliable pavement layers to match the soil properties and traffic volume to bring out the utmost efficiency of the pavement in its periodical life. The road construction work is aiming to get more economical and reliable pavement by reducing the periodical

N. S. Solanki · D. Pabbisetty · R. Majety · P. T. Ravichandran (✉) · G. Priyanga
Department of Civil Engineering, Faculty of Engineering and Technology,
SRM Institute of Science and Technology, Kattankulathur 603203, Tamil Nadu, India
e-mail: ptrsrm@gmail.com

© Springer Nature Singapore Pte Ltd. 2019
M. A. Bhaskar et al. (eds.), *International Conference on Intelligent Computing and Applications*, Advances in Intelligent Systems and Computing 846,
https://doi.org/10.1007/978-981-13-2182-5_29

303

maintenance and reduction in thickness of pavements. In order to reduce the thickness of pavement the soil should have more CBR value. But problematic soils have less CBR value, which is usually improved using various soil stabilization techniques. Here we have mainly concentrated using Geosynthetic reinforcements, they are used for the functions like partition, moisture barrier, cushion, drainage, confinement and reinforcement in many civil structures. The Geosynthetic are helpful in strengthening the soil characteristics and to construct the pavement in an economical way without compromising on the quality and minimization in periodical maintenance. It renders another way to meet the desire towards the greener construction by minimizing the formation of low lying areas and minimizing the excavation of earth materials, leading to cost benefits.

The service life of the pavements is affected by many pernicious factors which includes traffic density, temperature changes, sub-grade conditions, drainage conditions, and design life. Pavement design mainly deals with the combination of various construction materials and the different layers used to ensure maximum strength. The pavement, though well designed and constructed, they require long-term maintenance. For this, Geosynthetic material can be incorporated in the pavement design to improve the long-term maintenance, they are the most cost effective tools for safeguarding the pavement. Many researches have been conducted with the objective to evaluate the cost effectiveness and thickness reduction by using the Geosynthetic. Past studies on Geosynthetic fibers showed that there is an increase of CBR value while the best combination is used. these Geosynthetic are mainly made up of polymer material such as poly propylene, polyester, high density polyethylene (HDPE), Poly vinyl chloride (PVC), Nylon, synthetic rubber, natural fibers [1], etc. Many types of Geosynthetic have been used for the improvement purpose such as Geotextile [2–4], Geogrid [5], Geocomposite, Geonets, Geofibres, Geocells and mesh matting. The Geosynthetic [6, 7] acts as a seepage barrier, increases the bearing capacity of soil and acts as a tension member support.

2 Experimental Program

This project comprises with the design of low- or medium-volume traffic flexible pavement using Geosynthetic materials on a poor soil sub-grade. The details of the materials used and methodology followed are given below.

2.1 Materials Used

The soil sample is collected from Navallur, Chennai at 0.6 m below the ground level. The soil is sun-dried for 3 days and then pulverized for further use. Atterberg's limit [8] tests are conducted and from the result the soil is categorized as

CH, i.e., clay having high compressibility [9]. From the grain size analysis [10], it is to be calculated that the soil consists of 38% of sand and 62% of silt and clay. The specific gravity [11] of the soil was determined using density bottle and the value was found to be 2.43. The swell potential [12] of the soil was found using free swell index and the value obtained is 65% having the high swell characteristics. The water content [13] of the soil obtained by the proctor test was 16% and Maximum dry density [13] (MDD) is 18.1 kN/m³. Based on the test results, CBR [14] value of soil is 3.28%. So, as per IS:2720 (part-XVI) classification table, if CBR of soil is in between 2 and 4% then the soil is classified as very poor soil. The Unconfined compressive strength [15] of soil obtained by experimental study is 132.5 kN/m². The Geogrid used is a high strength polypropylene. The thickness of the Geogrid at joint is 3.1 mm. The Geotextile used is non-woven type having a thickness of 0.81 mm.

2.2 Methodology

The California Bearing Ratio (CBR) is the principle method for the design of flexible pavement. By using this method, placing of Geo-grids and Geotextile at different thickness like T/2, T/3, T/4 each layer individually and also in multilayers as shown in Fig. 1 the optimum combination is determined.

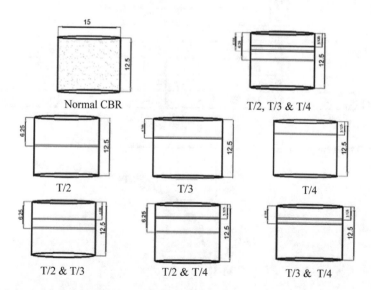

Fig. 1 Placement of geogrid/geotextile in the CBR mould

3   Results and Discussion

The pavement design is done with reinforced and unreinforced soil and the values obtained from the CBR tests are tabulated in Table 1.

3.1   Design of Flexible Pavement with Unreinforced Soil

The CBR of the sub-grade soil was calculated as 3.28% from the graph shown in Fig. 2 and designed as per IRC: 37-2012 [16], the minimum CBR of sub-grade or selected soil should be 8% for traffic more than 450 CVPD.

Initial traffic data in CV/day,

$A = P(1 + r)^\wedge x = 430(1 + 0.05)^\wedge 1 = 450$ CV/day

Cumulative no. of standard axles in msa = 6.2 msa

The pavement thickness is taken from plate 1 and 6.2 msa traffic. The obtained total thickness of pavement is 678 mm.

Table 1 CBR values of soil with reinforcement

Placement of geosynthetic material		CBR value (%)	
		Geotextile	Geogrid
Single layer	T/2	4.53	5.26
	T/3	5.33	5.48
	T/4	5.47	5.55
Double layer	T/2, T/3	5.25	6.21
	T/2, T/4	5.40	6.42
	T/3, T/4	4.82	5.47
Triple layer	T/2, T/3, T/4	4.01	5.10

Fig. 2 CBR for unreinforced soil

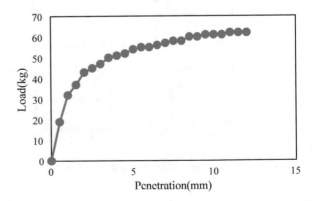

3.2 Design of Flexible Pavement with Reinforced Soil

Geogrid. The CBR value of 6.42% obtained for soil with reinforcement using Geogrid at thickness T/2 and T/4 from where the load is applied is shown in Fig. 3. The pavement thickness is taken from plate 1 and 6.2 msa traffic. The obtained total thickness of pavement is 541 mm.

Geotextile. The CBR value of 5.4% obtained for soil with reinforcement with Geotextiles at thickness T/2 and T/4 from where the load is applied is shown in Fig. 4. The pavement thickness is taken from plate 1 and 6.2 msa traffic. The obtained total thickness of pavement is 581 mm.

Quantity Estimation and Cost Analysis. The detailed estimate for the quantity of material required for the construction of pavement is done for per km length. Therefore, from the calculations we conclude that the cost [17] required for the design of pavement with unreinforced soil is more than the reinforced soil by 4% hence it results in the cost benefit for the pavement construction. Approximately the cost of pavement for unreinforced soil is Rs. 8,469,573 and for reinforced soil with Geogrid is Rs. 8,136,224.

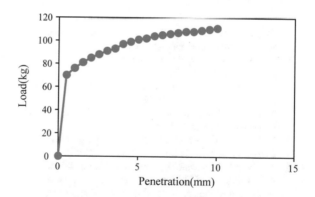

Fig. 3 CBR at thickness T/2 and T/4 (Geogrid)

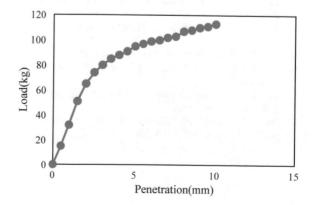

Fig. 4 CBR at thickness T/2 and T/4 (Geotextile)

4 Conclusion

From the experimental results conducted, it is obtained that the CBR value obtained using the Geogrid for reinforcement of soil is more than the geotextile used for the soil reinforcement. There is 85% increment obtained in the CBR value of soil using the Geogrid in soil as compared to untreated soil. The increase in CBR value results in the reduction in thickness of the pavement layers. The total pavement thickness found using the untreated soil is 688 mm while using the Geogrid the total pavement thickness reduced to 541 mm. There is 21% reduction in pavement thickness while comparing reinforced and unreinforced soil. In this way, decrement in pavement thickness helps to reduce the cost of construction of pavement by 4%. Hence, the cost required for the pavement construction using the geosynthetic is less than the pavement design without Geosynthetic. Therefore, it provides an innovative way to reduce the thickness of pavement and provides a solution to reduce the environmental problems by reducing the formation of low lying areas.

References

1. Sanyal.T. Jute.: Geotextile in erosion Control & strengthening of sub-grades. Jute Manufactures Development Council.
2. Bhavesh Joshi., Dr. R.P Arora.: Pavement Design By Using Geotextile. IJCIET, Volume 6 (11), pp. 39–44 (2015).
3. Meccai K. A., Hasan, E.: Geotextiles in transportation application. Second Gulf Conference on Roads, Abu Dhabi (2004).
4. Yang S.H.: Effectives of using Geotextiles in flexible pavements: life-cycle cost analysis. Blacksburg, Virginia (2006).
5. Richardson., G. N.: An Installation of geogrids and their role in paved roads and associates. Geosynthetic Association, Washington (1998).
6. Zornberg J.G.: Advances in the Use of Geosynthetic in Pavement Design. Geosynthetic India, Indian Institute of Technology Madras, Chennai, India, vol. 1, pp. 3–21 (2011).
7. M Rama Krishna., B Naga Malleswara Rao., Evaluation of CBR Using Geosynthetic in Soil Layers. IJRET, vol. 4(5) (2015).
8. IS: 2720 (Part V),"Methods of tests for soils- Determination of Atterberg limits", 1985, BIS, New Delhi.
9. IS: 1498, "Classification and identification of soils for general Engineering purposes", 1970, BIS, New Delhi.
10. IS: 2720 (Part IV),"Methods of tests for soils, Grain size analysis", 1985, BIS, New Delhi.
11. IS: 2720 (Part III), "Methods of Test for Soil, Determination of specific gravity", 1980, BIS, New Delhi.
12. IS: 2720 (Part XI), "Methods of Test for Soil, Determination of Free Swell Index of Soils", 1977, BIS, New Delhi.
13. IS: 2720 (Part XVII), "Methods of Test for Soil, Determination of moisture content - Dry density relationship", 1980, BIS, New Delhi.

14. IS: 2720 (Part XVI), "Methods of Test for Soil, Laboratory Determination of CBR", 1987, BIS, New Delhi.
15. IS: 2720 (Part X), "Methods of Test for Soil, Laboratory Determination of Unconfined compressive strength", 1991, BIS, New Delhi.
16. IRC: 37 (2012), "Tentative Guidelines for the Design of Flexible Pavements".
17. Standard Schedule of rates, 2016–2017, Government of Tamil Nadu- Public Works Department.

Optimal Reactive Power Dispatch Using Directional Bat Algorithm

CH. Sravanthi and D. Karthikaikannan

Abstract Power system operation and control plays a vital role in Electrical Power Systems (EPS). It becomes essential to control the severity of the problems like optimal reactive power dispatch (ORPD) erected during decentralization. Over the years various optimization mechanisms which are nature-inspired meta-heuristic have been introduced to handle ORPD. The Objective of this paper dealing with Optimal Reactive Power Dispatch (ORPD) is to minimize active power transmission loss using Directional Bat methodology (DBA) while varying decision variables such as voltage limits of generator terminals, tap settings of online tap changing transformer and reactive power output of shunt VAR compensators. DBA has been tested on IEEE 14—Bus and IEEE 57—Bus systems. The acquired results through DBAT methodology to solve ORPD problem is compared with various optimization methodologies (Particle swarm optimization, Genetic algorithm, etc.) to illustrate the effectiveness of the proposed approach in handling the optimal reactive power dispatch (ORPD) problem.

Keywords Optimal Reactive Power Dispatch (ORPD) · Directional Bat Algorithm (DBA)

1 Introduction

Electricity plays a major role in our daily life. The present electrical power systems have undergone many modifications in past few decades in order to serve quality power supply in reasonable cost to the consumers. On this basis, in the place of manual operations everything in possible extent was made digitized for effectiveness in dealing with power systems. During decentralization of power systems

CH. Sravanthi (✉) · D. Karthikaikannan
SASTRA University, Thanjavur 613401, Tamil Nadu, India
e-mail: sravanthi2438@gmail.com

D. Karthikaikannan
e-mail: gdkkannan@yahoo.co.in

© Springer Nature Singapore Pte Ltd. 2019
M. A. Bhaskar et al. (eds.), *International Conference on Intelligent Computing and Applications*, Advances in Intelligent Systems and Computing 846,
https://doi.org/10.1007/978-981-13-2182-5_30

many problems were raised which are related to security, reliability, and economic operation of power systems. Taking these as objectives many optimization techniques have emerged to deal with the problems raised in power systems while decentralization. ORPD is one of the major terms we always deal with in power systems and the major objective of ORPD is minimizing of active power transmission loss in network while satisfying various constraints.

To solve the ORPD many optimization techniques are emanated which are nature-inspired meta-heuristic methodologies such as Genetic algorithm (GA) [1, 2], particle swarm optimization (PSO) [3–5], Bat algorithm (BA) [6], Moth-flame optimization (MFO) [7], etc.

In this paper, the nature-inspired meta-heuristic technique known to be directional bat algorithm (DBA) [8] which is a modification of novel bat algorithm (NBA) [9] is used to solve ORPD problem. Results were compared with various optimization techniques for two test cases: IEEE 14 and 57 bus test systems.

2 Optimal Reactive Power Dispatch

Mathematical formulation of the objective function (1), equality (2) and inequality constraints are termed as follows:

$$P_{\text{loss}} = \min \left[\sum_{k=1}^{N_l} G_k \left(V_i^2 + V_j^2 - 2V_i V_j \cos \alpha_{ij} \right) \right] \tag{1}$$

Here, N_1 represents number of transmission lines, P_{loss} represents power loss, G_k represents kth bus conductance, V_i and V_j represents the voltages at ith and jth buses respectively.

2.1 Equality Constraints

$$P_{Gi} - P_{Di} = V_i \sum_{j=1}^{N_b} V_j \left[G_{ij} \cos \left(\delta_{ij} \right) + B_{ij} \sin \left(\delta_{ij} \right) \right] \tag{2}$$

$$Q_{Gi} - Q_{Di} = V_i \sum_{j=1}^{N_b} V_j \left[G_{ij} \sin \left(\delta_{ij} \right) + B_{ij} \cos \left(\delta_{ij} \right) \right] \tag{3}$$

Here, N_b denotes number of buses, P_{Gi} and Q_{Gi} denotes active and reactive power generation, P_{Di} and Q_{Di} are active and reactive load demand, G_{ij} and B_{ij} represents conductance and susceptance respectively.

2.2 Inequality Constraints

Generator Constraints.

$$\left.\begin{array}{l} V_{Gi}^{\min} \leq V_{Gi} \leq V_{Gi}^{\max} \\ Q_{Gi}^{\min} \leq Q_{Gi} \leq Q_{Gi}^{\max} \end{array}\right\} \quad i = 1 \ldots NG \tag{4}$$

Here, NG denotes number of generators; V_{Gi} and Q_{Gi} denotes generator voltages and generator reactive power outputs.

Transformer Constraints.

$$T_i^{\min} \leq T_i \leq T_i^{\max} \quad i = 1 \ldots NT \tag{5}$$

Here, NT denotes number of transformers; T denotes tap setting of the ith transformer.

Reactive Power Compensator Constraints.

$$Q_{Ci}^{\min} \leq Q_{Ci} \leq Q_{Ci}^{\max} \quad i = 1 \ldots NC \tag{6}$$

Here, NC denotes number of reactive power compensators. Q_{Ci} represents VAR injection of ith reactive power compensator.

2.3 Fitness Function

To restrict the state variables V_{load} and Q_g, penalty terms were included. The objective function including penalty terms are generalized as fitness function as follows:

$$F = P_{\text{loss}} + \lambda_v \sum_{N_V^{\lim}} \left(V_i - V_i^{\lim}\right)^2 + \lambda_Q \sum_{N_Q^{\lim}} \left(Q_{Gi} - Q_{Gi}^{\lim}\right)^2 \tag{7}$$

Here,

$$V_i^{\lim} = \begin{cases} V_i^{\min} & \text{if } V_i < V_i^{\min} \\ V_i^{\max} & \text{if } V_i > V_i^{\max} \end{cases}, \quad Q_{Gi}^{\lim} = \begin{cases} Q_{Gi}^{\min} & \text{if } Q_{Gi} < Q_{Gi}^{\min} \\ Q_{Gi}^{\max} & \text{if } Q_{Gi} > Q_{Gi}^{\max} \end{cases}$$

3 Directional Bat Algorithm

Directional Bat Algorithm (DBA) is an optimization technique based on echolocation behavior of bats. During the search of prey, bats exhibit ultrasonic waves into the environment, by using the Doppler Effect a bat can distinguish a difference between barrier and the prey. Directional bat algorithm is a modified form of Bat algorithm (BA) which includes additional constraints helps in obtaining better results.

A randomly selected bat k, emit pulses in two different possible directions, initially it chooses the direction where a bat holding a best possible solution. Later, it selects another random bat i. By means of Doppler Effect and echolocation, a bat identifies existence of food among the selected bats or with neighborhood. Depends on fitness value, the best position of the randomly selected bat is determined using fitness function. If the randomly selected bat exhibits a value with better fitness chosen as an actual one where the food exists otherwise it consider no food in the surroundings of the randomly selected bat. If the collision occurs it chooses the direction of the bat which contains excess of food.

The DBA has undergone four modifications from standard bat algorithm (BA) [8] as follows:

$$\begin{cases} X_i^{t+1} = X_i^t + (X^* - X_i^t)f_1 + (X_k^t - X_i^t)f_2 & (\text{if } F(X_k^t) < F(X_i^t)) \\ X_i^{t+1} = X_i^t + (X^* - X_i^t)f_1 & \text{otherwise} \end{cases} \tag{8}$$

Here X^* is the best solution, X_k^t is the direction of randomly selected bat (k must not be equal to i).

$$\begin{cases} f_1 = f_{\min} + (f_{\max} - f_{\min})\text{rand1} \\ f_2 = f_{\min} + (f_{\max} - f_{\min})\text{rand2} \end{cases} \tag{9}$$

Here f_1, f_2 are frequencies of two pulses and rand1, rand2 are vectors considered from uniform distribution (between 0 and 1).

Equations 8 and 9 are considered as the first modification.

The second modification is of local search mechanism. As in standard bat algorithm, bats allowed to move in random new positions from their current position using a local random walk. The equation of the random moves was modified as follows:

$$X_i^{t+1} = X_i^t + \langle A^t \rangle \varepsilon w_i^t \tag{10}$$

Here $\varepsilon \in [-1, 1]$ is a random vector, $\langle A^t \rangle$ is the average value of loudness of all bats, as the iterative process proceeds w_i = Scale regulating parameter for search process.

$$w_i^t = \left(\frac{w_{i0} - w_{i\infty}}{1 - t_m}\right)(t - t_m) + w_{i\infty} \tag{11}$$

Here, t = current iteration and t_m = maximum number of iterations.

$$w_{i0} = (Ub_i - Lb_i)/4 \tag{12}$$

$$w_{i\infty} = W_{i0}/100 \tag{13}$$

Here, Ub_i and Lb_i are upper and lower boundaries respectively.

$$A^t = \left(\frac{A_0 - A_\infty}{1 - t_m}\right)(t - t_m) + A_\infty \tag{14}$$

$$r^t = \left(\frac{r_0 - r_\infty}{1 - t_m}\right)(t - t_m) + r_\infty \tag{15}$$

The Eqs. 14 and 15 are considered as the third modification.

It is logical to assume when a bat approach its prey they would increase their pulse emission rate and decrease loudness. For better performance of algorithm rate of pulse emission and loudness is considered as follows: "$A_0 = 0.9$, $A_\infty = 0.6$ $r_0 = 0.1$ and $r_\infty = 0.7$".

Bat algorithm allows bats to update their loudness and rate of pulse emission. If the present solution produces a better solution than the previous one, then current solution is updated to new solution is considered as fourth modification.

The algorithm for directional bat algorithm is given below. This idea is used to implement optimal reactive dispatch problem.

Algorithm

1. Define objective function.
2. Initialization of bat population x_i within random boundaries Lb_i and Ub_i where (i = 1, 2, 3....n).
3. For every individual, run load flow method such as Newton raphson to obtain required operating points.
4. Evaluate the fitness function $F_i(x_i)$ according to (7).
5. Initialize loudness A_i, pulse emission rates r_i and scale regulating parameter w_i.
6. while ($t \le t_m$)
7. Select a bat k randomly which is different from bat i
8. Generate frequencies according to (9).
9. Update solutions considered as locations according to (8).
10. if ($rand(0,1) > r_i$)
11. Generate a local solution around the most selected solution according to (10).
12. Update w_i as per (11).

13. end if
14. if (rand(0,1) < A_i & $F(x_i^{t+1}) < F(x_i^t)$
15. Update the new solutions.
16. Decrease A_i as per (14).
17. Increase r_i as per (15).
18. end if
19. if $(F(x_i^{t+1}) < F(x^*))$
20. Update the obtained best solution x^*.
21. end if
22. end of while loop
23. Post-processing of output.

4   Results and Discussions

The results acquired for DBAT methodology based ORPD has been compared with base case and other optimization techniques for two test cases: IEEE 14 and 57 bus test systems (Table 1).

The minimum and maximum load bus voltage limits for the two test systems are considered to be as 0.95 and 1.1 p.u respectively.

Initially, test case 1 was considered as IEEE 14 bus test system. It is designed based on [10]. This test system having five generator buses (bus-1 is considered to be as slack bus, 2, 3, 6, and 8 buses are considered as PV buses), Shunt compensation is placed at 9 and 14 buses and tap changing transformers at (4–7, 4–9 and 5–6) buses with total 10 control variables.

The initial transmission line loss is taken as 0.1349 p.u. The system loads are P_{load} = 2.59 p.u, Q_{load} = 0.735 p.u.   Total   generation   is   considered   as $\sum P_G$ = 2.7239 p.u, $\sum Q_G$ = 0.8244 p.u. (Tables 2 and 3).

IEEE 57-bus test system was considered to be the second test case. It is designed based on [5]. This test system having seven generator buses (bus-1 is considered to be as slack bus, 2, 3, 6, 8, 9, and 12 are considered as PV buses), Shunt compensation is placed at 18, 25 and 53 buses and tap changing transformers at 15 branches with total 25 control variables.

The initial transmission line loss is taken as 0.28462 p.u. The system loads are P_{load} = 12.508 p.u, Q_{load} = 3.364 p.u.   Total   generation   is   considered   as $\sum P_G$ = 12.7926 p.u, $\sum Q_G$ = 3.4545 p.u. (Tables 4 and 5; Figs. 1 and 2).

Table 1 Limits of decision variables for IEEE 14 and 57 bus test systems	V_G^{max}	V_G^{min}	T_i^{max}	T_i^{min}	Q_C^{max}	Q_C^{min}
	1.1	0.95	1.1	0.9	0.3	0.0

Table 2 P_{loss} for the control variable setting before and after optimization of IEEE 14 bus test system

Control variable	Algorithms	
	Base case[10]	DBA
Generator voltage		
V_1 p.u	1.06	1.0273
V_2 p.u	1.045	1.056
V_3 p.u	1.01	1.0416
V_6 p.u	1.07	1.0730
V_8 p.u	1.09	1.0326
Transformer tap ratio		
T_{4-7}	0.9467	1.0195
T_{4-9}	0.9524	1.020
T_{5-6}	0.9091	0.9962
Capacitor banks MVAR		
Q_{c-9}	18	16.0657
Q_{c-14}	18	18.3201
P_{loss}, MW	13.49	12.1772

Table 3 P_{loss} comparison for IEEE 14 bus test system with various optimization techniques

Optimization technique	Power loss (MW)
Evolutionary programming [2]	13.3462
Particle swarm optimization [10]	13.327
Interior point method [10]	13.246
Differential Evolutionary computation [10]	13.239
Directional bat algorithm	12.1772

Table 4 P_{loss} for the control variable setting before and after optimization of IEEE 57 bus test system

Control variable	Algorithms	
	Base case	DBA
Generator voltage		
V_1 p.u	1.04	1.0495
V_2 p.u	1.01	1.0481
V_3 p.u	0.985	1.0486
V_6 p.u	0.98	1.0473
V_8 p.u	1.005	1.0591
V_9 p.u	0.98	1.0470
V_{12} p.u	1.015	1.0599
Transformer tap ratio		
T_{4-18}	0.97	1.0595
T_{24-25}	0.978	1.0546
T_{21-20}	1.043	1.045
T_{24-26}	1.043	1.0462
T_{7-29}	0.967	1.0252
T_{34-32}	0.975	1.0569
T_{11-41}	0.955	0.9694
T_{15-45}	0.955	0.9296
T_{14-46}	0.9	0.9391
T_{10-51}	0.93	0.9376
T_{13-49}	0.895	0.9450
T_{11-43}	0.958	1.0968
T_{40-56}	0.958	0.9521
T_{39-57}	0.98	1.0531
T_{9-55}	0.94	1.0010
Capacitor banks MVAR		
Q_{c-18}	0	13.369
Q_{c-25}	0	8.5606
Q_{c-53}	0	18.581
P_{loss}, MW	28.462	22.9636

Table 5 Comparison of P_{loss} for IEEE 57 bus test system with various optimization techniques

Optimization technique	Power loss (MW)
Adaptive genetic algorithm [5]	24.564
Biogeography based optimization [5]	24.544
Seeker optimization algorithm [5]	24.265
Gravitational Search algorithm [5]	23.461
Directional bat algorithm	22.9636

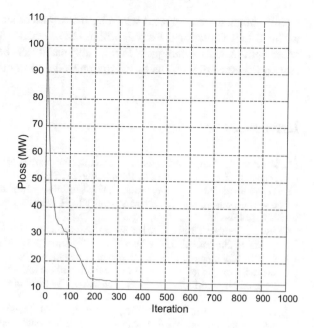

Fig. 1 P_{loss} convergence graph for an IEEE 14 bus test system

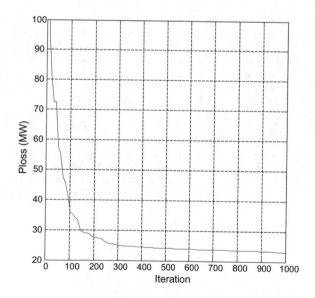

Fig. 2 P_{loss} convergence graph for an IEEE 57 bus test system

5 Conclusion

In this paper, the implementation of solution to obtain an optimal result for the ORPD problem using Directional Bat methodology is carried out. The effectiveness of the developed program was tested for IEEE 14 and 57 bus test systems. The results obtained are compared with a base case and other optimization techniques. It is found that Directional Bat Algorithm (DBA) perks up ORPD problem over all other optimization techniques. Further the result can be enhanced by using more real-world constraints so as to get optimum and accurate solution.

References

1. Devraj. D, "Improved genetic algorithm for multi-objective reactive power dispatch problem", Eur Trans Electrical Power, 569–81, 2007.
2. P. Subbaraj, P.N. Rajnarayanan, "Optimal reactive power dispatch using self-adaptive real coded genetic algorithm", Elsevier publications, Electrical Power Systems research 79, 374–381, 2009.
3. Esmin, Lamber-Torbes, G & De-Souza, "A hybrid particle swarm optimization applied to loss power minimization", IEEE Trans Power Systems, 20(2):859–66, 2005.
4. Mehdi Mehdinejad, Behnam Mohammadi-Ivatloo, "Solution of optimal reactive power dispatch of power systems using hybrid particle swarm optimization and imperialistic competitive algorithms", Elsevier publication, Electrical Power and Energy Systems 104–116, 2017.
5. Rudra Pratap Singh, V. Mukherjee, S.P. Ghoshal, "Optimal reactive power dispatch by particle swarm optimization with an aging leader and challengers", Elsevier publications, Applied soft computing 29, 298–309, 2015.
6. M. Mokhtarifard, H. Mokhtarifard, S. MoLaei, "Solution of reactive power dispatch of power systems using bat search algorithm", International journal on Technical and physical problems of Engineering", Issue 23, volume 7, No. 2, pages 65–70, June 2015.
7. Rebecca Ng Shin Mei, Zuriani Mustaffa, Mohd Herwan Sulaiman, Hamdan Daniyal, "Optimal reactive power dispatch solution by loss minimization using moth-flame optimization technique", May 2017, Elsevier publication, Applied Soft Computing 59, 210–222, 2017.
8. Xin-She Yang, Rabia Khelif, Mohamed Benouret, Asma Chakri, "New directional bat algorithm for continuous optimization problems", Oct 2016, Elsevier publication, Expert Systems With Applications 69, 159–175, 2017.
9. Xian-Bing Meng, X.Z. Gao, Yu Liu, Hengzhen Zhang, "A novel bat algorithm with habitat selection and Doppler effect in echoes for optimization", Elsevier publication, April 2015, Expert Systems with Applications 42, 6350–6364, 2015.
10. M. Varadarajan, K.S. Swarup, "Differential evolutionary algorithm for optimal reactive power dispatch", Elsevier publications, Electrical Power and Energy Systems 30, 435–441, 2008.

Study of Optimal Power Flow on IEEE Test Systems with TCSC Using Power World Simulator

B. Monisha, D. Karthikaikannan and K. Balamurugan

Abstract The Optimal Power Flow (OPF) is a static, nonlinear, and non-convex optimization problem, which finds a set of variables in optimal from the state of network, load data, and system parameters. In this paper, the values of optimal are computed by taking objective function of minimum generation cost subjected to number of equality and inequality constraints. Here, we consider the FACTS device TCSC (Thyrsitor-Controlled Series Capacitor) in the test network and study the change of objective function. The simulation of OPF on IEEE 6 and 30 bus systems with and without TCSC device are carried out using power world simulator. Simulation results show that after the installation of FACTS device, the over cost of generation is reduced.

Keywords OPF · Optimal solution · Generation loss

1 Introduction

Generation and distribution of power are accomplished with maximum efficiency but with minimum cost. This takes into each power plant for real and reactive power in order to reduce the cost of total operating system network. In other words, the different types of power like real and reactive generator is also allowed to vary the limits to reduce the generation cost of system this is called as Optimal Power Flow (OPF) [1]. In past, before opening the market, there was most realizing for power flow control by reallocating productions. The capability of steady state or

B. Monisha (✉) · D. Karthikaikannan
SASTRA University, Thanjavur, Tamil Nadu, India
e-mail: moni.bala94@gmail.com

D. Karthikaikannan
e-mail: karthikaikannan@eee.sastra.edu

K. Balamurugan
Dr. Magalingam College of Engineering and Technology, Pollachi, Tamil Nadu, India
e-mail: balapriya122@gmail.com

© Springer Nature Singapore Pte Ltd. 2019
M. A. Bhaskar et al. (eds.), *International Conference on Intelligent Computing and Applications*, Advances in Intelligent Systems and Computing 846,
https://doi.org/10.1007/978-981-13-2182-5_31

dynamic stability is studied and realized [2–4]. With the abilities, the transmission lines apparent impedance is also changed to control active power, reactive power, and voltage control FACTS device is used [5]. And, thus system loadability is increased [6, 7]. There are some devices in which improvement will be for load-ability that has been observed [8].

In other countries, power outages and disruptions were major issues which were due to power demand. Due to demand, the transmission system is operated nearer to stability limit of transmission line. The operation was done without adding new transmission line instead FACTS device is used to control all issues [9]. This device can regulate power flow and increase power transfer capability in transmission line [10, 11].

The FACTS device has some important issues like cost, optimal location, and size of device [12, 13]. Thus, this paper explains about the location of TCSC device mainly for transmission lines to increase power transfer capability. This paper presents different method total system real and reactive power loss minimization and real power flow minimization, respectively. IEEE 6 bus system and IEEE 30 bus system has been selected for implementation.

2 Problem Formulation

In OPF, to determine the objective, some of control variable is used to optimize. And, the objective is important for proper problem definition, quality based on accuracy of model. The main aim of the objective is to minimize the total pro-duction cost in generating unit.

General goals are:

1. From target schedule, the deviation is reduced.
2. The violation is minimized from control shifts.
3. Approximation of control shift is minimized.

2.1 *Objective Function of Fuel Cost Minimization*

The optimization problem for OPF problem
The objective function is expressed as

$$\text{Min}\, F(P_G) = f(x, u)$$

Total generation cost function is expressed as

$$F(P_G) = \sum_{i=1}^{N_g} (\alpha_i + \beta_i P_{Gi} + \gamma_i P_{Gi}^2)$$

where α, β and γ are fuel cost coefficient.

Subject to satisfaction of nonlinear equality constraints

$$g(s, u) = 0$$

And, nonlinear inequality constraints are

$$h(x, u) \leq 0$$

$$u^{\min} \leq u \leq u^{\max}$$

$$x^{\min} \leq x \leq x^{\max}$$

$F(P_G)$ = total cost function
$f(x, u)$ = scalar objective
$g(x, u)$ = nonlinear equality constraints
$h(x, u)$ = nonlinear inequality constraints
The vector x contains dependent variable consisting of

- Bus voltage magnitudes and phase angle.
- MVAr output of generators designated for bus voltage control.
- Fixed parameters such as the reference bus angle.
- Noncontrolled generator MW and MVAr outputs.

The vector u consists of control variables including:

- Real and reactive power generation.
- Phase shifter angles.
- Net interchange.
- Load MW and MVAr.
- DC transmission line flows.
- Control voltage settings.

The equality and inequality constraints are

- Limits on all control variables.
- Power flow equations.
- Generation/load balance.
- Branch flow limits.

2.2   Constraints for Objective Function of Fuel Cost Minimization

Representing a standard system with generator, load, and total system generation cost should be minimum. The network equality constraints are represented by the following load flow equation

$$P_i(V, \delta) - P_{Gi} + P_{Di} = 0$$

$$Q_i(V, \delta) - Q_{Gi} + Q_{Di} = 0$$

And, load balance equation is

$$\sum_{i=1}^{N_G}(P_{G_i}) - \sum_{i=1}^{N_D}(P_{D_i}) - P_L = 0$$

The inequality constraints representing the limits on all variables, line flow constraints are

$$V_{i\min} \leq V_i \leq V_{i\max}, \quad i = 1, \ldots, N,$$
$$P_{G_{i\min}} \leq P_{G_i} \leq P_{G_{i\max}}, \quad i = 1, \ldots, N_G$$
$$Q_{G_{i\min}} \leq Q_{G_i} \leq Q_{G_{i\max}}, \quad i = 1, \ldots, N_{Gq}$$
$$k_{vi}I_{1\max} \leq V_i - V_j \leq k_{vj}I_{1\max}, \quad i = 1, \ldots, N_1$$

$$-k_{\delta i}I_{1\max} \leq \delta_i - \delta_j \leq k_{\delta j}I_{1\max}, \quad i = 1, \ldots, N_1$$

$$S_{li} \leq S_{li} \leq S_{l_{i\max}}, \quad i = 1, \ldots N_1$$
$$T_{k\min} \leq T_k \leq T_{k\max}, \quad i = 1, \ldots N_1$$
$$\delta_{i\min} \leq \delta_i \leq \delta_{i\max}$$

3   Static Model of TCSC

In transmission line by using TCSC which is type of series compensator which increases the power transfer capacity by reducing series impedance of line. The capacitor and inductor are connected in parallel along with thyristors. The firing angle "α" ranges between $90°$ and $180°$ which acts in bidirectional [14].

The possibilities are

- $90° < \alpha < \alpha_{L\text{ lim}}$ for inductive region
- $\alpha_{C\text{ lim}} < \alpha < 180°$ for capacitive region (Fig. 1).

Fig. 1 Schematic diagram of
TCSC

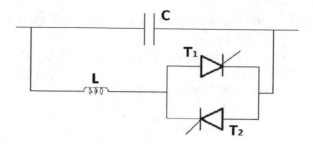

The impedance of TCSC is given by

$$X_{\text{TCSC}}(\alpha) = \frac{X_C X_l(\alpha)}{X_l(\alpha) - X_C}$$

To compensate inductive reactance in transmission line, where TCSC is inserted in line is a controllable reactance. This results in reducing transfer capability in transmission line, i.e., "X" get reduced.

Transmission lines power transfer is given by

$$P_e = \frac{Vi * Vj}{X} * \sin(\theta i - \theta j)$$

This equation is used to reduce transfer reactance, which increases and reactive power loss is effectively reduced.

4 Method for Optimal Location of Tcsc

4.1 Reduction of Total System Reactive Power Loss

This is totally based on reactive power loss ($\sum Q_L$) also with the control variable of TCSC.

The loss sensitivity with respect to Xij is

$$a_{ij} = \frac{\partial Q_L}{\partial X_{ij}}$$

$$= [V_i^2 + V_j^2 - 2V_i V_j \cos \delta_{ij}] \frac{r_{ij}^2 - x_{ij}^2}{(r_{ij}^2 + x_{ij}^2)^2}$$

By using $X_{\text{TCSC}} = X_C = 0$, we can find the sensitivity loss of each line.

4.2 Reduction of Total System Real Power Loss

This is totally based on real power loss ($\sum Q_L$) also with the control variable of TCSC.

The loss sensitivity with respect to X_{ij} is

$$d_{ij} = \frac{\partial P_L}{\partial X_{ij}}$$

$$= [V_i^2 + V_j^2 - 2V_iV_j \cos \delta_{ij}] \frac{(-2r_{ij} * x_{ij})}{(r_{ij}^2 + x_{ij}^2)^2}$$

4.3 Reduction of Real Power Flows

Control variable of TCSC is taken by real power flow, i.e., X_{ij}. When TCSC is placed between i and j, u_{ij} is loss sensitivity with respect to X_{ij}

$$u_{ij} = \frac{\partial P_{i-j}}{\partial x_{ij}}$$

$$= [-V_i^2 + V_iV_j \cos \delta_{ij}] \frac{(2r_{ij} * x_{ij})}{(r_{ij}^2 + x_{ij}^2)^2} + V_iV_j \sin \delta_{ij} \frac{(r_{ij}^2 - x_{ij}^2)}{(r_{ij}^2 + x_{ij}^2)^2}$$

5 Power World Simulator (PWS)

The Power World Corporation was found in 1996. The main goal of power world is to introduce a developed software package in such a way the Power World Simulator (PWS) is found and distributed to the market. This software was developed by Professor Thomas Overbye of the University of Illinois at Urbana–Champaign. The Simulator 5.0 was introduced first which has a normal menu structure. The developed version 17 has a new ribbon interface which replaces the old menu and toolbar interface. First, it was introduced in Microsoft office 2007 desktop application.

The changes was made in new case information menu from old cases, this menu help the users to have certain number of case information displays simultaneously in interface. This new explorer provides convenient direct access for the users in single window.

The remainder of this document will provide a brief introduction about new ribbon interface and model explore. Power World Corporation continues to produce software that is complex. At the same time, we invest in user interface improvement, ensuring that use of the software remains both efficient and enjoyable. In this, we are able to find different solution like contingency analysis, fault analysis, optimal power flow, available transfer capability tool, etc.

The main advantage of finding optimal power flow using this software is it removes transmission line overload and also calculate spot prices. OPF is available as an add-on to simulator.

The steps to be followed are given below:

Step 1: Sketch the single line diagram of given power system network with specified data.
Step 2: Run the optimal power flow and find out the power flow in each line.
Step 3: Finding the load, loss and generator value related to the bus, also the hourly cost is found without connecting FACTS device.
Step 4: Finding the load, loss and generator value related to the bus, also the hourly cost is found by connecting FACTS device in two mode (with one FACTS and two FACTS).
Step 5: comparison is made between different case and the conclusion is made.

6 System Description

6.1 Implementation of IEEE-6 Bus System

Case-I Without FACTS device:
The single line diagram of the IEEE-6 bus system is shown below with respected data of generator, load, and transmission line along the generator cost (Figs. 2 and 3; Tables 1, 2, 3, and 4).

Case-II With one FACTS devices:
The optimal location for FACTS device is found by the power of the system by calculating the maximum real power in the system the priority is given for location.

Fig. 2 Single line diagram of IEEE-6 bus system

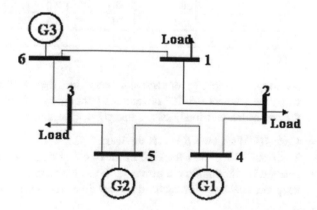

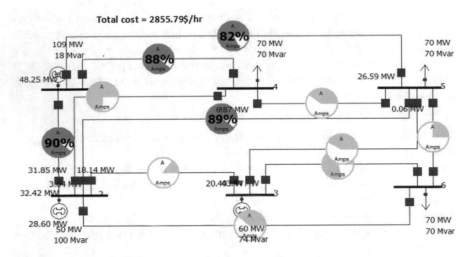

Fig. 3 Simulink for IEEE-6 bus system without FACTS devices

Table 1 Bus data for IEEE-6 bus system

Bus no.	Bus type	V (p.u)	Pd (MW)	Qd (MVAr)	Pg (MW)	Qg (MVAr)
1	Slack	1.05	0	0	0	0
2	PV	1.05	0	0	50	0
3	PV	1.07	0	0	60	0
4	PQ	1.0	70	70	0	0
5	PQ	1.0	70	70	0	0
6	PQ	1.0	70	70	0	0

Table 2 Generator data for IEEE-6 bus system

Bus no.	Pmin (MW)	Pmax (MW)	Qmin (MVar)	Qmax (MVar)	a_i	b_i	c_i
1	50	200	−20	100	0.0107	11.669	213.1
2	37.5	150	−20	100	0.0178	10.333	200
3	45	180	−15	100	0.0148	10.833	240

By keep on checking for each and every line, there will be particular line in which, when we connect FACTS device, the generator cost will reduce automatically with the same demand load, losses, and total generator (Fig. 4; Table 5).

Case-III With two FACTS devices:

The optimal location for FACTS device is found by the power of the system by calculating the maximum power in the system the priority is given for location. By keep on checking for each and every line, there will be particular line in which,

Table 3 Line data for IEEE-6 bus system

From bus	To bus	R (p.u)	X (p.u)	Half line charging suspectance (p.u)	Thermal limit (MVA)
1	2	0.1	0.2	0.02	40
1	4	0.05	0.2	0.02	60
1	5	0.08	0.3	0.03	50
2	3	0.05	0.25	0.03	40
2	4	0.05	0.1	0.01	70
2	5	0.10	0.3	0.02	30
2	6	0.07	0.2	0.025	90
3	5	0.12	0.26	0.025	70
3	6	0.02	0.1	0.01	80
4	5	0.2	0.4	0.04	20
5	6	0.1	0.3	0.03	40

Table 4 Simulation output

Description		Output
Generator (MW)	PG1	78.97
	PG2	66.47
	PG3	73.97
Total demand load (MW)		210
Losses (MW)		9.41
Total generator (MW)		219.41
Hourly cost ($/hr)		2855.79

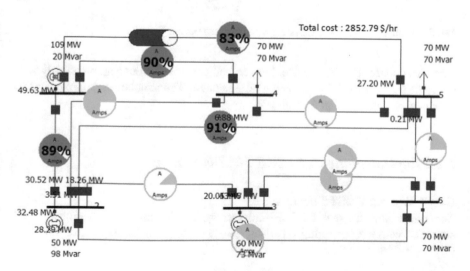

Fig. 4 Simulink for IEEE-6 Bus system with one FACTS device

330   B. Monisha et al.

Table 5 Simulation output

Description		Output
Generator (MW)	PG1	89.63
	PG2	72.02
	PG3	55.49
% line reactance		40
Total demand load (MW)		210
Losses (MW)		7.14
Total generator (MW)		217.14
Hourly cost ($/hr)		2852.79

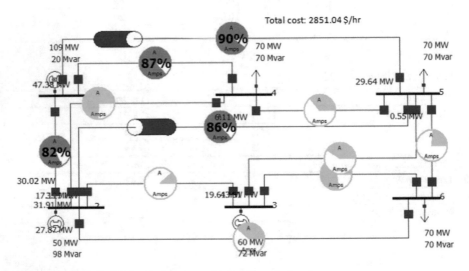

Fig. 5 Simulink for IEEE-6 bus system with two facts devices

when we connect FACTS device, the generator cost will reduce automatically with the same demand load, losses, and total generator. The simulation diagram for this case is given below (Fig. 5; Table 6).

6.2 Implementation of IEEE-30 Bus System

Case-I Without FACTS devices:
The single line diagram for IEEE-30 bus system is simulated in power world simulator with the correcting bus data, generator, load, and transmission line data (Fig. 6; Tables 7 and 8).

Table 6 Simulation output

Description		Output
Generator (MW)	PG1	90.63
	PG2	68
	PG3	57.49
First FACTS % line reactance		30
Second FACTS % line reactance		40
Total demand load (MW)		210
Losses (MW)		6.02
Total generator (MW)		216.12
Hourly cost ($/hr)		2851.04

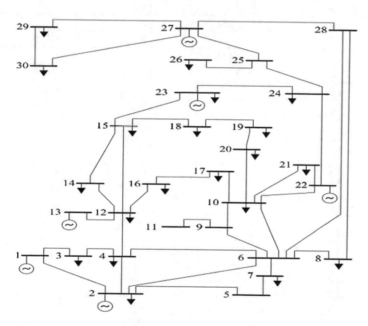

Fig. 6 Single line diagram of IEEE-30 bus system

Simulink for IEEE-30 bus system without FACTS devices:
The simulation diagram for IEEE-30 bus system is drawn by single line diagram and their respective data (Fig. 7; Table 9).

Case-II With one FACTS devices:
The optimal location for FACTS device is found by the power of the system by calculating the maximum power in the system the priority is given for location. By keep on checking for each and every line, there will be particular line in which, when we connect FACTS device, the generation cost will reduce automatically with the same demand load, losses, and total generator (Fig. 8; Table 10).

Table 7 Generator and load data for IEEE-30 bus system

Bus No	Generator		Load	
	Real power (p.u)	Reactive power (p.u)	Real power (p.u)	Reactive power (p.u)
1	1.3848	−0.0279	0	0
3	0	0	0.025	0.012
4	0	0	0.076	0.016
5	0	0.37	0.0943	0.019
6	0	0	0	0
7	0	0	0.228	0.109
8	0	0.373	0.3	0.3
9	0	0	0	0
10	0	0	0.058	0.02
11	0	0.162	0	0
12	0	0	0.112	0.075
13	0	0.106	0	0
14	0	0	0.062	0.016
15	0	0	0.082	0.025
16	0	0	0.035	0.018
17	0	0	0.09	0.058
18	0	0	0.032	0.009
19	0	0	0.095	0.024
20	0	0	0.022	0.007
21	0	0	0.175	0.112
22	0	0	0	0
23	0	0	0.032	0.016
24	0	0	0.087	0.067
25	0	0	0	0
26	0	0	0.0.35	0.023
27	0	0	0	0
28	0	0	0	0
29	0	0	0.024	0.009
30	0	0	0.106	0.109

Table 8 Transmission line data for IEEE-3 bus system

From bus	To bus	Line impedance		Half line charging susceptance	MVA rating
		Resistance	Reactance		
1	2	0.0192	0.0575	0.0264	130
1	3	0.0452	0.1652	0.0204	130
2	4	0.0570	0.1737	0.0184	65
3	4	0.0132	0.0379	0.0042	130
2	5	0.0472	0.1983	0.0209	130
2	6	0.0581	0.1763	0.0187	65
4	6	0.0119	0.0414	0.0045	90
5	7	0.0460	0.1160	0.0102	70
6	7	0.0267	0.0820	0.0085	130
6	8	0.0120	0.0420	0.0045	32
6	9	0	0.2080	0	65
6	10	0	0.5560	0	32
9	11	0	0.2080	0	65
9	10	0	0.1100	0	65
4	12	0	0.2560	0	32
12	13	0	0.1400	0	32
14	15	0.2210	0.1997	0	16
15	18	0.1073	0.2185	0	16
18	19	0.0639	0.1292	0	32
19	20	0.0304	0.0682	0	32
10	20	0.0936	0.2090	0	32
10	17	0.0324	0.0845	0	32
10	21	0.0348	0.0789	0	32
10	22	0.0727	0.1499	0	16
21	22	0.0116	0.0236	0	16
15	23	0.1000	0.2029	0	16
22	24	0.1150	0.1790	0	16
23	24	0.1320	0.2700	0	16
24	25	0.1885	0.3290	0	16
25	26	0.2544	0.3800	0	65
28	27	0.1093	0.087	0	16
28	27	0	0.3960	0	16
27	29	0.2198	0.4153	0	26
27	30	0.3202	0.0627	0	32
29	30	0.0239	0.4533	0	32
8	28	0.0636	0.2000	0.0214	32
6	28	0.0169	0.0588	0.0055	32

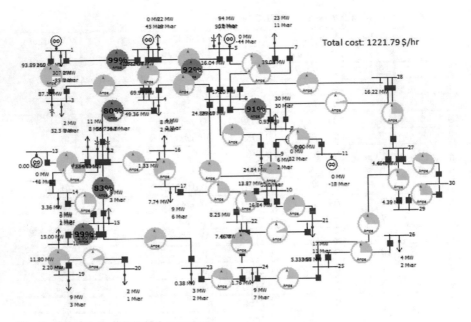

Fig. 7 Simulink for IEEE 30 bus system

Table 9 Simulation output

Description		Output
Generator (MW)	PG1	306
	PG2	1.18
	PG3	0
	PG4	0
	PG5	0
	PG6	0
% line reactance		40
Total demand load (MW)		283.02
Losses (MW)		24.16
Total generator (MW)		307.18
Hourly cost ($/hr)		1221.27

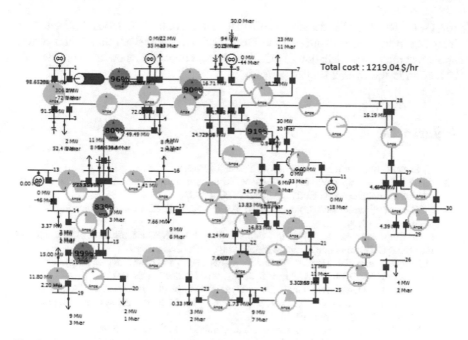

Total cost : 1219.04 $/hr

Fig. 8 Simulink for IEEE 30 bus system with one FACTS device

Table 10 Simulation output

Description		Output
Generator (MW)	PG1	304
	PG2	1.18
	PG3	0
	PG4	0
	PG5	0
	PG6	0
% line reactance		40
Total demand load (MW)		283.02
Losses (MW)		22.16
Total generator (MW)		305.18
Hourly cost ($/hr)		1219.04

7 Conclusion

The Optimal Power Flow (OPF) analysis is conducted on IEEE-6 bus system and IEEE-30 bus system with and without FACTS device using Power World Simulator (PWS). The minimization of overall generation cost is taken as objective function for OPF. The FACTS device TCSC (Thyristor-Controlled Series Capacitor) is included in the test network and the OPF results are observed. Simulation results show that after the installation of TCSC in the test networks, there is a considerable reduction in the overall cost of generation

References

1. H. Sadat, Power System Analysis, McGraw-Hill, Boston, 1999.
2. "Study of system operating impacts of FACTS technologies," GE, ECC Lnc., Carson Taylor Seminar, Final Report on EPRI RP 3022-25, Oct 1993.
3. FACTS Application Task Force, FACTS Applications: IEEE Power Engineering Society, 1996, pp 1, 1–4, 9.
4. N.G. Hingorani and L. Gyugyi, Understanding FACTS Concepts and Technology of Flexible AC Transmission Systems. Piscataway: IEEE Press, 1999.
5. D.J. Gotham and G.T. Heydt, "Power flow control and power flow studies for systems with FACTS devices, IEEE Trans. Power Systems, vol. 13, no. 1, Feb 1998.
6. J. Griffin, D. Atanackovic, and F.D. Galiana, "A study of the impact of FACTS on the secure-economic operation of power systems," in 12th Power systems computation Conference, Dresden, Aug. 19–23, 1996.
7. F.D. Galiana, K. Almeida, M. Toussaint, J. Griffin, and D. Atanackovic, "Assessment and control of the impact of FACTS devices on power system performance," IEEE Trans. Power Systems, vol 11, no. 4, Nov 1996.
8. S. Gerbex, R. Cherkaoui, and A. J. Germond, "optimal location of FACTS devices in a power system using genetic algorithms," in Proceedings for the 13th Power System Computation Conference, 1999, pp. 1252–1259.
9. Kamel T., Tawfik G. and Hasan, H.A., "Optimal number, location and parameter setting of mulitple TCSCs for security and system loadability enhancement", 10th International Multi-conference on Systems, Signals & Devices (SSD), 2013, pp 1–6.
10. Ghahremani, E. and Kamwa, I, " Optimal Placement of Multiple-Type FACTS Devices to Maximize Power System Loadability Using a Generic Graphical User Interface", Power Systems, IEEE Transactions, Vol. 28, Issue: 2, 2013, pp. 764–778.
11. Wartana I.M., Singh J.G., Ongsakul W., Agustini N.P., "Optimal Placement of a Series FACTS Controller in Java-Bali 24-bus Indonesian System for Maximizing System Loadability by Evolutionary Optimization Technique", Third International Conference on Intelligent Systems, Modelling and Simulation (ISMS), 2012, pp. 516–521.
12. Vanitila R. and Sudhakaran M., "Differential Evolution algorithm based Weighted Additive FGA approach for optimal power flow using muti-type FACTS devices", International Conference on Emerging Trends in Electrical Engineering and Energy Management(ICETEEEM), 2012, pp 198–204.
13. Chansareewittaya S and Jirapong P., "Power transfer capability enhancement with optimal maximum number of FACTS controllers using evolutionary programming", IECON 2011— 37th Annual Conference on Industrial Electronics, pp 4733–4738.
14. Nagalakshmi S. and Kamaraj N., "Evaluation of loadability limit of pool model with TCSC using optimal featured BPNN", ICONRAEeCE, 2011 International Conference, pp 454–458.

Maximum Power Tracking for PV Array System Using Fuzzy Logic Controller

M. Pushpavalli and N. M. Jothi Swaroopan

Abstract This paper is to design by adapting the Perturbation and Observation (P&O) algorithm of Maximum Power Point Tracking (MPPT) and intelligent controller to vary the duty ratio of KY boost converter using Photo Voltaic (PV) array as an energy input. This PV system extracts the energy from the sun, after the power converter module energy is fed to the battery for storage. The KY boost converter is used to boost the voltage to a certain level. The switches mainly work on MPPT conditions. This converter has higher voltage conversion ratio, low ripple, and continuous current. This combination makes the integration work perfectly under any transient disturbance condition occurs in the input side and it stays connected to the battery through charge controller. In the charge controller unit, PI controller is used to regulate the battery parameters. MPPT method-based Fuzzy Logic Controller (FLC) is implemented for PV panel by change in irradiance from (600 to 1000 W/M^2) with constant temperature at 25 °C.

Keywords Fuzzy logic controller (FLC) · Perturbation and observation (P&O) algorithm · Maximum Power Point Tracking (MPPT) · PV (Photovoltaic) array KY boost converter

1 Introduction

Humans are overloading the nature by increase in global warming, burning fossil fuels, emission of carbon dioxide, increase in fuel energy, and low economic backgrounds. The only solution for the mentioned problem is renewable energy. It gives inexpensive energy across the global, low operating cost, economic benefits,

M. Pushpavalli (✉)
Sathyabama Institute of Science and Technology, Chennai, India
e-mail: pushpa.murugan@gmail.com

N. M. Jothi Swaroopan
RMK Engineering College, Chennai, India
e-mail: nmjothi@yahoo.com

© Springer Nature Singapore Pte Ltd. 2019
M. A. Bhaskar et al. (eds.), *International Conference on Intelligent Computing and Applications*, Advances in Intelligent Systems and Computing 846,
https://doi.org/10.1007/978-981-13-2182-5_32

337

and inexhaustible energy and gives stable power values in the future. The PV is powerful, potential, cleanness, and so on. So, it is popular among all possible renewable resources [1]. Although the PV has more advantages but the characteristics are nonlinear, it is mainly dependent on insolation and temperature. The IV (Current–Voltage) and PV (Power–Voltage) curves mainly varies due to voltage and current which are changes due to temperature and insolation, respectively [2]. The PV energy is well suited to connect with standalone mode and grid-connected mode. In Power Generation System (PGS), grid connection mode is more suitable [3]. PV is a non-linear in nature, so that obtain the Maximum Power Point is essential. Many methods and controllers are available to track the maximum power [4]. Its necessary to find all the step sizes according to various operating point conditions in Perturbation and Observation (P&O) algorithm. For a nonlinear system, FLC techniques are applicable [5]. FLC-based MPPT tracking methods have become a good one [6]. A wide range of comparison among all MPPT techniques like P&O, (ICT) Incremental Conductance Techniques, Voltage and Current control, Pilot cell, and so on, among that P&O is more efficient because it is fit to various temperature and solar irradiation conditions [7]. Here, it is mainly focused to change the irradiation level and keep the temperature constant. Due to suspicions conditions of PV, nonlinear variables like weather forecasting conditions are also incorporated using fuzzy logic techniques [8]. Fluctuation of tracking the power has been based on the control of FLC [9]. Two MPPT methods discussed with Fuzzy Logic Controller and Neural Networks among that FLC improves the power [10]. A fuzzy feed-forward tracker for Coupled Inductor Interleaved Boost Converter gives better tracking performance than PI controller [11]. Adaptive Fuzzy Logic Control can supply to the grid with high power factor and less harmonics [12]. MPPT FLC increases the efficiency under variable weather conditions [13]. A steady-state performance obtained quickly with P&O algorithm implemented FLC [14]. The dual-axis tracking system using fuzzy logic increases the overall efficiency of the system [15]. Fuzzy logic controller-based genetic algorithms provides the better performance [16]. PI control used as charge controller and PV system work with fuzzy logic controller proves the dynamic response as well as charging the battery [17]. One converter is used to track the power and other converter used to regulate the battery, without any power fall continuously delivered it to the load [18]. Direct methods are used for tracking the MPPT which require less memory to calculate maximum power point [19]. The proposed KY boost converter has higher conversion ratio, low ripples, continuous inductor current compared to traditional boost converter [20]. KY boost converter has high stability because the poles and zeros lie on the left half of the s plane without requiring any bulk inductance. The new KY converter is proposed with synchronized boost converter and the main advantages are ZVS turn-on, continuous current, small in size, fewer ripples, less voltage, and current stress [21]. So, in this paper, it is mainly focused to design Perturb and Observe (P&O) algorithm with Fuzzy Logic Controller (FLC).

2 Proposed System

The block diagram of the proposed system given by Fig. 1 is consisting of a solar array as a main source, KY boost converter, a FLC, a battery, and a charge controller. FLC is mainly used to adjust the duty ratio of KY boost converter switches which depends on the MPPT conditions. Charge controller is used to charge the battery with a given rating.

2.1 PV Array

The solar panel is the energy source which results of a specified set of photovoltaic cells connected in the combination of parallel and series. PV cell directly converts the sun irradiation into electricity. Due to its nonlinear I-V characteristics, the combinations form an array with appropriate voltage and power levels. Figure 2 shows an equivalent circuit of solar cell.

The current drawn from photovoltaic array is given by

$$Ipv = Ipanel \ Ipv1 > Iph \tag{1}$$

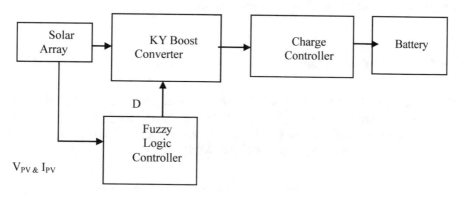

Fig. 1 Block diagram of the proposed system

Fig. 2 Equivalent circuit of solar cell

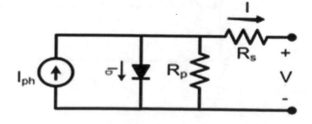

where

Ipv	Photovoltaic array current
Ipanel	Individual panel current
Iph	Individual phase current

The voltage drawn from photovoltaic array is given by

$$Vpv = \sum Vpanel \tag{2}$$

where

Vpv	Photovoltaic array voltage
Vpanel	Individual panel voltage

2.2 Characteristics of Solar Array

The VI and PV characteristics of the solar array are shown in Figs. 3 and 4. The PV array is maintained with constant temperature at different solar irradiation levels varied between 1000 W/M^2 and 600 W/M^2. KY boost converter is used to have Maximum Power Points of PV array. Due to variation in solar irradiation, the corresponding Maximum Power Points also varies.

2.3 Power Converter

A new proposed KY boost converter is incorporated with a usual combined rectified boost converter and a coupling inductor as shown in Fig. 5 [21]. This is a power converter which is used to connect between PV and load. The duty ratio of this power converter is varied by FLC with MPPT algorithm.

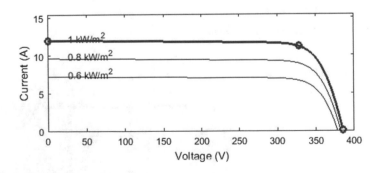

Fig. 3 Current-Voltage characteristics

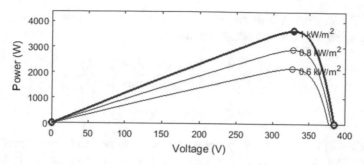

Fig. 4 Power–Voltage characteristics

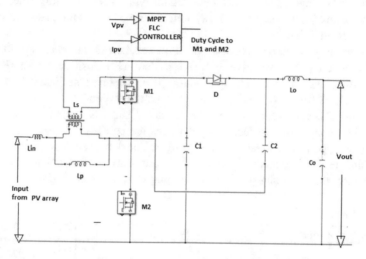

Fig. 5 Power converter for PV array

MPPT algorithm is adopted to increment or decrement the duty ratio by observing the instantaneous power Ppv(K) and the preceding power Ppv(K−1). If the Ppv(K) is greater than Ppv(K−1), the control of perturbation is in same direction, otherwise, it is inverted.

When $\frac{\Delta P}{\Delta V}$ is greater than zero, MPPT voltage is enlarged and the corresponding duty cycle is also incremented by the step size. When $\frac{\Delta P}{\Delta V}$ **less than zero**, MPPT voltage is decreased and the corresponding duty cycle is also decremented by the step size.

3 Fuzzy Logic MPPT Controller

In this proposed system, change in power perturbation by change in reference voltage as per the approach defined in P&O algorithm of MPPT using FLC. In many research papers, it concludes that FLC-based MPPT algorithm is more suitable for tracking MPPT power.

The FLC input variables are error e(K) and change in error ce(K) at sampled times K which explains the below equations

$$e(K) = \frac{\Delta P}{\Delta V} = \frac{Ppv(K) - Ppv(K-1)}{Vpv(K) - Vpv(K-1)} \tag{3}$$

$$ce(K) = e(K) - e(K-1) \tag{4}$$

where Ppv, Vpv are the instant panel power and voltage. Ppv(K−1), Vpv(K−1) are the preceding panel power and voltage. Load operating point at the instant K is situated on left or right expressed by e(K). Moving direction of the point is expressed by change in error ce(K). Mamdani type of fuzzy is implemented. The inputs are e and ce, the output of the system is duty cycle D and these control rules are indicated in Table 1.

Each input and output consists of five normalized membership functions: Negative Small (NS), Negative Big (NB), Zero (Z), Positive Small (PS), and Positive Big (PB). The membership function plots are shown in Figs. 6, 7, and 8. As shown in Fig. 9, the duty ratio D is varying between 0 and 1.

Input membership functions are error (e) and change of error (ce) and are normalized by an input scaling factor. Input values are between −1 and 1. Maximum–Minimum composition is based on these 25 rules. The duty cycle D has been generated using triangular membership function of defuzzification. The output membership function duty ratio D scaling is done by varying between 0 and 1 (Fig. 9).

Table 1 Fuzzy control rules

Error/change in error	NB	NS	Z	PS	PB
NB	Z	Z	PS	NS	NS
NS	Z	Z	Z	NS	NB
Z	PB	PS	Z	NS	NB
PS	PB	PS	Z	Z	Z
PB	PB	PS	NS	Z	Z

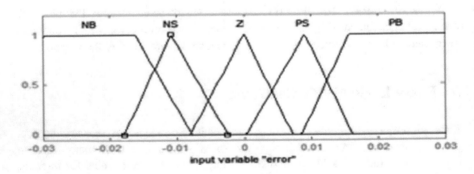

Fig. 6 Membership function plots for (i/p e-error)

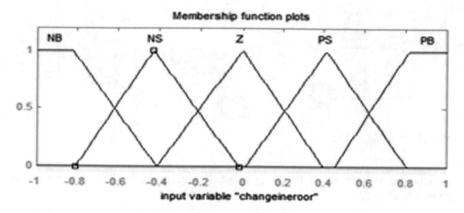

Fig. 7 I/p ce-change in error

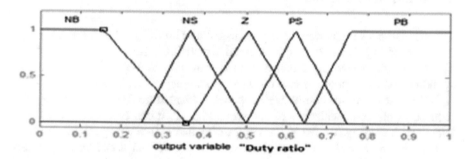

Fig. 8 Membership function plots for (c) o/p D-duty cycle

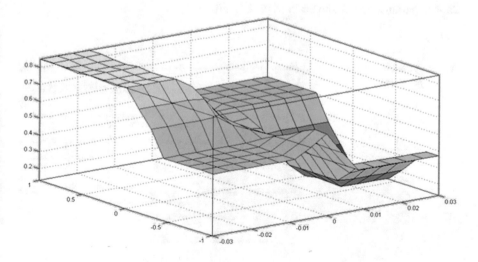

Fig. 9 3D surface viewer for 25 Fuzzy rules

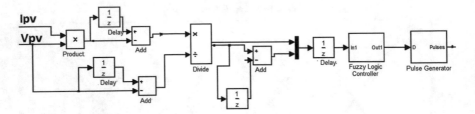

Fig. 10 MPPT-based FLC

The Fuzzy logic-based P&O algorithm is created in MATLAB Simulink as shown in Fig. 10.

4 Simulation Results

The simulations were performed using the PV system of Fig. 1 and the KY boost converter of Fig. 5. Some of the obtained simulation results are illustrated here. The PV panel change in irradiance ranges from 600 to 1000 W/m². The corresponding PV panel power is also changed as shown in Figs. 11 and 12 (Table 2).

KY boost converter boosts the PV panel voltage from 340 to 640 volts with reduced ripple and oscillations as shown in Fig. 12. It can be further stored in a battery with a voltage of 270 V. This output can be fed into voltage source inverter and it will be converted as three phase ac voltage. So that it can be passed into utility grid. Figure 13 represents PI charge controller for battery.

To regulate the output voltage and power of KY boost converter, charge controller is used as shown in Fig. 15. It is mainly used to charge 270 V battery. Battery parameters are shown in Figs. 14 and 15.

Fig. 11 Variable irradiation and PV panel power

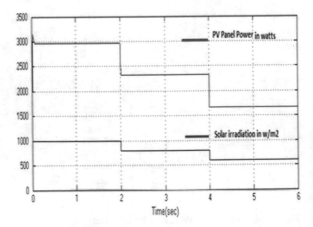

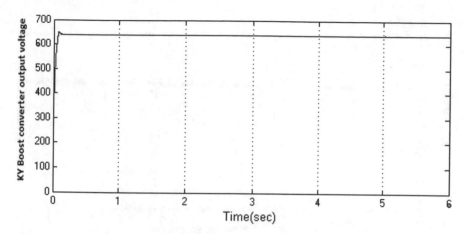

Fig. 12 KY boost converter output voltage

Table 2 Electrical Characteristics of PV

Parameters	Values
Voc—Open-circuit voltage	64.6 V
Isc—Short circuit current	6.14 A
Vmpp—Voltage at maximum power point	54.7 V
Impp—Current at maximum power point	5.76 A
Pmpp—Power of maximum power point	315.07 W
Series-connected modules per string	6
Parallel strings	2

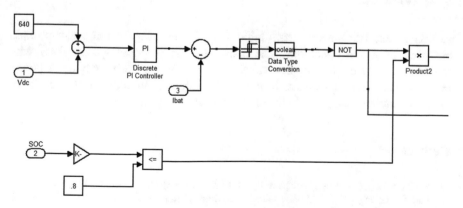

Fig. 13 PI charge controller for battery

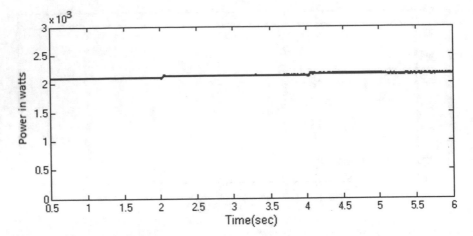

Fig. 14 Battery power regulated by charge controller

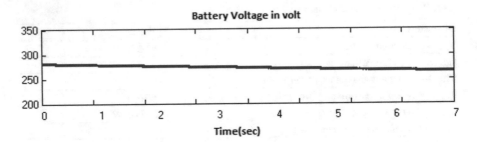

Fig. 15 Battery voltage

5 Conclusion

In this paper, P&O-based fuzzy logic MPPT controller is designed for rapid changing insolation conditions. The results of this proposed system show with reduced oscillations, less ripple, and faster convergence under maximum power point of steady state as well as in transient conditions. In addition, the results of MPPT controllers using FLC gives more power than usual methods.

References

1. S. Choudhury, P.K. Rout, "Adaptive Fuzzy Logic Based MPPT Control for PV System under Partial Shading Condition", International Journal of Renewable Energy Research Vol. 5, No. 4, 2015.
2. H. Bounechba, A. Bouzid, K. Nabti and H. Benalla, "Comparison of perturb & observe and fuzzy logic in maximum power point tracker for PV systems", Energy Procedia 50 (2014) 677–684.

3. Bader N. Alajmi, Khaled H. Ahmed, Stephen J. Finney, and Barry W. Williams. "Fuzzy-Logic-Control Approach of a modified Hill-Climbing Method for Maximum Power Point in Microgrid Standalone Photovoltaic System", IEEE Transactions on Power Electronics, Vol. 26, No. 4, April 2011.
4. T. Bogaraj, J. Kanakaraj, J. Chelladurai, "Modeling and simulation of stand-alone hybrid power system with fuzzy MPPT for remote load application", Archives of Electrical Engineering Vol. 64(3), pp. 487–504 (2015).
5. Po-Chen Cheng, Bo-Rei Peng, Yi-Hua Liu, Yu-Shan Cheng and Jia-Wei Huang, "Optimization of a Fuzzy-Logic-Control-Based MPPT Algorithm Using the Particle Swarm Optimization Technique", Energies 2015, 8, 5338–5360.
6. Shiau, J.K.; Wei, Y.C.; Lee, M.Y. Fuzzy Controller for a Voltage-Regulated Solar-Powered MPPT System for Hybrid Power System Applications. Energies 2015, 8, 3292–3312.
7. L. Fangrui, K. Yong, Z. Yu, and D. Shanxu, "Comparison of P&O and hill climbing MPPT methods for grid-connected PV converter," in Proc. 3rd IEEE Ind. Electron. Appl. Conf., 2008, pp. 804–807.
8. B Wichert and Wb Lawrance, "Application of Intelligent Control Methods to the Management of Modular Hybrid Energy Systems", Proceedings of Solar '97—Australian and New Zealand Solar Energy Society.
9. Won, Chung-Yuen, Duk-Heon Kim, Sei-Chan Kim, Won-Sam Kim, and Hack-Sung Kim. "A new maximum power point tracker of photovoltaic arrays using fuzzy controller." In Power Electronics Specialists Conference, PESC'94 Record., 25th Annual IEEE, vol. 1, pp. 396–403. IEEE, 1994.
10. Esram, Trishan, and Patrick L. Chapman. "Comparison of photovoltaic array maximum power point tracking techniques." IEEE Transactions on energy conversion 22, no. 2 (2007): 439–449.
11. Veerachary, Mummadi, Tomonobu Senjyu, and Katsumi Uezato. "Neural-network-based maximum-power-point tracking of coupled-inductor interleaved-boost-converter-supplied PV system using fuzzy controller." IEEE Transactions on Industrial Electronics 50, no. 4 (2003): 749–758.
12. Patcharaprakiti, Nopporn, and Suttichai Premrudeepreechacharn. "Maximum power point tracking using adaptive fuzzy logic control for grid-connected photovoltaic system." In Power Engineering Society Winter Meeting, 2002. IEEE, vol. 1, pp. 372–377. IEEE, 2002.
13. Algazar, Mohamed M., Hamdy Abd El-Halim, and Mohamed Ezzat El Kotb Salem. "Maximum power point tracking using fuzzy logic control." International Journal of Electrical Power & Energy Systems 39, no. 1 (2012): 21–28.
14. Cheikh, MS Aït, C. Larbes, GF Tchoketch Kebir, and A. Zerguerras. "Maximum power point tracking using a fuzzy logic control scheme." Revue des energies Renouvelables 10, no. 3 (2007): 387–395.
15. Al Nabulsi, Ahmad, and Rached Dhaouadi. "Efficiency optimization of a DSP-based standalone PV system using fuzzy logic and dual-MPPT control." IEEE Transactions on Industrial Informatics 8, no. 3 (2012): 573–584.
16. Larbes, C., SM. Ait Cheikh, T. Obeidi, and A. Zerguerras. "Genetic algorithms optimized fuzzy logic control for the maximum power point tracking in photovoltaic system." Renewable Energy 34, no. 10 (2009): 2093–2100.
17. Unal Yilmaza, Ali Kircaya and Selim Borekci "PV system fuzzy logic MPPT method and PI control as a charge controller" Volume 81, Part 1, January 2018, Pages 994–1001.
18. Lalouni, S., D. Rekioua, T. Rekioua, and Ernest Matagne. "Fuzzy logic control of stand-alone photovoltaic system with battery storage." Journal of Power Sources 193, no. 2 (2009): 899–907.

19. Salas, V., E. Olias, A. Barrado, and A. Lazaro. "Review of the maximum power point tracking algorithms for stand-alone photovoltaic systems." Solar energy materials and solar cells 90, no. 11 (2006): 1555–1578.

20. A Novel Negative-Output KY Boost Converter," Industrial Electronics(ISIE) 2009, ISBN No: 978-1-4244-4349-9, 272–274.

21. Kuo-Ing Hwu† and Wen-Zhuang Jiang, "A KY Converter Integrated with a SR Boost Converter and a Coupled Inductor" Journal of Power Electronics, Vol. 17, No. 3, pp. 621–631, May 2017.

Improvement of Low Voltage Ride Through Capability of Grid-Connected DFIG WTs Using Fuzzy Logic Controller

M. Kuzhali, S. Joyal Isac and S. Poongothai

Abstract Due to its advancements and enhanced usability, Doubly Fed Induction Generator find its application in almost the Wind Turbines and in all wind Power Generation. Grid-connected wind turbine will generally experience faulty condition; it will lead to fault ride through of the turbine which in turn reduces the reliability of the system. To improve the low voltage ride through capability of wind turbine, a control strategy is to be framed. The existing crowbar method for DFIG is to protect Rotor Side Converter and Grid Side Converter but makes the system weaker by absorbing more reactive power from the system. The proposed method works with Fuzzy Logic controller to limit the deviations of steady-state values during and after faults in the grid and thereby reduces the need of crowbar for the system. The designed fault detection and confrontation system manages to attenuate the disturbance during fault which would supply the reactive power to the grid. Simulation result shows that the controller successfully limits the rotor overcurrent. The MATLAB/SIMULINK TOOL is used to model the controller and test the validity of the control over the Grid-connected wind turbine over weak ac signal.

Keywords Fuzzy control · DFIG · LVRT

1 Introduction

The grid-connected Doubly Fed Induction Generator (DFIG) contains wound rotor and stator windings. The stator windings are connected to grid directly and rotor winding and the grid are connected via two back-to-back converters. This converter provides a optimized control of active and reactive power with cost effective and smaller in size. Although DFIG absorbs more reactive power from the grid when

M. Kuzhali (✉) · S. Joyal Isac
Saveetha Engineering College, Chennai, India
e-mail: kuzhali@saveetha.ac.in

S. Poongothai
Velammal Engineering College, Chennai, India

© Springer Nature Singapore Pte Ltd. 2019
M. A. Bhaskar et al. (eds.), *International Conference on Intelligent Computing and Applications*, Advances in Intelligent Systems and Computing 846,
https://doi.org/10.1007/978-981-13-2182-5_33

subjected to fault [1]. Whenever fault occurs, at the stator windings, due to the magnetic coupling, it results in change in stator flux of the DFIG which provides an overcurrent to the rotor Windings [2].

Due to this current, the rotor side converter is affected and that leads to large fluctuations of DC-LINK voltage. Crowbar circuit consists of bank of resistors and with the help of power electronic devices, it is connected to the rotor windings. Whenever the fault occurs, Rotor Side Converters is disabled temporarily and crowbar circuit is connected to the rotor windings. There is considerable loss of output power and hence this type of protection is not much effective. When the large transients and faults occur in the grid crowbar, protection is deactivated and DFIG is disconnected from the grid.

Nowadays, wind generator contributes substantial amount of power over the total power generation in India. As per grid code requirements, the WT should have LVRT capability during grid faults resulting in more than 85% voltage drop. This clearly shows that they should provide supply power to the grid during and after grid faults [3].

In [4], proposes a control by flux linkage tracking control method. The control system does not depend on the system parameters and for that decoupling of DC and negative sequence components are not necessary. Even though the proposed method manages to enlarge the control action, but it has many drawbacks that could not be avoided. Properly designed fuzzy controller (FC) is better than a traditional proportional-integral (PI) controller. There are many research works are carried out to prove that the FC manages to limit the rotor current during the fault, with the elimination of external devices. But in these, work grid code requirements are not described properly. This control scheme can be applied to the small voltage dip. Fuzzy logic controller acts as a protective device when the DFIG is subjected to external fault when compared to traditional PI controller, result FC satisfies the grid code requirements.

Based on the various analysis and results studied from works so far, this paper interested to improve [6] safety of the DFIG without the presence of added hardware. This endeavor is applied to DFIG during asymmetrical fault [5]. The proposed method contributes to the most optimum coordination of the two converters. The objective is to improve system stability and reduce the disturbances to the system. In order to come across from the difficulties, the controller was proposed based on Fuzzy Logic Controller (FLC). By using the concept FC model, the rotor overcurrent and dc link overvoltages are effectively diminished. In addition, the grid code requirement fulfilled such as FRT proficiency of the DFIG and reactive power supplied to the machine.

2 Modeling of DFIG

Figure 1 shows the generalized schematic arrangement of doubly fed induction generator. As per grid code requirement, the DFIG has been modeled in such a way that the low-speed WT shaft should be linked to high-speed generator shaft through

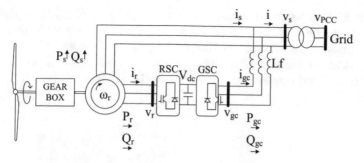

Fig. 1 General schematic diagram of DFIG

gear box. The DFIG is a variable speed wind generator, it consists of two windings rotor and stator side, where the stator windings are linked to a grid output without any intermediary and rotor windings are linked via two back-to-back converters, namely RSC and GSC. The ac/dc/ac converter is an IGBT-based PWM converter.

2.1 Modeling of Wind Turbine

The efficient construction of Wind Turbine Rotor plays a major role to accomplish maximum wind power generation. To achieve an optimum wind power generation design of wind turbine blades, diameter of rotor, blade pitch angle, the transmission system and gear box, blade chord, and tower length should be determined. The maximum power obtained from the wind is directly depends on the speed of the rotor and maximum power can be expressed as

$$P_{\max} = \frac{1}{2}\rho C C_{\mathrm{p}} \pi R^2 V_{\mathrm{w}}^3 \tag{1}$$

At the optimum operating point, the maximum value of power can be written as

$$P_{\max} = \frac{1}{2} C_{\mathrm{opt}} \pi \left(R^5 | \lambda_{\mathrm{opt}}^3 \right) \tag{2}$$

$P_{\max}$	Maximum power
R	Radius of the rotor
λ	Tip speed ratio
ω	Rotor speed
V_{w}	Wind speed.

The maximum power extraction is the function of tip speed ratio and pitch angle, so that the specific value of TSR (λ) should be maintained. Figure 2 shows typical characteristics of the wind turbine using the C_p versus λ curve.

Turbine speed versus turbine output power is shown in ABCD curve in pitch angle β maintained constant up to rated speed (D) as shown in Fig. 3.

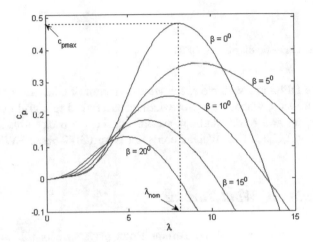

Fig. 2 C_p versus λ curve

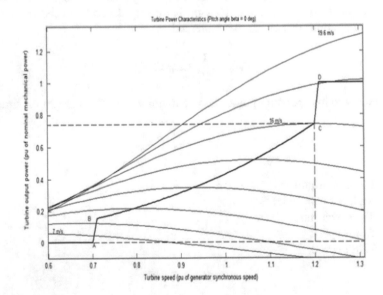

Fig. 3 ABCD curve

2.2　Mathematical Modeling of DFIG

The studied system represented by the following equations, the mechanical power of the generator (stator and rotor) are calculated as follows:

$$P_{\text{stator}} = T_{\text{m}} * \omega_{\text{s}} \tag{1}$$

$$P_{\text{rotor}} = T_{\text{m}} * \omega_{\text{r}} \tag{2}$$

For ideal generator, the relationship between the total mechanical torque is calculated using

$$T_{\text{m}} = T_{\text{sm}} \text{ and } P_{\text{m}} = P_{\text{s}} + P_{\text{r}} \tag{3}$$

$$P_{\text{r}} = P_{\text{m}} - P_{\text{s}} = T_{\text{m}}\omega_{\text{r}} + T_{\text{em}}\omega_{\text{s}} = -sP_{\text{s}} \tag{4}$$

where $s = (\omega_{\text{s}} - \omega_{\text{r}})/\omega_{\text{s}}$.

Based on the dynamic operating characteristics of DFIG, the voltage equations are given by

$$V_{\text{qs}} = -R_{\text{s}}i_{\text{qs}} + \omega_{\text{s}}\lambda_{\text{ds}} + \frac{\mathrm{d}\lambda_{\text{qs}}}{\mathrm{d}t} \tag{5}$$

$$V_{\text{dr}} = R_{\text{r}}i_{\text{dr}} - (\omega_{\text{s}} - \omega_{\text{r}})\lambda_{\text{qr}} + \frac{\mathrm{d}\lambda_{\text{dr}}}{\mathrm{d}t} \tag{6}$$

$$V_{\text{qr}} = R_{\text{r}}i_{\text{qr}} + (\omega_{\text{s}} - \omega_{\text{r}})\lambda_{\text{dr}} + \frac{\mathrm{d}\lambda_{\text{qr}}}{\mathrm{d}t} \tag{7}$$

where

$V_{\text{qs}}, V_{\text{ds}}, i_{\text{qs}}, i_{\text{ds}}$	d–q-axis of stator components
$\lambda_{\text{qs}}, \lambda_{\text{ds}}$	direct and quadrature axis stator flux linkage
$V_{\text{qr}}, V_{\text{dr}}, i_{\text{qr}}, i_{\text{dr}}$	d–q-axis rotor components
$\lambda_{\text{qr}}, \lambda_{\text{dr}}$	direct and quadrature axis rotor flux linkages
$R_{\text{r}}, R_{\text{s}}$	rotor and stator resistances of generator per phase.

Optimized control system for the DFIG has been separated into two converters RSC and GSC as shown in Figs. 4 and 5. These two converters are generally used to obtain regulated voltage by controlling real and reactive power. In order to reach enhanced controlling of active and reactive power, the park transformation of current components is accomplished using a reference frame oriented technique where q-axis current component controls the active power in stator side.

The efficiency of the system is influenced by the load characteristics of the entire system, thus the Maximum Power Point Tracking (MPPT) Technique is employed to determine the feasible operating point of the system that can be given to the reference value of the active power. The input of the power controller is the

354

Fig. 4 Controller—rotor

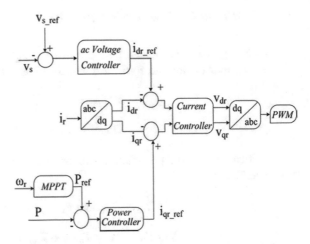

Fig. 5 Controller—grid

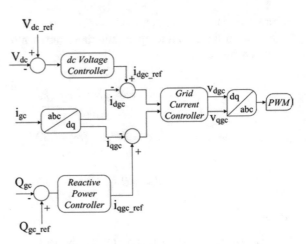

difference between P_r and $P_{r\text{-ref}}$, the error is given to the power controller. The observed value of the q-axis rotor current is related with the actual value of i_{qr} and the error is driven to the current controller. This output value has been considered as reference voltage for q-axis component v_{qr}.

When the DFIG is subjected to fault and connected with weak power system, the RSC can be activated to control active and reactive power. In case, the DFIG is linked with strong power system, the control action is not involved. In this work, the DFIG is linked to weak ac grid and unconventional performance of the system is studied and appropriate control action has been taken instead of reactive power control. The computed voltage from the generator terminals is compared with its reference value and error signal for the d-axis current. The current controller provides a reference voltage for the d-axis rotor terminals in which input of the controller is obtained by comparing the error signal from voltage regulator and the *d*-axis current. The desired output signal v_{dr} and v_{qr} are transformed into ABC components. These components are driven by the PWM module.

The dc link voltage is kept as constant with the help of GSC. In this paper, the reactive power is considered to be neutral by setting $Q_{gc\_ref} = 0$. The control system of GSC is shown in Fig. 5. The reference frame-oriented vector control is applied for three-phase quantities into dq transformation. The dc voltage and the reactive power can be controlled by *d*-axis reference frame current and *q*-axis reference current, respectively.

3 Optimized Control System

The optimized control system intends to have enhanced LVRT capability of the DFIG independently. RSC is modified with fine-tune fuzzy controller. In other words, the modified RSC makes it an optimized control system. Figure 6 shows optimized control system which generated PWM signal from the current voltage and power values. Depending on the deviation of the percentage of voltage control, block is activated (or) else set to be ideal. Only 10% of voltage sag is allowed from the reference value. The abstraction of the control strategy is achieved by considering two major deviations in RSC. By the means of the drastic decrement in rotor overcurrent and dc link overvoltage will lead to successful protection of DFIG. Due to sudden changes in the system, dc voltage and rotor current will go beyond their maximum values. During the time of transient, the existing quantity of energy that

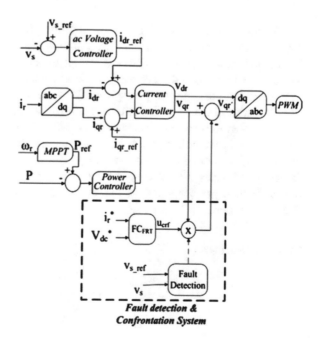

Fig. 6 Optimized control system

is to be appropriately pumped through converter and connected to the grid, in order to fetch the value of the rotor current and dc voltage back to their desired value.

DC voltage can be attenuated or eliminated by minimizing the rotor current but the rotor current cannot be minimized after a certain value. Fault ride through proficiency of the DFIG can be achieved only on taking rectifying signals into account and also the value of dc voltage should be considered.

3.1 Fuzzy Controller

Fuzzy logic controller controls the speed and power of DFIG. It is known for its precision and it can be implemented in simple manner. Mamdani-type fuzzy controller is used due to its robustness of control. Rotor current and dc voltage are input signals of fuzzy controller and output signal is command signal generator which will be compared with the quadrature axis rotor voltage and this signal from the fault detection. The error between the actual and the reference value of current is minimized by calculating an index in Fuzzy Controller, FC_{FRT}. The input values of FC_{FRT}, V'_{dc}, i'_r are given by Eqs. 8 and 9. (Table 1)

$$V'_{dc} = \frac{V_{dc} - V_{dc\,ss}}{V_{dc\_mv} - V_{dc\_ss}} \tag{8}$$

$$i'_r = \frac{i_r - i_{r\,ss}}{i_{r\_mv} - i_{r\_ss}} \tag{9}$$

The actual values of V'_{dc}, i'_r are processed through fuzzy control system. They are made as fuzzy values when it is processed through fuzzy interference system. Triangular membership function will be apt for both voltage and current as their values will increase and reach peak value at a time. The modulation index (U_{crf}) calculated from the fuzzy expert system finds the deviation from the settled value. The deviations include both positive and negative deviations. Positive deviations shall be taken into account leaving out negative deviations. The fuzzification process is done in Mamdani system. The rules framed in fuzzy set between two inputs and the index is classified into five subsets for the three different values [7] classification of inputs. Neglecting of negative values of U_{crf} will lead to easy defuzzification process when we can obtain the actual value to be applied to the controller system.

Table 1 FC_{FRT} inputs

i'_r	V'_{dc}		
	S	M	B
S	OK	SP	BP
M	SN	OK	SP
B	BN	SN	OK

4 Simulation Result

MATLAB STIMULATION TOOL is used for stimulation work. The twice fed induction generator (1.5 MW) is modeled which supplies electrical power to a weak power system. A three-phase fault is introduced to an electrical system at $t = 1.2$ s resulting in voltage sag of 85% dip in normal voltage. The simulation output is compared for the above electrical system, the proposed control system moderates the fault periods and the response during and after fault periods seems to be good which shows that this control system can make the DFIG to successfully overcome the fault.

By the rapid recovery system of the proposed control system, the need for large amount of reactive power from the DFIG during faulty conditions can be reduced. DFIG along with the proposed control system supplies the required amount of reactive power to the grid which maintains the ac voltage of the grid. Apart from good ride through capability of DFIG, Rotor Side Converter is continuously being active throughout the fault period and after the fault period, thereby supplies reactive power to the grid, which is useful for balancing the voltage caused by three-phase fault

Table 2 Parameters of the ac grid

Rated voltage	25 kV
Rated frequency	60 Hz
Short circuit ratio at the PPC	2.23

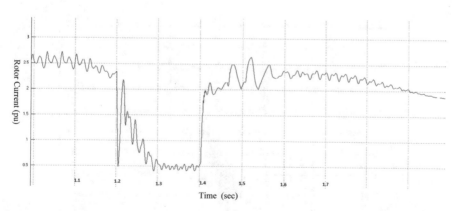

Fig. 7 Rotor current with controller

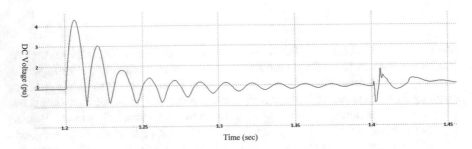

Fig. 8 DC voltage with controller

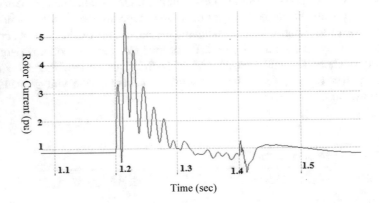

Fig. 9 Rotor current without controller

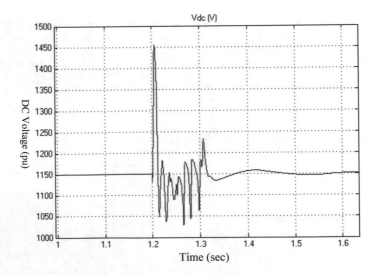

Fig. 10 DC voltage without controller

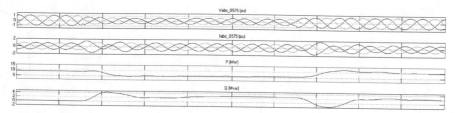

Fig. 11 Response of the system with controller

5 Conclusion

Improved LVRT capability grid-connected DFIG has been implemented with the absence of external hardware. This paper explains the DFIG could feed the electrical system during the fault and after the fault period. It can be achieved through fuzzy controller by effectively maintaining the current flowing through the rotor and capacitor voltages within the limit. This new enhanced system makes our wind power generation system more stable with greater efficiency.

References

1. J. Lopez, P. Sanchis, X. Roboam, and L. Marroyo, "Dynamic behavior of the doubly fed induction generator during three-phase voltage dips," IEEE Trans. Energy Convers., vol. 22, no. 3, pp. 709–717, Sep. 2007.
2. F. Lima, A. Luna, P. Rodriguez, E. Watanabe, and F. Blaabjerg, "Rotor voltage dynamics in the doubly fed induction generator during grid faults," IEEE Trans. Power Electron., vol. 25, no. 1, pp. 118–130, Jan. 2010.
3. D. Xiang, L. Ran, P. J. Tavner, and S. Yang, "Control of a doubly fed induction generator in a wind turbine during grid fault ride-through," IEEE Trans. Energy Convers., vol. 21, no. 3, pp. 652–662, Sep. 2006.
4. S. Xiao, G. Yang, H. Zhou, and H. Geng, "An LVRT control strategy based on flux linkage tracking for DFIG-based wecs," IEEE Trans. Ind Electron., vol. 60, no. 7, pp. 2820–2832, Jul. 2013.
5. L. Xu, "Coordinated control of DFIG's rotor and grid-side converters during network unbalance," IEEE Trans. Power Electron., vol. 23, no. 3, pp. 1041–1049, May 2008.
6. S. Hu, X. Lin, Y. Kang, and X. Zou, "An improved low-voltage ride through control strategy of doubly fed induction generator during grid faults," IEEE Trans. Power Electron., vol. 26, no. 12, pp. 3653–3665, Dec. 2011.
7. S. Joyal Isac, K. Suresh Kumar, "Optimal capacitor placement in radial distribution system to minimize the loss using fuzzy logic control and hybrid particle swarm optimization." Power electronics and renewable energy systems, 2015 – Springer, pp. 1319–1329.

Design, Development and Testing of a Downsized Offshore Wind Mill Floating Model

K. Subha Sharmini

Abstract The principal goal of this article is to gauge the challenges that are faced in the process of design, development, and deployment of the downsized floating model of an offshore wind turbine. A 100:1 scale model of a ballast/spar buoy buoyant wind turbine is designed and built. The model is then subjected to testing, i.e., preliminary validation check for the design is done with the help of a locally created water flume tank environment.

Keywords Built · Ballast · Design · Floater · Offshore · Small scale
Spar buoy · Testing

1 Introduction

Since medieval days, human beings, the social animals have been interdependent on various earth-based energy resources for their survival. Apart from having food and shelter as basic needs, power has also become one of their mandatory requirements. The power resources are generally classified as renewable/sustainable and nonrenewable/conventional sources. Although coal, oil, and nuclear fuel are the major supply resources, they are scarce in availability and will be depleted in the near future. Hence, all the countries are on the lookout for suitable alternative sources of energy in the form of wind, solar, biomass, etc.

As per the statistics indexed in Fig. 1, India due to its geographical location enjoys having wind and coal as its prime resources of energy. Wind as a natural resource, when churned in an effective manner, produces an enormous yield. Wind energy is converted into useful electrical energy with the help of wind energy conversion systems (WECS). The question that now arises is whether to locate the WECS offshore/onshore? The answer is given in Table 1.

K. S. Sharmini (✉)
SRM IST, Kattankulathur, Chennai, India
e-mail: subha_kannan@srmuniv.edu.in

© Springer Nature Singapore Pte Ltd. 2019
M. A. Bhaskar et al. (eds.), *International Conference on Intelligent Computing
and Applications*, Advances in Intelligent Systems and Computing 846,
https://doi.org/10.1007/978-981-13-2182-5_34

INSTALLED CAPACITY, MW

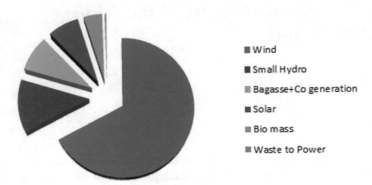

- Wind
- Small Hydro
- Bagasse+Co generation
- Solar
- Bio mass
- Waste to Power

Fig. 1 Installed power capacity in India

Table 1 Offshore and onshore wind turbines—comparison table

Parameters	Onshore	Offshore
Wind speed	Moderate	High
Area	Restriction	No such restriction
Visual impact	Bad impact	Zero impact
Noise	Affects surroundings	Minimal effects
Efficiency	High	Higher

2 Offshore Wind Turbines

Professor William E. Heronemus of University of Massachusetts was the visionary man behind large-scale offshore floating wind turbines [1]. Such systems can either have a fixed bottom or floating bottom. The present day fixed bottom technology-based wind turbines are deployed only up to a water depth of 20 m. But when it comes to deeper water, floating bottom becomes the suitable choice. Thousands of marine and offshore oil rigs that have been already set up in a widespread manner all around the world, are the fore runners for the successful demonstration of these floating structures [2–8].

3 Components and Classification

The anatomy of any wind energy conversion systems [9] (WECS) is sectionalized into

- Top mass including wind turbine (Blades + hub), gear mechanism, and nacelle (Pitch/Yaw controller, generator).
- Tower.
- Bottom mass—Platform or floating platform in case of offshore wind turbines.

3.1 Components

The choice of the blade number decides the power generation, cost, and efficiency. The number may range 2–3 and the length of the blades may vary generally between 30 and 50 m. The design and materials chosen adjudicate the weight of the blades. Blades in conjunction with the hub form the rotor. The gearbox, generator, and the controller are housed inside the nacelle. The gearbox adjusts the speed to torque ratio between the turbine and the generator. The controller keeps a check on the blade pitch and nacelle yaw. The control system aims to augment power productivity [10].

Induction generators or permanent magnet alternators are commonly employed for converting mechanical energy (turbine rotation) into electrical energy. Towers are tall standing structures that support the top mass and if it is to be an offshore type, it acts as the connector between top mass and floating base. The elevations of the towers are between 60 and 100 m. Steel is the preferable material and lattice is the popular shape.

The presence of the floating body is the one that distinguishes the offshore wind turbines from its onshore counterpart. The mandatory requirement of the floater is that it should have enough buoyancy to support the entire structure in open sea. Hence, various categories of offshore wind turbine plants are excogitated solely based on the platform.

3.2 Classification

Shallow water turbines use concrete gravity bases or driven monopiles for their foundation. But deep water wind turbines use a floating structure that provides enough buoyancy to support the weight of the turbine. The structure must also restrain pitch, roll and heave motions within acceptable limits [11]. Given below is a brief vignette on the prevalent types of deepwater floating wind turbine.

Ballast/Spar Buoy. Ballast type has a central buoyancy tank below which ballast weights are hung to achieve stability. This arrangement creates the high inertial resistance, necessary righting moment and draft against roll, pitch, and heave motion [4, 12]. Figure 2 represents a typical spar buoy structure. Its floating base constitutes cylindrical tanks (steel/concrete) replete with ballast. This assures that the structure stands upright and afloat in sea as the center of buoyancy becomes greater than the center of gravity. Grout, solid iron ore, concrete, or gravel can be used for ballast. Secure anchorage is provided with mooring. Catenary spread mooring system or single vertical tendon with swivel connection-based mooring is used. Although swivel arrangement permits turning of wind turbine with wind, it may cause the floater to drift.

Tension Leg Platform/Mooring Line Type. The design derives its name because of a floater held in position with the help of pretensioned vertical tethers

Fig. 2 Spar buoy type

(tendons) for its righting stability. A template foundation or pile-driven anchors or suction caissons may be used for anchorage [4, 12]. The TLP type has a sluggish dynamic response when compared with its other counterparts (Fig. 3).

Buoyancy/Barge/Pontoon Type. Barge floaters utilize weighted water plane area/distributed buoyancy to achieve righting moment/stability. As illustrated in Fig. 4, it consists of a raft-like base on which one or more than one wind turbines may rest. Tethering/mooring is achieved by means of catenary anchor chains. However, the major obstacle of this type is its high degree of sensitivity to roll and pitch motion. Placid locations are condign for the setting up of such types.

4   Design of the Downsized Prototype

The strong and weak points of the aforementioned types were analyzed. On the basis of which it was settled to fix on designing a spar buoy type of wind floater. The design calculations and procedure accord with the article [10] except for some minor differences to suit the local conditions and to trim money spent out of pocket. Froude scaling was used to the downsizing of the upscale model (NREL baseline wind turbine) [10]. Length measurements are reduced by 100:1 and weights and forces by 106:1 and 1010:1 for moments of inertia. The chief parts of the spar buoy base are cylindrical tank (platform) with top and bottom covers (Fig. 5) and ballast. The measures of these components are as recorded in Table 2. The decisions and design procedures are as follows.

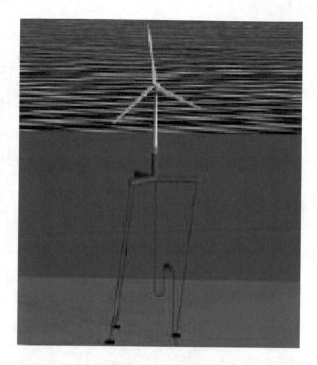

Fig. 3 Tension-leg platform type

Fig. 4 Pontoon type

4.1 Platform

Platform is a cylindrical tank with both the ends closed. The tower of the wind turbine is attached to the top cover. PVC is selected to be the material for tank as it has many credible characteristics like high buoyancy, flexibility, water tightness, safety, and nominal cost. Ballast is introduced into this tank.

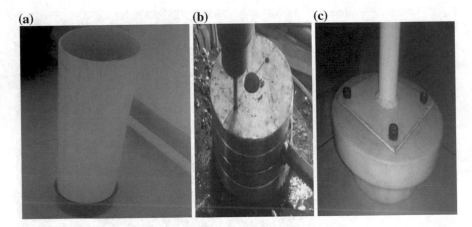

Fig. 5 Platform, steel ballast, and nylon top cover with tower attached

Table 2 Design specifications

Part	Outer diameter (cm)	Inner diameter (cm)	Height (cm)	Weight (kg)
Platform	14	13.3	60	1.51
Nylon top cover	17	13.9	–	0.23
Ballast	13	2.5	8	5.25

4.2   Tank Cover

Tank cover is used to cover the open ends of the platform/cylindrical tank. The upper tank cover forms the link between platform and tower of the top mass. Upper cover was purposely designed in such a way that bending and buckling of walls are less likely to occur once the tower is assembled and attached to the top of the cover. A round nylon block (Fig. 5c) was shaped to fit into the platform. The tower is then riveted on the other side of the cover with bolt and nut. Nylon block is strong as well as it has an air locking feature which makes it waterproof and thus fulfill our requirement.

4.3   Ballast

Ballast has a conspicuous task of providing the required balance for the structure to stay afloat. According to [13] Sungho Lee's thesis report, the center of gravity of the spar buoy should be approximately 39.68 cm below the surface of the tank, or 20.32 cm above the bottom of the tank. A better ballast configuration should be explored, such that it will result in more space between the ballast and the center of gravity for possible component adjustments and positioning changes. Thus, only steel plates comply with all requirements for the ballast. The reason is the higher density of steel (than aluminum). The number of steel plates can be adjusted according to the requirement of the ballast height. The design has four number of steel plates located at the base of the tank (Fig. 5b).

5 Testing

5.1 Field Test

To validate the prototype, the field study was conducted at different local conditions. Data such as wind speed, shaft speed, and voltage output was observed as indicated in Fig. 6.

5.2 Floatation Test

The floatation check of the developed model was done in a nearby swimming pool having a water depth of 1.54 m. This preliminary testing was carried out in three stages.

- In the first stage, the base without ballast was placed in the pool.
- In the second stage, the steel plates ballast was introduced into the tank.
- As the last step, the top mass of wind turbine fixed onto the upper cover was used to seal the spar buoy platform (Fig. 7).

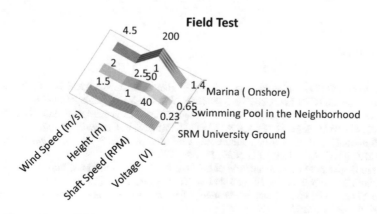

Fig. 6 Observations—field test

Fig. 7 Floatation test

When the spar buoy was thus placed in the pool, water did not enter into the base assembly. Hence, the steel plates placed in the ballast remained moisture free. This shows that the platform is waterproof. Hence, floatation test for the prototype proved to be a success.

6 Conclusion

A 100:1 reduced scale semisubmersible spar buoy-type floating wind turbine was designed and it had undergone floating test which was proven to be successful. Minor refinements (coating of ballast with noncorrosive material and platform inside with ABS cement) are desired for long-term survivability of the designed prototype. Means of tethering for better anchorage and stability should and will be provided in time ahead.

References

1. Heronemus W E: Pollution free energy from offshore winds. In: 8th ANNUAL CONFERENCE AND EXPOSITION MARINE TECHNOLOGY SOCIETY, Washington (1972).
2. S Butterfield, W Musial and J Jonkman.: Engineering challenges for offshore wind turbines. In: COPENHAGEN OFFSHORE WIND CONFERENCE, Denmark (2005).
3. S Butterfield and W Musial.: Future for offshore wind energy in the United States. In: ENERGY OCEAN PROCEEDINGS, Florida (2005).
4. Hasan Bagbanci.: Dynamic analysis of offshore floating wind turbines. In: DISSERTATION, Instituto Superior Tecnico, Universidad de Tecnica de Lisboa (2011).
5. Lee Sungho.: A non—linear wave model for extreme and fatigue responses of offshore floating wind turbines.

6. Zhang Yong Li and Li Jie.: Seabed instability simplified model and application in offshore wind turbine. In: 2nd INTERNATIONAL CONFERENCE ON POWER ELECTRONICS AND INTELLIGENT TRANSPORTATION SYSTEM, (2009).
7. G Betti, M Farina, A. Marzorati, R Sacttolini and G. A Guagliardi.: Modeling and control of a floating wind turbine with spar buoy platform. In: IEEE CONFERENCE PROCEEDINGS, (2012).
8. Tore Bakka, Hamid Reza Karimi, Soren Christiansen.: Linear parameter varying modeling and control of an offshore wind turbine with constrained information. IET Control Theory and Applications 8(1), 22–29 (2014).
9. Gary L Johnson.: Wind energy systems. 2nd edn. Kansas State University, (2006).
10. S Christiansen, S M Tabatabaeipour and T Knudsen.: Wave disturbance reduction of a floating wind turbine using a reference model based predictive control. In: AMERICAN CONTROL CONFERENCE, Washington (2013).
11. Musial W D, Butterfield and Boone A.: Feasibility of floating platform systems for wind turbines. In: 23rd ASME WIND ENERGY SYMPOSIUM PROCEEDINGS, Nevada (2004).
12. J M Jonkman: Dynamics modeling and load analysis of an offshore floating wind turbine. In: NREL TECHNICAL REPORT, Colorado (2007).
13. Justin Frye, Nathan Horvath and Arnold Ndegwa.: Design of scale model floating turbine: Spar buoy. In: MQP REPORT, Worcester Polytechnic Institute (2011).

Fuzzy Logic Controller-Based Bridgeless Cuk Converter for Power Factor Correction

R. Meena Devi, M. Kavitha, V. Geetha and M. Preethi Pauline Mary

Abstract The Fuzzy logic controller-based single-phase bridgeless Cuk PFC converter with positive output voltage is proposed in this paper. This converter is used for both buck and boost operations. In case of low voltage output applications, it acts in buck mode which converts high input voltage to low output voltage. Whereas for applications involving high output voltage, it operates in boost mode which converts low input voltage into high output voltage. The conduction losses are reduced and efficiency of this system can be improved due to the absence of bridge diodes. Since bridge diodes are absent, the rectifier has a single switch, thus reducing the complexity of the circuit. The current loop circuit is not required, as discontinuous mode of operation is involved in rectifier. The simulation is put into execution by using MATLAB/Simulink software. The results of both open and closed loop were executed and compared. In closed loop, Fuzzy Logic Controller (FLC) provides reduced output voltage ripple with less settling time and better power factor.

Keywords Cuk · Bridgeless · PFC · FLC · Positive output voltage

R. Meena Devi · M. Kavitha · V. Geetha · M. Preethi Pauline Mary (✉)
Sathyabama Institute of Science and Technology, Chennai, India
e-mail: preethi.drives@gmail.com

R. Meena Devi
e-mail: devimalathi2010@gmail.com

M. Kavitha
e-mail: kaveem@gmail.com

V. Geetha
e-mail: geethasendray28@gmail.com

© Springer Nature Singapore Pte Ltd. 2019
M. A. Bhaskar et al. (eds.), *International Conference on Intelligent Computing and Applications*, Advances in Intelligent Systems and Computing 846,
https://doi.org/10.1007/978-981-13-2182-5_35

371

1 Introduction

The rectifier converts ac input to dc output. The use of transformers, bridge rectifiers, and capacitors can achieve this easily but there will be serious distortions in
the input current. In order to prevent this, power factor corrector is used [1, 2].
Therefore, power factor correctors are required for ac-dc conversion. Due to its
simplicity, the converters are widely used, but for PFC applications, the buck
converters are used. Due to discontinuous input current, the buck converter may fail
to hold the control when output voltage is high compared to input voltage [3, 4].
A passive LC filter is used to filter the input of buck converter [5, 6]. At zero
crossing of the input line voltage, the current cannot follow the input voltage and
this increases total harmonic distortion [7, 8]. In addition to this, power factor is
also reduced. For such conditions, the Cuk converters are used next to bridge
rectifier. This converter is used to shape the input current while operating in DCM
and if a buck–boost converter operates in DCM, it has drawbacks such as high
current stress and discontinuous input current. The input current is continuous in
Cuk and SEPIC converter when operating in DCM and the output voltage is less
than the input voltage. Cuk produces the efficient and reliable output than the
SEPIC converter [9, 10]. The disadvantage of SEPIC converters is discontinuous
current at the output, which may lead to increased output ripple [11–16].

In this paper, bridgeless Cuk PFC converter under open loop and closed loop are
discussed. Section 2 deals with existing Cuk converter with negative output voltage. Section 3 describes the modes of operation of bridgeless Cuk PFC converter
with output positive voltage. Simulation results of Cuk converter under open and
closed loop using FLC is discussed in Sect. 4. Section 5 describes the conclusion.

2 Existing Bridgeless Cuk Converter with Negative Output Voltage

Figure 1 displays the bridgeless cuk converter with negative output voltage. In PFC
applications, the Cuk converters has the following merits—Inrush currents can be
protected, transformer isolation is easily implemented, ripple current is low, and
electromagnetic interference is less with DCM topologies. The Cuk converters are
used in an application that needs low input and output current both at source and
sink side as it produces continuous current in both input and output with reduced
ripple. Due to the increased current stress on the components of the circuit, the
conduction losses also increase in DCM mode practically. It is limited to operate in
DCM rather than CCM because of the cost of components, since it requires extra
components. Since we are using bridgeless Cuk converter, the cost of the circuit is
decreased due to reduced components in the circuit.

An inverse amplifier can be used to convert negative voltage to positive voltage.
The addition of the amplifier increases the cost of the circuit. To reduce the cost and

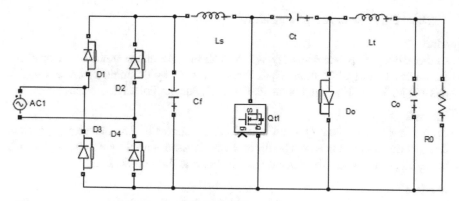

Fig. 1 Existing Cuk PFC with negative output voltage

complexity of the circuit, the polarity of all the components are reversed. So, the polarity of the circuit is reversed and the polarity of the output voltage is also reversed. Bridgeless Cuk converter with positive and negative output voltage is simulated and controlled using FLC converter and discussed.

3　Bridgeless Cuk Converter with Positive Output Voltage

After the change of polarities, the whole components in the circuit are reversed, so the output is also reversed. The whole circuit is transformed as shown in the Fig. 2

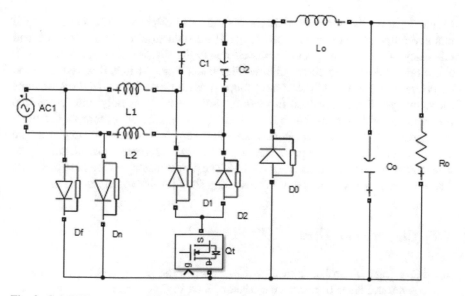

Fig. 2 Cuk PFC converter with positive output voltage

There are three modes of operation.

Mode 1
In this mode, the power switch Q_t is ON. The inductor at the input L_2 accumulates the energy linearly. The diode D_f is forward biased by the inductor current i_{L2}. In this mode, $V_{L0} = V_{ac}$. The output inductor i_{Lo} current linearly increases..

Mode 2
In this mode, the switch Q_t is off and diode D_o conducts. The inductor i_{L2} current delivers the energy linearly. The diode D_f is forward biased by current i_{L2}. In this mode $V_{Lo} = V_o$. Thus, the current i_{Lo} decreases and diode D_o is off.

Mode 3
In this mode, diode D_f is ON and provides path for the inductor current i_{L2}. Here, $V_{L2} = V_{Lo}$, therefore L_2 and L_o acts as constant current source. The capacitor C_2 is charged by i_{L2}. The capacitor C_o at the output releases energy to the load. This mode is called as freewheeling mode.

4 Results and Discussion

The Cuk PFC converter which is bridgeless is simulated using MATLAB software under open and closed loop. Fuzzy logic controller is used in closed loop.

4.1 Open Loop

Figure 3 shows the simulation circuit diagram of bridgeless Cuk converter with positive output voltage under open loop. The conduction losses are reduced and efficiency of this system can be improved due to the absence of bridge diodes [17]. Since bridge diodes are absent, the rectifier has a single switch, thus reducing the complexity of the circuit. The current loop circuit is not required, as discontinuous mode of operation is involved in rectifier. Since it involves only one switch, the power factor of the circuit can be improved with simple PWM technique. The input ac voltage of 90 V is supplied. Both input voltage and input current are in phase. The waveform is displayed in Fig. 4. The output voltage waveform is shown in Fig. 5. The voltage at the load side is around 48 V. The power factor was found to be 0.984. The ripple voltage is around 4 V. The settling time for this system is 0.3 s.

4.2 Closed Loop Using FLC

The fuzzy logic controller is designed to improve the power factor of bridgeless Cuk converter. Figure 6 shows the bridgeless Cuk PFC converter under closed loop using FLC.

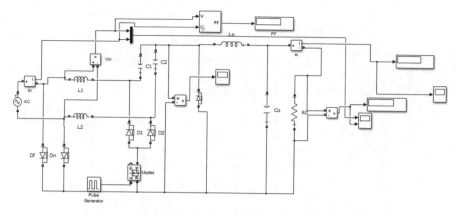

Fig. 3 Simulation circuit diagram of open loop bridgeless Cuk PFC converter

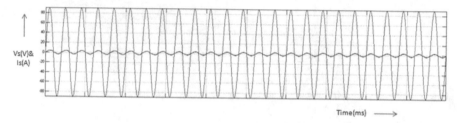

Fig. 4 Waveform of input voltage and current of the cuk converter under open loop

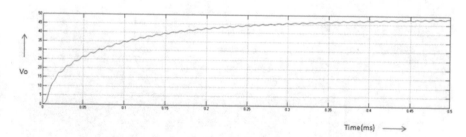

Fig. 5 Waveform of output voltage of the Cuk converter under open loop

Fuzzy logic controller requires linguistic rules. Detailed knowledge of the process to be controlled is required for framing the rules. FLC has four major blocks. In Fuzzification, crisp inputs are converted into fuzzy input. Knowledge base is the combination of rule base and database. Interference Engine fuzzy input to required fuzzy output using IF..THEN rules. Defuzzification converts fuzzy to crisp output. Figure 7 shows the inputs and output of FLC.

The membership functions of the input decide the output of the system. In the circuit, a feedback is given with desired membership functions at the input. Two inputs are given with error(e) and change in error(ce). Both the inputs are compared

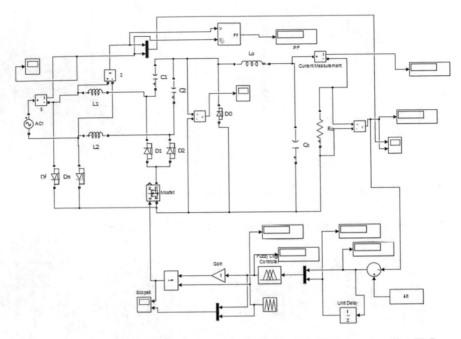

Fig. 6 Simulation circuit diagram of closed loop bridgeless Cuk PFC converter using FLC

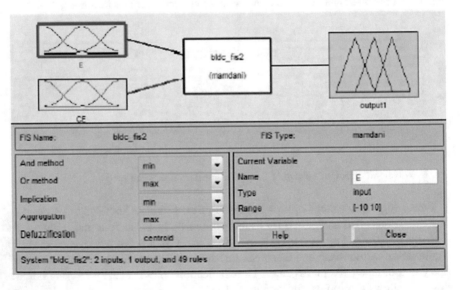

Fig. 7 Inputs and output of FLC

Table 1 Fuzzy control rules

e	ce						
	NB	KM	KS	ZE	PS	PM	PB
NB	KB	KB	KB	NB	NM	NS	ZE
NM	KB	KB	KB	NM	NS	ZE	PS
NS	KB	KB	KM	NS	ZE	PS	PM
ZE	KB	KM	ZE	PS	PM	PB	NM
PS	KM	KS	ZE	PS	PM	PB	PB
PM	KS	ZE	PS	PM	PB	PB	PB
PB	ZE	PS	PM	PB	PB	PB	PB

and the output is given based on the rules table of fuzzy logic. The fuzzy rules are shown in Table 1.

Thus the fuzzy logic controller is implemented in this paper to improve the power factor of the bridgeless cuk power factor converter.e Fig. 8 displays the input voltage and input current waveform of the bridgeless cuk converter under closed loop using FLC. The power factor was found to be 0.998. Figure 9 shows the waveform of the dc output voltage of the proposed Bridgeless Cuk Converter using fuzzy logic control. The output voltage is 48 V and the voltage ripple is 2 V. The settling time for this system is around 0.03 s. Specification is given in Table 2.

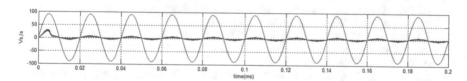

Fig. 8 Waveform of input voltage and current of the cuk converter under closed loop using FLC

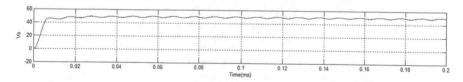

Fig. 9 Waveform of output voltage of the Cuk converter under closed loop using FLC

Table 2 Specifications

	Specifications
Input voltage	90 V ac, rms
Output voltage	48 V dc
Inductor L_1, L_2	1 mH
Inductor L_o	22 μH
Capacitor C_1, C_2	1 μF
Capacitor C_o	10 mF

5 Conclusion

The bridgeless Cuk PFC converter is simulated under open loop and closed loop. The input ac voltage is 90 V. By using bridgeless cuk PF converter, the output DC voltage is obtained as 48 V. The power factor of the circuit is 0.984 under open loop and 0.998 under closed loop using Fuzzy logic Controller. The output DC voltage with fewer ripples is obtained by using FLC. The settling time of the output voltage is even much lesser in FLC.

References

1. A. Fardoun, E. H. Ismail, and A. J. Sabzali, "New efficient bridgeless Cuk rectifiers for PFC applications," IEEE Trans. Power Electron., vol. 27, no. 7, pp. 3292–3301, Jul. 2012.
2. T. Ching-Jung and C. Chern-Lin, "A novel ZVT PWM Cuk power factor corrector," IEEE Trans. Ind. Electron., vol. 46, no. 4, pp. 780–787, Aug. 1999.
3. M. Mahdavi and H. Faarzanehfard, "Bridgeless CUK power factor correction rectifier with reduced conduction losses," IET Power Electron., vol. 5, no. 9, pp. 1733–1740, Nov. 2012.
4. L. Huber, L. Gang, and M. M. Jovanovic, "Design-oriented analysis and performance evaluation of buck PFC front end," IEEE Trans. Power Electron., vol. 25, no. 1, pp. 85–94, Jan. 2010.
5. A. J. Sabzali, E. H. Ismail, M. A. Al-Saffar, and A. A. Fardoun, "New bridgeless DCM Sepic and Cuk PFC rectifiers with low conduction and switching losses," IEEE Trans. Ind. Appl., vol. 47, no. 2, pp. 873–881, Mar./Apr. 2011.
6. Y. S. Roh, Y. J. Moon, J. G. Gong, and C. Yoo, "Active power factor correction (PFC) circuit with resistor-free zero-current detection," IEEE Trans. Power Electron., vol. 26, no. 2, pp. 630–637, Feb. 2011.
7. P. F. de Melo, R. Gules, E. F. R. Romaneli, and R. C. Annunziato, "A modified SEPIC converter for high-power-factor rectifier and universal input voltage applications," IEEE Trans. Power Electron., vol. 25, no. 2, pp. 310–321, Feb. 2010.
8. T. Ching-Jung and C. Chern-Lin, "A novel ZVT PWM Cuk powerfactor corrector," IEEE Trans. Ind. Electron., vol. 46, no. 4, pp. 780–787, Aug. 1999.
9. Y. Jang and M. M. Jovanovic, "Bridgeless high-power-factor buck converter," IEEE Trans. Power Electron., vol. 26, no. 2, pp. 602–611, Feb. 2011.
10. R. Oruganti and M. Palaniapan, "Inductor voltage control of buck-type single-phase AC-DC converter," IEEE Trans. Power Electron., vol. 15, no. 2, pp. 411–416, Mar. 2010.
11. R. Meena Devi, "Variable Sampling Effect for BLDC Motors using Fuzzy PI Controller" Indian Journal of Science and Technology, Vol 8(35), https://doi.org/10.17485/ijst/2015/v8i35/68960, December 2015.
12. M. Kavitha, Dr. V. Sivachidambaranathan, "PV based high voltage gain Quadratic DC-DC Converter integrated with coupled inductor", International Conference on Computation of Power, Energy Information and Communication (ICCPEIC'16), on 21st April 2016.
13. R. Meena Devi, Dr. L. PREMALATHA," A New PFC converter using bridgeless single-ended primary induction converter(SEPIC)" International Conference on Control, Instrumentation, Communication and Computational Technologies (ICCICCT) pp: 344–347.
14. D. S. L. Simonetti, J. Sebastian, and J. Uceda, "The discontinuous conduction mode Sepic and Cuk power factor pre regulators: Analysis and design," IEEE Trans. Ind. Electron., vol. 44, no. 5, pp. 630–637, Oct. 1997.

15. Kunapuli sahiti and V.Geetha," Simulation of series resonant inverter using pulse density modulation" ARPN journal of engineering and applied sciences vol. 10, no. 7, April 2015.
16. L. Petersen, "Input-current-shaper based on a modified SEPIC converter with low voltage stress," in Proc. 32nd IEEE Power Electron. Spec. Conf., Jun. 2001, vol. 2, pp. 666–671.
17. Hong-Tzer Yang, Hsin-Wei Chiang, Chung-Yu Chen," Implementation of Bridgeless Cuk Power Facto Corrector with Positive output voltage", IEEE Transactions on Industry Application, Vol. 51(4),Aug 2015.

Measurement of Electromagnetic Properties of Plastics and Composites Using Rectangular Waveguide

Zaharaddeen S. Iro, C. Subramani and S. S. Dash

Abstract Rectangular waveguides are the most common type of waveguides used in transporting microwave signals, and they are still being used to date in many applications. Waveguides can be used in high power systems, millimeter wave systems, and in some precision test applications. In this work, a model of a rectangular waveguide was created using FEKO software, using the model some plastic samples like polyethylene and acrylic were put into the rectangular waveguide and there S-parameters were measured. The dimensions of the modeled rectangular waveguide model were used in fabricating an actual rectangular waveguide. The same plastic samples used in the model were used in the constructed rectangular waveguide and their S-parameters were measured. The simulated and measured S-parameters were compared and there was a good agreement between the results.

Keywords Rectangular waveguide · FEKO · Acrylic · Carbon fiber-reinforced plastic · S-parameters

1 Introduction

All materials have distinctive electrical properties dependent on the dielectric properties. A precise measurement of the dielectric properties can offer valuable information about the material, which can then be used to accurately qualify the material for its proposed application for further solid designs or to observe the manufacturing process for an enhanced quality control. The dielectric properties of materials are design parameters needed for many electronics applications. For instance, the loss of a cable insulator, the impedance of a substrate, or the frequency of a dielectric resonator can be calculated from its dielectric properties. Further

Z. S. Iro (✉) · C. Subramani · S. S. Dash
Department of Electrical and Electronics Engineering, SRM Institute of Science and Technology, Kattankulathur, Chennai 603203, India
e-mail: zaharaddeensiro@gmail.com

© Springer Nature Singapore Pte Ltd. 2019
M. A. Bhaskar et al. (eds.), *International Conference on Intelligent Computing and Applications*, Advances in Intelligent Systems and Computing 846,
https://doi.org/10.1007/978-981-13-2182-5_36

recent applications in various fields such as industrial microwave processing of food, rubber, ceramics, plastic, and carbon fiber-reinforced plastics have benefitted from knowledge of dielectric properties measurements [1].

A transmission line refers to a device that is intended to guide electrical energy from one point to another. In other words, it is a means of transferring the output RF energy of a transmitter to an antenna. It becomes impossible to transfer this energy using a normal electrical wire at high frequencies as the losses will be excessive. The transmission line has a single purpose for both the transmitter and the antenna which is to transfer energy between them with the least possible power loss [2]. Some examples of transmission lines are Coaxial line, Two-wire, and Microstrip line.

Coaxial line is used at frequencies below 10 GHz, coaxial cables are used mostly to connect RF components and their operation is practical at that frequency. At frequencies beyond that, the losses become too excessive. The power rating is usually of the order of 1 kw at 100 MHz, but only 200 W at 2 GHz, being limited primarily because the coaxial conductors and the dielectric between the conductors get heated up [3]. Coaxial cables consist of a core wire that is enclosed by a nonconductive material, which is then surrounded by a covered shielding mostly made out of braided wires. The nonconductive material is there to keep the core and the shielding apart. Finally, the coax is protected by an outer shielding which is normally a PVC material. The inner conductor carries the RF signal and the outer shield is there to keep the RF signal from radiating to the atmosphere and to stop outside signals from interfering with the signal carried by the core [4].

Waveguides are metallic enclosed transmission lines used at microwave frequencies, usually to connect transmitters and receivers with antennas. Amongst the advantages of waveguides includes: being completely shielded (excellent isolation between adjacent transmission lines can be obtained), at microwave frequencies it transmits extremely high peak power and it has a very low loss (so low it is negligible) [5]. Common examples of waveguides have rectangular handling capability and circular cross sections.

Rectangular waveguides are the most common type of waveguides used in transporting microwave signals, and they are still being used to date in many applications. Waveguides can be used in high power systems, millimeter wave systems, and in some precision test applications [6]. A rectangular waveguide can support propagation of both TE and TM modes. The dominant mode in a rectangular waveguide is TE_{10} mode, therefore it has the lowest cut-off frequency. Generally, the conducting walls are made of brass or aluminum and the dielectric region is air. Since the wall thickness is greater than several skin depth, it is not considered to be part of the analysis [7].

Carbon fiber-reinforced plastic (CFRP) is made up of high stiffness, high strength carbon fibers reinforced by a reasonably compliant, and low strength polymer matrix. When compared with other structural materials or composite materials, it has relatively high specific stiffness and low specific weight [8]. For the past 50 years, the usage of CFRP has increased in a number of applications such as load-bearing structures in aerospace, automotive, naval, and construction industries. In some rare cases, the high specific conductivity of CFRP has led to its increased usage in the electronics industry [9, 10]. Even though CFRP composites were commercially available since in the 1960s, making good use of their electronic and electromagnetic properties has been a recent development. Both electrical and mechanical properties of CFRP are highly dependent on the fiber orientation [11]. There has been a lot of research done on the conductivity and electromagnetic properties of CFRP to exploit its electrical behavior [12–14].

Nicholson–Ross–Weir (NRW) uses the samples S-parameters measured from the vector network analyser (VNA) for its calculations. There are different types of conversion techniques. NRW is mostly used for these types of conversions. To obtain the reflection and transmission coefficients, it needs all four (S_{11}, S_{21}, S_{12}, and S_{22}) or pair (S_{11}, S_{21}) of the measured S-parameters from the sample. The technique has an advantage of being noniterative and can be used on coaxial line and rectangular waveguide cells. At frequencies equivalent to integer multiples of one-half wavelength in the sample, the technique diverges for low loss materials [15, 16].

2 Experimental

Dielectric properties are determined from the S-parameters measured, the S-parameters of the sample is measured when the sample is inserted into the rectangular waveguide and it is connected to voltage network analyzer (VNA). The measured S-parameters are processed and then using NRW the respective permittivity and permeability of the sample can be obtained (Fig. 1).

The operating frequency used for the project is 5.05 GHz, from there a rectangular waveguide specification sheet designed by Cobham Defense Electronic Systems was used in order to know the type of rectangular waveguide to use. The recommended operating range for TE_{10} mode of WR159 rectangular waveguide is between 4.90 and 7.05 GHz, since the operating frequency we are using is within that range it was selected for the experiment. It has a cut-off frequency of

Fig. 1 Measurement setup

3.705 GHz, with a cut-off wavelength of 8.092 cm. The internal dimensions which were critical were given as 40×20 mm, it is customary to label the large dimension as **a** which is the width and **b** as the height of the waveguide.

2.1 FEKO Model

After obtaining the dimensions of the rectangular waveguide, cut-off frequency and guided wavelength. The next thing done was using FEKO to model a rectangular waveguide so that we can have a better idea on how the rectangular waveguide works before constructing it. All the units used in FEKO are in mm. The variables introduced were probe length, back-short and guided wavelength. The rectangular waveguide was created using a cuboid. The medium was set to free space. And since it was an enclosed rectangular waveguide a section of it was cutoff so that we can see what is happening inside. The probes used were represented as line and voltage sources were added to them. Figure 2 shows a cross section of the rectangular waveguide.

The distance of the probe from the wall is known as the back-short distance. The first value of the back-short distance used was $\lambda_g/8$.

2.2 Hardware

For the hardware design, the most important parameters were the internal width and height, then the back-short distance (Fig. 3).

2.3 Samples

The polyethylene and acrylic samples were gotten from the lab and they were made to fit tight into the rectangular waveguide in order to avoid air gaps. Sample 1 is

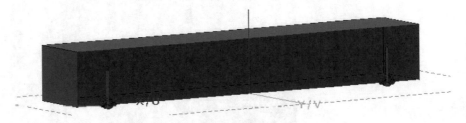

Fig. 2 Rectangular waveguide cross section in FEKO

Fig. 3 Constructed rectangular waveguide

(a)   **(b)**

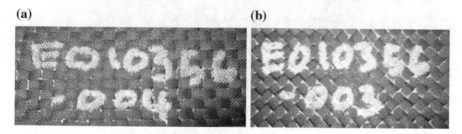

Fig. 4 (a) and (b) 45° and 90° CFRP respectively

polyethylene sample, while $S2$ is the acrylic sample. Sample 3 and Sample 4 is 45° and 90° CFRP, respectively, as shown in Fig. 4.

2.4 Probes

Using the probe length of 12 mm. The samples were put into the rectangular waveguide and their respective S-parameters values were measured. The S-parameters obtained were S_{11}, S_{21}, S_{12}, and S_{22} each one of the S-parameters had magnitude and phase values, respectively (Fig. 5).

3 Results and Discussions

3.1 Attenuation Curve

An attenuation curve plotted to see the losses due to the walls of the rectangular waveguide (WR-159). It can be seen that attenuation losses increases rapidly as the operating frequency approaches the cut-off frequency. At frequencies greater than cutoff, it possesses a broad minimum (Fig. 6).

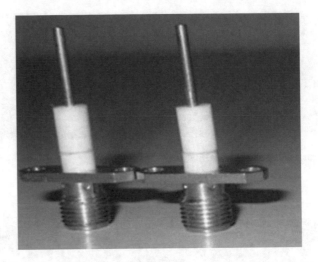

Fig. 5 Probes

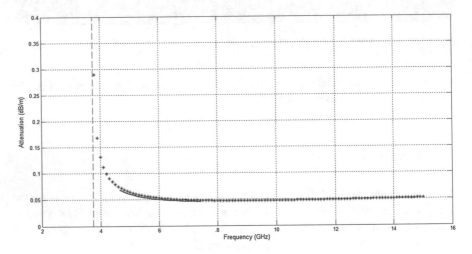

Fig. 6 Attenuation curve for rectangular waveguide

3.2   S-parameter Graphs

Both the magnitude and phase for the measured S-parameters of the Polyethylene sample and the Acrylic sample are in good agreement with the corresponding magnitude and phase for the simulated S-parameters (Figs. 7, 8, 9, 10, 11, 12, 13, and 14).

The 45° and 90° CFRP S-parameters were measured.

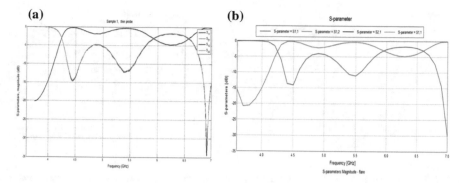

Fig. 7 **a** Measured *S*-parameters of sample 1 magnitude, **b** simulated *S*-parameters of sample 1 magnitude

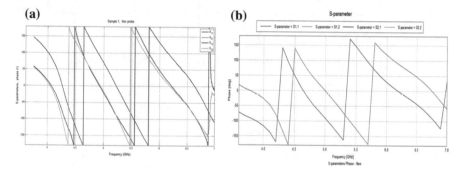

Fig. 8 **a** Measured *S*-parameters of sample 1 phase, **b** simulated *S*-parameters of sample 1 phase

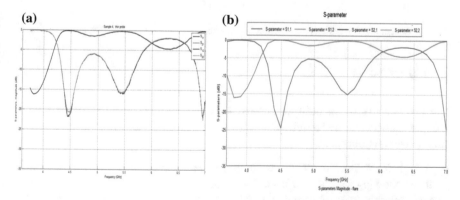

Fig. 9 **a** Measured *S*-parameters of sample 2 magnitude, **b** simulated *S*-parameters of sample 2 magnitude

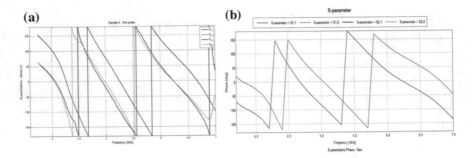

Fig. 10 **a** Measured *S*-parameters of sample 2 phase, **b** simulated *S*-parameters of sample 2 phase

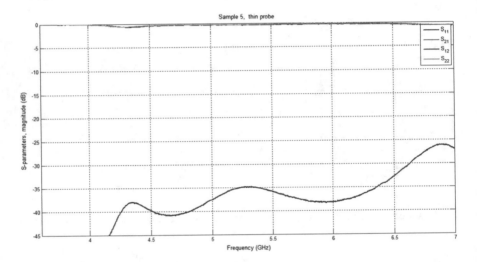

Fig. 11 Measured *S*-parameters of sample 3 magnitude

4 Conclusion

The test cell designed, built and tested here did not calibrate the surface of the samples under test as required by Nicholson-Ross-Weir method, but alot of knowledge was obtained.

- A high level of agreement was found between experiment and FEKO simulations, showing that despite been a Method of Moments software, FEKO can be used for this type of rectangular waveguide working confidence in its further use.

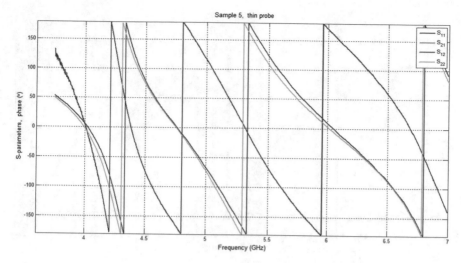

Fig. 12 Measured *S*-parameters of sample 3 phase

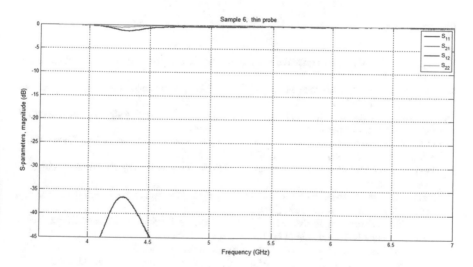

Fig. 13 Measured *S*-parameters of sample 4 magnitude

- The novel construction method using a trough and lid was sound and convenient.
- The full list of equipment and procedure to calibrate to a sample face as required by the Nicholson–Ross–Weir method is now understood and defined.

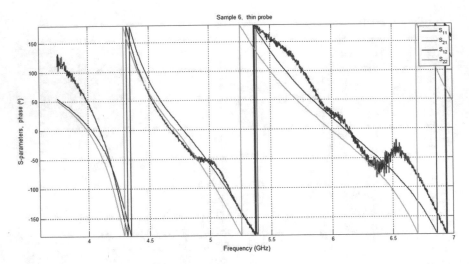

Fig. 14 Measured S-parameters of sample 4 phase

References

1. Agilent Basics of Measuring the dielectric properties of materials, Agilent Technologies, Inc., USA, 2006.
2. M. Solari, http://www.techlearner.com/2000. [Online]. Available: http://www.techlearner.com/Apps/TransandGuides.pdf. [Accessed 26 August 2012].
3. Sophocles J. Orfanidis, Waveguides, Electromagnetic waves and Antenna, Piscataway NJ, ECE Department Rutgers University, 362–396 (2008).
4. Carlo Fonda and Marco Zennaro, Transmission Lines in Radio Laboratory Handbook - Vol 1, Trieste, Radio Communications Unit-ICTP, 10–36 (2004).
5. January 2008. [Online]. Available: http://www.microwaves101.com/encyclopedia/waveguide.cfm.
6. David M. Pozar: Transmission Lines and Waveguides. Microwave Engineering, New Jersey, John Wiley & Sons, Inc, 91–160 (2005).
7. Peter Rizzi: Microwave Transmission Lines. Microwave Engineering Passive Circuits, New Jersey, Prentice Hall, 200–298 (1987).
8. Paul J. Callus, Kamran Ghorbani and Amir Galehdar: A Novel Method of Conductivity Measurements for Carbon-Fiber Monopole Antenna. IEEE TRANSACTIONS ON ANTENNAS AND PROPAGATION (6), 2120–2126 (2011).
9. Harvey M. Flower: High Performance Materials in Aerospace. London (1995).
10. Xuifeng Wang, Yonglan Wang and Zhihao Jin: Electrical conductivity characterization and variation of carbon fiber reinforced cement composite. JOURNAL OF MATERIALS SCIENCE 37, 223–227 (2002).
11. Christopher L. Holloway, Maria S. Sarto and Martin Johansson: Analyzing Carbon-Fiber Composite Materials With Equivalent-Layer Models. IEEE TRANSACTIONS ON ELECTROMAGNETIC COMPATIBILITY 47(4), 833–844 (2005).
12. J B Park, T K Hwang and Y D Doh: Experimental and numerical study of the electrical anisotropy in unidirectional carbon-fiber-reinforced polymer composites. SMART MATERIALS AND STRUCTURES(16), 57–66 (2006).

13. Alexander Markov, Bodo Fiedler and Karl Schulte: Electrical conductivity of carbon black/ fibres filled glass-fibre-reinforced thermoplastic composites. Composites(37), 139–1395 (2005).

14. Zdenka Rimska, Vojteuch and Josef Spacek: AC Conductivity of Carbon Fiber-Polymer Matrix Composites at the Percolation Threshold. POLYMER COMPOSITES 23(1), 95–103 (2002).

15. Abdel-Hakim Boughriet, Christian Legrand and Alain Chapoton: Non iterative Stable Transmission/Reflection Method for Low-Loss Material Complex Permittivity Determination. IEEE TRANSACTIONS ON MICROWAVE THEORY AND TECHNIQUES 45(1), 52–57 (1997).

16. ROHDE & SHWARZ: Measurement of Dielectric Material Properties. ROHDE & SHWARZ (2006).

An Investigation on Strength Properties and Microstructural Properties of Pozzolanic Mortar Mixes

L. Krishnaraj, P. T. Ravichandran and P. R. Kannan Rajkumar

Abstract This study concentrates the replacement of cement with Raw Fly Ash (RFA) and Micro Silica (MS) were investigated. The Ordinary Portland Cement is partially replaced at 15, 30, 45, and 60% with RFA and MS used to mortar preparation. The mortar specimens are prepared for evaluation of the mechanical strength of mortar and are tested at the curing days of 3, 7, 14, and 28 of ages. It is found that the strength developed with increase in replacement of cement with RFA up to an optimum value, beyond which the strength start decreasing. SEM, FTIR, and X-Ray diffraction tests were conducted on OPC, RFA, and MS samples to analyze the microstructural properties. It was observed that inert surface of RFA was transformed to more reactive one by ball milling process. The particle shape size, shape, and texture of OPC, RFA, and MS were determined by using scanning electron microscopy analysis.

Keywords Fly ash mortar · Micro silica · Compressive strength
Split tensile strength

1 Introduction

Cement is the most common materials used for the construction. It is generally used as the binding material for the preparation of concrete and mortar. Increased cement production causes CO_2 emissions and as a consequence of the greenhouse effect [1–4]. Fly ash, Silica Fume, slag cement, GGBS, etc., are some commonly used secondary cementitious materials (SCMs). In this, Fly ash exhibits better durability properties and it is easily available [5–8].

The generation of Portland cement consumes a great deal of common assets and energy and radiates CO_2, SO_2, and NOx. These gases can detrimentally affect

L. Krishnaraj · P. T. Ravichandran (✉) · P. R. Kannan Rajkumar
Department of Civil Engineering, SRM Institute of Science and Technology,
Kattankulathur, Chennai 603203, Tamil Nadu, India
e-mail: ptrsrm@gmail.com

© Springer Nature Singapore Pte Ltd. 2019
M. A. Bhaskar et al. (eds.), *International Conference on Intelligent Computing
and Applications*, Advances in Intelligent Systems and Computing846,
https://doi.org/10.1007/978-981-13-2182-5_37

nature bringing about acid rain and adding to the greenhouse impact. Consequently, it is crucial that the construction industry responds to artificially use resources, particularly industrial byproducts. Fly ash has been generally utilized as a substitute for Portland concrete in numerous applications as a result of its favorable on fresh, hardened and durability properties. One clear detriment in the utilization of most fly ash for cement substitution reasons for existing is that the substitution of concrete, particularly in high volumes (>40%), diminishes rate of early strength development of the concrete [9, 10]. Alkali treatment included the separating of the glassy phases in a raised alkali environment to quicken the reaction. Sulfate initiation depends on the capacity of sulfates to respond with aluminum oxide in the glassy phase of fly ash to form sulfates (AFt) that adds to quality at early ages. Another technique regularly utilized by industry to enhance early age qualities is to utilize substance admixtures. Numerous usually accessible admixtures can possibly quicken the hydration of PC systems, while likewise being high range water reducers and superplasticizer in the meantime [11, 12].

Micro Silica or silica fumes are available in fine powder material and are being used in the experiment to study the sulfate attacks and to know the behavior of the modified characteristics of mortar with the addition of Silica. Micro Silica is very fine pozzolanic material composed of ultrafine amorphous glassy sphere [13].

2 Experimental Program

2.1 Materials Used

The binder materials are consist of 43 grade of OPC as per ASTM C 150 [14] with a specific gravity of 3.15 and secondary cementitious materials like RFA and Micro Silica. For the curing of specimens saturated lime water was used. The C Class fly ash as per ASTM C 618 standards is collected from Neyveli Lignite Corporation. The sand from three different grades of standard Ennore sand with the size range of 2–90 mm was utilized as fine aggregate for preparing standard mortar cubes. Each grade of sand was taken equally for the mix (33%). Micro Silica used in the study is powder in form and it is light pink in color with a density of 2.2 gm/cm^3. For improving the workability of the mortar mix, Conplast SP430 is used as the superplasticizer.

2.2 Mix Ratios Adopted

The sand to binder ratio is 2.75:1 used to prepare the mortar cubes. The mix ratio is taken from the IS 10262-2009. The basic binder materials are OPC, RFA and MS samples. The OPC material was partially replaced with 15, 30, 45, and 60% of MS and RFA. All replacement was made by mass of the OPC. Two series of mortar samples were prepared, one with tap water to binder ratio of 0.485 and the other with a water-binding ratio of 0.4 with 1% Super Plasticizer based on flow spread value 110 ± 5 as per ASTMC 230.

2.3 Material Preparation and Testing

For compressive strength testing cubes of size $50 \times 50 \times 50$ mm steel molds are prepared and for shrinkage tests mortars cubes of size $250 \times 25 \times 25$ mm are prepared. The molds are cleaned, fasteners are tightened, greased. Hand compaction was done using 20 mm diameter tamping rod. Mortar is placed in 3 layers with 25 numbers of strokes applied for each layer and compacted well [15–18]. It is De-molded after 24 h and kept in the water tank for curing with water replaced after every 7 days. The mortar cubes strength were analyzed at curing ages of 3, 7, 14, and 28 curing days as per ASTM C109, IS:1727, and IS:4037 and the split tensile strength in accordance with ASTMC 496.

3 Results and Discussions

3.1 Physical Properties

The basic material properties of OPC, RFA, and sand were compared in Table 1. Micro Silica has a specific gravity of 2.2. Micro Silica was found to have very fine particles than the OPC and RFA and particle size of the RFA was bigger than MS but smaller than OPC.

3.2 Particle Size Analysis

The Particle size distribution of OPC, RFA, and MS samples are discussed in Fig. 1. It is seen that OPC has 13.6% particles retained on 45 micron and its specific

Table 1 Physical characteristics of materials

Sample type	Specific gravity	45 micron sieve retained (%)	Baine's fineness (cm^2/g)	Median particle size (μm)
OPC	3.15	86.32	2630	23.28
RFA	2.62	87.21	3717	21.35
Sand	2.64	0	0	2 mm–0.09

Fig. 1 Particle size distribution of OPC, RFA, and MS samples

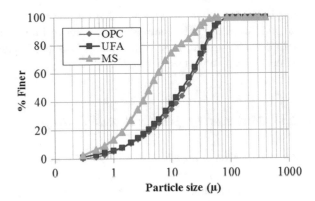

gravity is 3.15 and median particle size is 23.28 μm. The weight of particles retained on sieve No. 325 for RFA is 12.7%. The specific gravity and median particle sizes of RFA are 2.62 and 21.35 μm [19–22].

Much difference was not shown in the particle size distribution of OPC and RFA which might be due to the homogeneity in the particle characteristics, which can be greatly reduced by the ball milling process. Grinding process causes break up of large spheres and decreases the particle toughness [23, 24]. The main effect of grinding is that coarse particles are crushed on smaller particles by grinding process. The smaller reactive particles, which on the other hand, can increase the specific surface area and act as good filet materials.

3.3   Compressive Strength of Mortar

The compressive strength of different percentages (15, 30, 45, and 60%) of fly ash combination tested at various time interval of period. Figures 2 and 3 shows the compressive strength of specimens prepared by replacement of cement with RFA and MS, respectively. RFA and MS replaced specimens have the higher strength than the OPC specimens. The introduction of finer particles (RFA) into the mix reduces the cement voids which results in higher compressive strength [25, 26].

The compressive strength result was higher for replacement of OPC with 15% RFA and replacement of OPC with 15% MS. Higher percentage of replacement gave lower compressive strengths.

3.4   Split Tensile Strength of Mortar

The split tensile behavior of blended cylinder with different percentages like 15, 30, 45, and 60% of RFA and MS combination were tested and given in Fig 4. It is clear

Fig. 2 Compressive strength of OPC replaced with different percentages of RFA

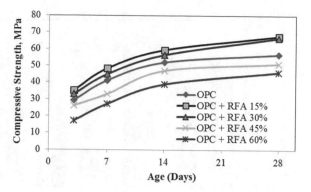

Fig. 3 Compressive strength of OPC replaced with different percentages of MS

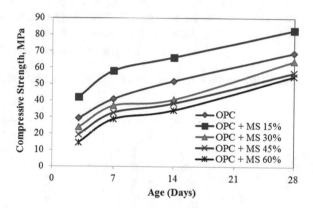

that the tensile strength MS mortar samples are higher than the RFA mortar samples and OPC mortar sample. The 28 day split tensile strength of the standard cement mortar with 0.4 w/b + 1% SP is found to be higher than the standard mortar split tensile strength with 0.485 w/b. For replacement percentage of OPC with 30% RFA gave split tensile of 3.75, 4.93 N/mm^2 for both w/b ratios, respectively. Split tensile strength results of the RFA and MS added specimens with super-plasticizer are shown in Fig. 5. The strength increases gradually with respect to the curing ages [27].

3.5 Morphological Studies

To analyze the microstructure of OPC, RFA samples was conducted the Scanning Electron Microscopy (SEM) and results are shown in Fig. 6. From the images show the SEM micrographs of OPC with 30 and 5 μm magnitude. The spherical structure and the size of OPC were measured. The SEM images show the OPC, RFA particle

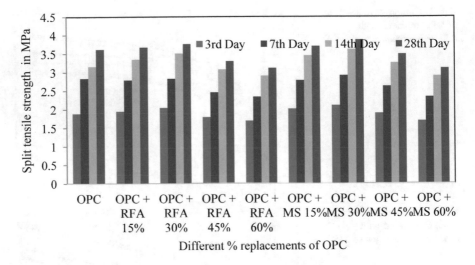

Fig. 4 Split tensile strength of OPC, RFA, and MS mortar samples without SPL

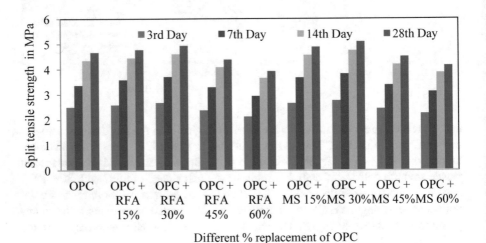

Fig. 5 Split tensile strength of OPC, RFA, and MS mortar samples with 1% SPL

sizes are 3.82 μm, 651 nm, respectively. The OPC and RFA sample images are shows particle nonspherical shape and irregular and of the particles But the RFA sample shows the spherical shape of particles compared with OPC sample SEM images [28].

(a)   **(b)**

(c)   **(d)**

(e)   **(f)**

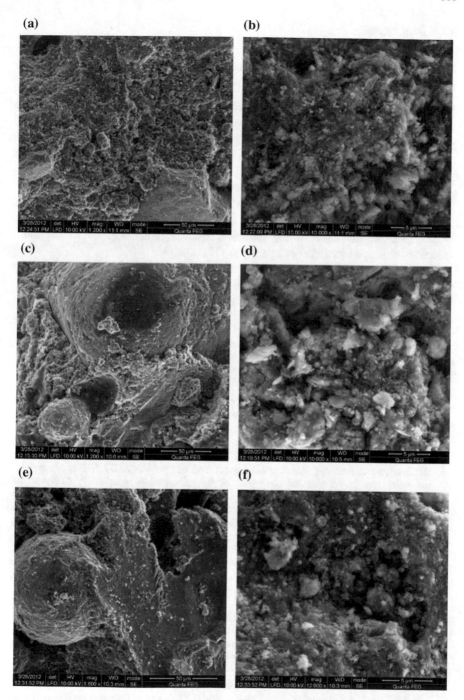

Fig. 6 SEM images of particles shapes of OPC, RFA, and MS samples

4 Conclusions

The following points the can be concluded from the RFA, MS blended mortar specimens of experimental investigation:

- Replacement of cement with 15% of RFA and 15% MS has shown compressive strength of 16.2 and 16.5%, respectively.
- The 28 day split tensile strength of the standard cement mortar with 0.4 w/b + 1% SP is found to be higher than mortar specimens with 0.485 w/b. For replacement percentage of OPC with 30% RFA, split tensile strength of 3.75, 4.93 N/mm^2 for both w/b ratios, respectively, was achieved.
- The SEM images shows that OPC, RFA particle sizes are 3.82 μm, 651 nm, respectively. The OPC and RFA samples images show irregular in shape and nonspherical in shape of the particles. But the RFA sample shows more spherical-shaped particles compared to OPC samples.
- The reduction in crystallite size from 68.48 to 46.50 nm, increases the unstructured realms. The unstructured realm enhances compatibility with various polymeric matrices.

References

1. C.S. Poon, L. Lam and Y.L. Wong, (2000), "A study on high strength concrete prepared with large volumes of low calcium fly ash", Cement and Concrete Research, 447–455.
2. Deepak Ravikumar, Sulapha Peethamparan and Narayanan Neithalath, (2010), "Structure and strength of NaOH activated concretes containing fly ash or GGBFS as the sole binder", Cement &Concrete Composites; 32:399–410.
3. DimitriosPanias, Ioanna P. Giannopoulou and Theodora Perraki, (2007), "Effect of synthesis parameters on the mechanical properties of fly ash-based geopolymers", Colloids and Surfaces, vol-301, 246–254.
4. L. Krishnaraj, P. T. Ravichandiran, R. Annadurai, P.R. Kannan Rajkumar, Study on Micro Structural Behavior and Strength Characteristics of Ultra Fine Fly Ash as a Secondary Cementitious Material with Portland cement, Int. Jour. Chem. Tech. Res. (2015), 7 (2), 555–563.
5. L. Krishnaraj, P.T. Ravichandiran, P.R. Kannan Rajkumar, Investigation on Effectiveness of the Top down Nanotechnology in Mechanical Activation of High Calcium Fly Ash in Mortar. Ind. Jour. Sci. Tech., (2016), 9, 23.
6. Djwantoro Hardjito, Shaw Shen and Fung, (2010), "Fly Ash-Based Geopolymer Mortar Incorporating Bottom Ash", Modern applied science, vol-4.
7. Hardjito Djwantaro, Chua chung cheak and Carrie Ho Lee Ing, (2008), "Strength and setting times of low calcium fly ash-based geopolymer mortar", Journal of Modern Applied Science.
8. Hua Xu van and Deventer JSJ, (1999), "The geopolymerization of alumino-silicate minerals", Int J Miner Process, 59:247–66.
9. Jeevaka Somaratna, Deepak Ravikumar and Narayanan Neithalath, (2010) "Response of alkali activated fly ash mortars to microwave curing", Cement and Concrete Research, vol-40, 1688–1696.

10. L. Krishnaraj, P. T. Ravichandran, P. R. Kannan Rajkumar and P. Keerthy Govind, "Effectiveness of Alkali Activators on Nano Structured Fly ash in Geopolymer Mortar', Indian Journal Of Science And Technology, ISSN: 0974-5645, Vol 9(33), 1-7, September 2016.

11. P. R. Kannan Rajkumar, P. T. Ravichandran, J. K. Ravi and L. Krishnaraj, Investigation on the Compatibility of Cement Paste with SNF and PCE based Superplasticizers., Indian Journal of Science and Technology, ISSN: 0974-5645, Vol 9 (34), pp 1-5, September 2016.

12. M. Ahmaruzzaman, (2010), "A review on the utilization of fly ash", Progress in energy and combustion science, 327–363. K. Thomas Paul,

13. Satpathy, I. Manna, K. K. Chakraborty and G. B. Nando, (2007), "Preparation and Characterization of Nano structured Materials from Fly Ash: A Waste from Thermal Power Stations, by High Energy Ball Milling", Nanoscale Res Lett, vol-2, 397–404.

14. ASTM C109: Standard Test Method for Compressive Strength of Hydraulic Cement Mortars (Using 2-in. or [50-mm] Cube Specimens).

15. L. Krishnaraj, Madhusudhan N. and P. T. Ravichandran, Experimental Study Of Ultra Fine Particles In Mechanical And Durability Properties Of Fly Ash Cement Composite Mortar, ARPN Journal of Engineering and Applied Sciences, VOL. 12, NO. 7, APRIL 2017.

16. L. Krishnaraj and P.T. Ravichandran, Impact of Ball Milled Fly Ash Nano Particles on the Strength and Microstructural Characteristics of Cement Composite Mortars, International Journal of Control Theory and Applications, Volume 10, Number 12, 2017.

17. Manjit Singh and MridulGarg, (1999), "Cementitious binder from fly ash and other industrial wastes", Cement and Concrete Research, vol-29,309–314.

18. Nguyen Van Chanh, Bui Dang Trung and Dang Van Tuan, (2008), "Recent research geopolymerconcrete", The 3rd ACF International Conference, 235–241.

19. Ravindra N. Thakur and SomnathGhosh, (2009), "Effect of mix composition on compressive strength and microstructure of fly ash based geopolymer composites", ARPN journal of engineering and applied sciences, vol-4, 68–72.

20. L. Krishnaraj, Yeddula Bharath Simha Reddy, N. Madhusudhan and P.T. Ravichandran, Effect Of Energetically Modified Fly Ash On The Durability Properties Of Cement Mortar, Rasayan J. Chem., 10(2), 423–428(2017).

21. L. Krishnaraj, R. Suba Lakshmi and P.T. Ravichandran, Utilisation Of Rmc Waste With Chemical Admixtures To Manufacturing Of Sustainable Building Components, Rasayan J. Chem., 10(2), 592–599(2017).

22. T. Bakharev, (2005), "Resistance of geopolymer materials to acid attack", Cement and Concrete Research, 658–670. Peng J, Huang L, Zhao Y, Chen P, Zeng L, Zheng W. (2013). "Modeling of carbon dioxide measurement on cement plants". Adv Mater Res, 610–613.

23. Smith songpiriyakij, Seksum Chutubtim. (2001). "Study of ground coarse fly ashes with different finenesses from various sources as pozzolanic materials". Cement & concrete composites, 23, 335–343.

24. Steve W.M. Supit, Faiz U.A. Shaikh. (2013). "Effect of ultra fine fly ash on mechanical properties of high volume fly Ash mortar". Construction and building materials, 278–286.

25. Suresh Thokchom, Partha Ghosh, Somnath Ghosh. (2009). "Effect of water absorption, porosity and sorptivity on durability of Geopolymer mortars". ARPN Journal of Engineering and Applied Sciences, volume 4, No.7.

26. Suresh Thokchom, Dr. Partha Ghosh, Dr. Somnath Ghosh. (2009). "Resistance of fly ash based geopolymer mortars in sulfuric acid". International Journal of Recent Trends in Engineering, Vol. 1, No. 6. Turhan Bilir, Osman Gencel, Ilker Bekir Topcu. (2001). "Properties of mortars with fly ash as fine aggregate". Construction and building materials.

27. N. Bouzoubaa, M.H. Zhang, A. Bilodeau, and V.M. Malhotra. (1998). "Laboratory-produced high-volume fly ash blended cements: physical properties and compressive strength of mortars". Cement and Concrete Research, Vol. 28, No. 11, pp. 1555–1569.

28. Norsuzailina M. Sutan, Sinin Hamdan, Habibur Rahman Sobuz, Vincent Laja, Md. Saiful Islam. (2011). "Porosity and Strength of Pozzolan Modified Cement Systems". Concrete Research Letters. Vol 2, No 4.

Author Index

© Springer Nature Singapore Pte Ltd. 2019
M. A. Bhaskar et al. (eds.), *International Conference on Intelligent Computing
and Applications*, Advances in Intelligent Systems and Computing 846,
https://doi.org/10.1007/978-981-13-2182-5

Printed in the United States
By Bookmasters